Introduction
to Control System
Analysis and Design

SECOND EDITION

Introduction to Control System Analysis and Design

FRANCIS J. HALE

Professor of Mechanical and Aerospace Engineering
North Carolina State University

PRENTICE HALL, Englewood Cliffs, New Jersey 07632

Library of Congress Cataloging-in-Publication Data

Hale, Francis J. (date)
   Introduction to control system analysis and design.

   Bibliography: p.
   Includes index.
   1. Automatic control.   2. Control theory.   I. Title.
TJ213.H29 1988   629.8   87-14494
ISBN 0-13-479767-1

Editorial/Production Supervision: Evalyn Schoppet
Cover Design: 20/20 Services, Inc.

 © 1988 by Prentice-Hall, Inc.
A Division of Simon & Schuster
Englewood Cliffs, NJ 07632

Printed in the United States of America

10 9 8 7 6 5 4 3 2 1

ISBN 0-13-479767-1

Prentice-Hall International (UK) Limited, *London*
Prentice-Hall of Australia Pty. Limited, *Sydney*
Prentice-Hall of Canada Inc., *Toronto*
Prentice-Hall Hispanoamericana, S. A., *Mexico*
Prentice-Hall of India Private Limited, *New Delhi*
Prentice-Hall of Japan, Inc., *Tokyo*
Simon & Schuster Asia Pte. Ltd., *Singapore*
Editora Prentice-Hall do Brasil, Ltda., *Rio de Janeiro*

Contents

Preface

This introductory book was developed as a teaching text for a one-semester course that has been taught for many years at North Carolina State University. Initially the enrollment consisted primarily of mechanical and aerospace engineering undergraduates, but over the years has broadened to include undergraduates from other engineering curricula as well as graduate students who have encountered dynamic analysis and control problems in the course of their research or design. The course, and thus the book, may be terminal (the student's only formal exposure to control theory) or may serve as a preparation for specialized applications courses and for more advanced control theory courses.

This book has four general objectives. The first is to remove the mystique that many students and engineers seem to associate with control theory and to show that control applications are all around us. The second is to impart a sound understanding, both physical and mathematical, of the fundamentals of control problems and of their solutions. The third is to provide the student/engineer with techniques that can be easily and personally applied to the analysis and design of control systems. The fourth is to show the relationships (and differences) between the techniques of classical and modern control theory.

This text concentrates on what is known as classical control theory and limits the treatment of modern control theory to description and discussion in the final chapters. There is no intention whatsoever of disparaging modern control theory; I happen to believe that the classical approach should precede the modern approach, both in the classroom and in practice. One reason for this belief is the fact that the level of mathematics required and the obscuration of the physical nature of the problem are considerably less in the classical approach than in the modern approach.

Furthermore, a multivariable nonlinear system can frequently be reduced by linearization and decoupling approximations to a set of single-variable linear systems that can be handled with the classical techniques. These classical techniques provide analytical and physical relationships among the system parameters and inputs that are not easily attainable with modern control theory. The approximate solutions obtained with classical control theory are often sufficient in themselves; in fact, most working control systems are still being designed by a combination of classical and experimental methods. For complex and sophisticated problems, the approximate solutions aid in scaling analog simulations and in checking computer programs and numerical solutions. In addition, the classical techniques can readily be used by the designer, who can rapidly sketch root locus, Bode, and Nyquist diagrams to determine the major effects of parameter and design changes, and then use the computer and more advanced techniques for the detailed analysis and design.

For an introductory book such as this, any claim to merit is based on organization, coverage, and style. The first chapter presents an overview of the control problem and establishes a frame of reference for the following chapters, which are in the order that I have found to be most effective. The sequence can obviously be altered at the discretion of the instructor; similarly, sections and topics can be omitted*, supplemented, emphasized, or deemphasized. A one-semester course should include the material of the first nine chapters. Chapter 10 is strongly recommended if time permits; as a minimum, the concept of nonlinear controllers and the destabilizing effects of inherent nonlinearities should be discussed. Chapters 11 and 12 are beyond the normal scope of the course; students, however, should be encouraged to read these chapters on their own, if only to increase their awareness of other areas of control theory.

The topic coverage strives to be complete but not necessarily exhaustive. For example, specialized topics such as Nichols charts are discussed only briefly in order to indicate their use, to broaden the control vocabulary, and to provide an entry into the supporting literature such as the books listed as references at the end of this text. I apologize for any inadvertent omissions and for the deliberate omission of the many papers that are the foundation of control theory.

Since components and the derivation of the differential equations describing such pieces of hardware are functions of the reader's area of specialization, they are essentially ignored. The standard mechanical and electrical elements are used to illustrate procedures for obtaining the general form of transfer functions. Applications are also omitted, not by choice but by necessity.

As desirable as they may be, worthwhile applications not only require large amounts of space and time but also break the continuity and flow of ideas. If the application is simplified so as to minimize the disruption, it is no longer representative of problems actually encountered. Furthermore, no one application will satisfy all of the readers with their diverse interests. I strongly believe that the applications of control theory should be covered in specialized follow-on courses, such as vehicle stability and control, thermal control, vibration control,

---

* The sections that I feel can be omitted or treated briefly are preceded by asterisks.

pneumatic and hydraulic systems, etc. It is not possible to do justice to both theory and its applications in a single one-semester course.

This book should also be suitable for self-teaching and study. It requires only an understanding of the nature of differential equations. Familiarity with the Laplace transformation and complex variables is desirable but not necessary. Mathematical derivations and proofs are included only when they might contribute to an understanding of the control problem and are generally less than rigorous.

None of the problems in this book requires computer solutions, since I want the user to understand the "why" of the various techniques before going to the computer. Access to a computer with appropriate software is desirable for verification of sketches, for obtaining detailed plots and time responses, and for controller design. Suitable software with varying degrees of flexibility and capability is available; PROGRAM CC from Systems Technology, Inc., was very useful in preparing this edition.

In summary, this book represents an effort to present in a readable form, suitable for both teaching and learning, an integrated introduction to the important and widely-applicable areas of dynamic analysis and control with the major objectives of understanding and enjoyment.

ACKNOWLEDGMENTS

I should like to express my appreciation to my two colleagues at North Carolina State University, Clare Maday and Fred Smetana, for their comments and suggestions, and to Clare for his review of the revised manuscript.

FRANCIS J. HALE

Introduction
to Control System
Analysis and Design

1

Introduction

1-1 THE WORLD OF CONTROL

Control theory deals with the dynamic response of a system to commands or disturbances. If a system is stable, it will eventually settle down to an *equilibrium* or *steady-state condition;* the behavior of the system prior to this settling down is called the *transient response.* An unstable system, on the other hand, will never settle down. Mathematically, its transient response will continue indefinitely; physically, it will continue until constrained by some physical limitation such as amplifier or motor saturation or by damage to or destruction of the system itself.

The laws of nature are such that everything in the known universe is controlled, with the degree of control varying widely. (Compare, for example, regulation of the temperature of the human body with that of the earth's atmosphere.) People, however, tend to think of nature as uncontrolled and manifest their egoism by regarding the word *control* as descriptive of human applications only. A further distinction is made between *manual control* with a human operator as an element of the system and *automatic control* without an explicit human-machine relationship.

Let us consider the flow of traffic in a typical city. In the absence of police officers and traffic lights, traffic would be described as uncontrolled even though there would be an inherent control of the flow by such factors as the density, type, and speed of vehicles on various streets; number of intersections and lanes; courtesy and training of drivers; etc. If police officers were placed at the intersections, traffic would then be manually controlled. Replace the police officers with traffic lights, and control would then be described as automatic. If the operating sequence of the lights is programmed or timed for one set of traffic conditions, the control is called *open loop* because it does not consider the actual traffic conditions; for example, crosstown traffic would be stopped at an intersection even though there were no automobiles on the main thoroughfare. A police officer, however, would observe this condition and allow the crosstown traffic to proceed without delay. This type of control is an example of *closed-loop* control, where the actual system response is compared with the desired response and appropriate control action is taken. With the addition of sensors and computing devices, the automatic control by traffic lights can be changed from open loop to closed loop, but the change obviously entails additional expense and complexity.

Now for some terminology—troublesome at times but essential to a working knowledge of any field or discipline. The *plant* is whatever is to be controlled. In our traffic example, the plant is the network of roads plus the vehicles; the *plant parameters* would be such things as the number of intersections and lanes, speed limits, traffic density, etc. *System* or *control system* generally denotes the plant plus all the elements necessary for control, including all inputs and disturbances. With manual traffic control, the system would be the plant plus the police officers.

A *controlled variable* is a specific characteristic of a plant that we wish to control. In our example, a controlled variable might be the number of vehicles per hour that can pass through the plant, i.e., enter the downtown area. The terms *output* and *controlled variable* are often used interchangeably even though the output cannot always be directly controlled.

A *subsystem* is defined as a system that is contained within a larger system. An individual vehicle could be considered a subsystem of our traffic control system. The vehicle itself could be the plant; the vehicle plus the driver, the subsystem; and the speed of the automobile, the controlled variable.

The plant can be anything for which a suitable mathematical model can be obtained. It might be a physical subsystem, such as an airplane, a tracking radar, a milling machine, a clothes dryer, or a robotic device. It may be a process, such as a chemical or nuclear reaction, crystallization, water purification, or heat transfer. It could be biological, ecological, or demographic: the flow of blood in the human body, the elk herds in Wyoming, or the population growth in underdeveloped countries. It may even be political, social, economic, or any combination thereof: the political attitudes of the South, the morality of the under-thirty population group, or the welfare system of the United States.

The application of control theory has essentially two phases: *dynamic analysis* and *control system design*. The analysis phase is concerned with determination of the response of a plant to commands, disturbances, and changes in the plant parameters. If the dynamic response is satisfactory, there need be no second phase. If the response is unsatisfactory and modification of the plant is unacceptable, a design phase is necessary to select the control elements (the *controller*) needed to improve the dynamic performance to acceptable levels. A word of caution is in order. Control can be expensive in many ways and should be added with care and wisdom. Control for the sake of control or to create an image of impressive technology should not be tolerated.

Control systems are used by humans to extend their physical capabilities, to compensate for their physical limitations, to relieve them of routine or tedious tasks, or to save money. In a modern aircraft, for example, the power boost controls amplify the force applied by the pilot to move the control surfaces against large aerodynamic forces. The reaction time of a human pilot is too slow to enable him or her to fly an aircraft with a lightly damped Dutch roll mode without the addition of a yaw damper system. An autopilot (flight control system) relieves the pilot of the task of continuously operating the controls to maintain the desired heading, altitude, and attitude. Freed of this routine task, the pilot can perform other tasks, such as navigation and/or communications, thus reducing the number of crew required and consequently the operating costs of the aircraft.

Control theory itself has two categories: *classical* and *modern*. *Classical control theory,* which had its start during World War II, can be characterized by the transfer function concept with analysis and design principally in the Laplace and frequency domains. *Modern control theory* has arisen with the advent of high-speed digital computers and can be characterized by the state variables concept with emphasis on matrix algebra and with analysis and design principally in the time domain. As might be expected, each approach has its advantages and disadvantages as well as its proponents and detractors. The "compleat control system designer" should be familiar with both approaches; he or she should exploit the advantages of each, both individually and in combination, to execute the analysis and design.

As compared to the modern approach, the classical approach has the tutorial advantage of placing less emphasis on mathematical techniques and more emphasis on physical understanding. Furthermore, in many design situations the classical approach is not only simpler but may be completely adequate. In those more complex cases where it is not adequate, the classical approach solution may aid in applying the modern approach and may provide a check on the more complete and exact design. For these reasons, the major portion of this book is devoted to the classical approach.

1-2 TYPES OF CONTROL SYSTEMS

Control systems are classified in terms that describe either the system itself or its variables. These descriptive terms are mainly of the either-or form. In the preceding section, for example, the traffic control system was either open loop or closed loop.

An *open-loop system* is shown in Fig. 1-2-1a and is characterized by the input entering directly into the control elements unaffected by the output; the output is related to the input solely by the characteristics of the plant and control elements. In the *closed-loop system* of Fig. 1-2-1b, however, the input is modified by the actual output before entering the control elements. Since the output is fed back in a functional form determined by the nature of the feedback

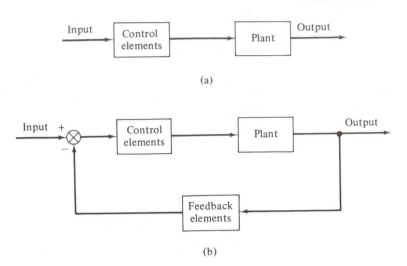

Figure 1–2–1. Control systems: (a) open loop (b) closed loop.

elements and then subtracted from the input, a closed-loop system is often referred to as a *negative feedback system* or simply as a *feedback system.*

A system is *linear* if *all* of its elements are linear, and *nonlinear* if *any* element is nonlinear. Linearity and nonlinearity are defined and discussed in Sec. 2–1.

Lumped parameter systems are those for which the physical characteristics are assumed to be concentrated in one or more "lumps" and thus independent of any spatial distribution. In effect, bodies are assumed rigid and treated as point masses; springs are massless and electrical leads resistanceless, or suitable corrections are made to the system mass or resistance; temperatures are uniform; etc. In *distributed parameter systems,* the continuous spatial distribution of a physical characteristic is taken into account. Bodies are elastic, springs have a distributed mass, electrical leads have a distributed resistance, and temperatures vary across a body. Lumped parameter systems are described by ordinary differential equations, while distributed parameter systems are described by partial differential equations.

A *stationary* or *time-invariant system* is one whose parameters do not vary with time. The output of a stationary system is independent of the time at which an input is applied, and the coefficients of the describing differential equations are constants. A *nonstationary* or *time-variant system* is a system with one or more parameters that vary with time. The time at which an input is applied must be known, and the coefficients of the differential equations are time-dependent.

A system or variable is *deterministic* if its future behavior is both predictable and repeatable within reasonable limits. If not, the system or variable is called *stochastic* or *random.* Analyses of stochastic systems and of deterministic systems with stochastic inputs are based on probability theory.

A *single-variable system* is defined as one with only one output for one reference or command input and is often referred to as a *single input–single*

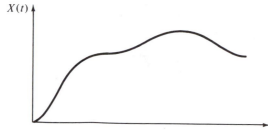

t **Figure 1–2–2.** A continuous variable.

output (SISO) system. A *multivariable (MIMO) system* has any number of inputs and outputs. A linear multivariable system may be, and often is, treated as comprising the appropriate number of single-variable systems.

A *continuous-variable* system is one for which *all* the system variables are continuous functions of time, as in Fig. 1–2–2; the describing equations are differential equations. A *discrete-variable system* has one or more variables known only at particular instants of time, as in Fig. 1–2–3; the equations are difference equations. If the time intervals are controlled, the system is termed a *sampled-data system.* Discrete variables occur naturally, as from a scanning radar that obtains position data once per scan or a data channel that transmits many pieces of information in turn, or they occur deliberately; for example, to provide an input to a digital computer which accepts only discrete data. A discrete variable will obviously approach a continuous variable as the sampling interval is decreased. *Discontinuous variables,* such as shown in Fig. 1–2–4, occur in "on-off" or "bang-bang" control systems and are treated separately in a subsequent chapter.

The techniques of the classical approach are directly applicable only to systems with all of the following characteristics: closed loop, linear, lumped parameter, stationary, deterministic, single variable, and continuous variable. The first nine chapters will be devoted exclusively to such systems, which often are referred to as *servomechanisms* or *servos.* In subsequent chapters, other types of systems will be discussed briefly.

The basic elements of a closed-loop control system are shown in Fig. 1–2–5a. More detail and some symbols are shown in Fig. 1–2–5b. The separation

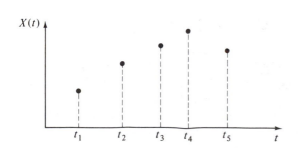

Figure 1–2–3. A discrete variable.

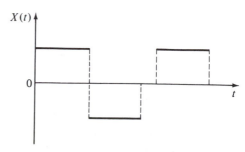

Figure 1–2–4. A discontinuous variable.

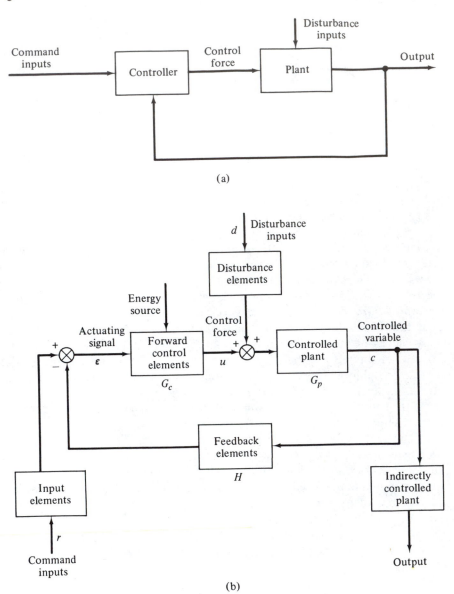

Figure 1-2-5. Closed-loop system schematics: (a) generalized (b) detailed.

between the plant and controller is not always obvious; for example, is the drive motor for a tracking antenna part of the plant or part of the controller? The separation is made to emphasize the fact that disturbance (undesirable) inputs primarily affect the plant and may enter the system at a different point than a command input.* The controlled variable and the output are usually (but not necessarily) identical; a central heating system controls the temperature only at

* A command input is sometimes referred to as a *reference input*—thus the symbol *r*.

the thermostat, with an output objective of controlling the temperature throughout the building. The input elements convert the command input into suitable form for entry into the system; in the simplest case, an input element may be a potentiometer for changing a mechanical rotation into a voltage.

Control systems may also be described as optimal, adaptive, self-adaptive, or process control. These are generic descriptions, and systems so described will also have varying combinations of the specific classifications of the preceding paragraphs. An *optimal* (or *optimum*) *control system* is directly designed to maximize or minimize a selected performance index; optimal design and modern control theory are practically synonymous. An *adaptive control system* is a nonstationary system whose controller parameters are varied, either continuously or discretely, so as to maintain proper control of the system output. If the system itself detects changes in the plant parameters and makes appropriate changes in the controller, it is called a *self-adaptive* or *learning system*. *Process control* is merely a descriptive term applied to the control of chemical or similar physical processes.

1–3 THE TRANSFER FUNCTION CONCEPT

If the input-output relationship of the linear system of Fig. 1–3–1a is known, the characteristics of the system itself are also known. The input-output relationship in the Laplace domain is called the *transfer function* (TF). By definition, the transfer function of a component or system is the ratio of the transformed output to the transformed input:

$$TF(s) = \frac{\text{output}(s)}{\text{input}(s)} = \frac{c(s)}{r(s)} \tag{1–3–1}$$

This relationship is shown in Fig. 1–3–1b.

(a)   (b)

Figure 1–3–1.   Input-output relationships: (a) general (b) transfer function.

This definition of the transfer function requires the system to be linear and stationary, with continuous variables and with zero initial conditions. The transfer function is most useful when the system is also lumped parameter and when transport lags* are absent or neglected. Under these conditions the transfer function itself can be expressed as a ratio of two polynomials in the complex Laplace variable s, or

$$TF(s) = \frac{N(s)}{D(s)} \tag{1–3–2}$$

* Transport lag is treated in Sec. 9–7.

For physical systems, $N(s)$ will be of lower order than $D(s)$ since nature integrates rather than differentiates. It will be shown later that a frequency transfer function (FTF) for use in the frequency domain can be obtained by replacing the Laplace variable s in the transfer function by $j\omega_f$.

In Eq. (1–3–2) the denominator $D(s)$ of the transfer function is called the *characteristic function* since it contains all the physical characteristics of the system. The *characteristic equation* is formed by setting $D(s)$ equal to zero. The roots of the characteristic equation determine the stability of the system and the general nature of the transient response to any input. The numerator polynomial $N(s)$ is a function of how the input enters the system. Consequently, $N(s)$ does not affect the absolute stability or the number or nature of the transient modes. It does, however, along with the specific input, determine the magnitude and sign of each transient mode and thus establishes the shape of the transient response as well as the steady-state value of the output.

The transfer function can be obtained in several ways. One method is purely mathematical and consists of taking the Laplace transform of the differential equations describing the component or system and then solving for the transfer function; nonzero initial conditions, when they occur, are treated as additional inputs. A second method is experimental. A known input (sinusoids and steps are commonly used) is applied to the system, the output is measured, and the transfer function is constructed from these data. A third method is a variation of the second in that the transfer function is obtained from operating data and curves. The transfer function for a subsystem or complete system is often obtained by proper combination of the known transfer functions of the individual elements. This combination or *reduction* process is termed *block diagram algebra* and is discussed in the next section.

For multivariable systems, a transfer function can be written for each input-output combination, as shown in Fig. 1–3–2. If there are m inputs each affecting n outputs, mn transfer functions will be required to describe the system. These transfer functions will all have identical denominators and thus identical characteristic equations and transient modes. The numerators, however, will all be different; consequently, the magnitudes of the transient modes will differ, as will the steady-state values. The variations in the relative importance of the individual transient modes may allow, as a first approximation, the neglect of specific transient modes in certain transfer functions and even the neglect of certain transfer functions themselves.

If the number of inputs and outputs is large and the interactions among

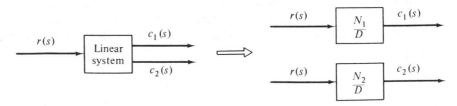

Figure 1–3–2. Transfer function representation of a multivariable system.

them strong, it may be more convenient to use a single transfer function matrix rather than many single input–single output transfer functions. Transfer function matrices may be used directly by an extension of the techniques used with transfer functions or changed into a state space model to be treated by modern control theory techniques. The point at which to switch from transfer functions to transfer function matrices or from transfer function matrices to state variables is often a matter of judgment and personal taste. It is sometimes desirable to use two or more approaches, exploiting the advantages of each.

1–4 THE BLOCK DIAGRAM

One advantage of the transfer function concept is the representation of a linear component or system by a single box or block as shown in Fig. 1–4–1.

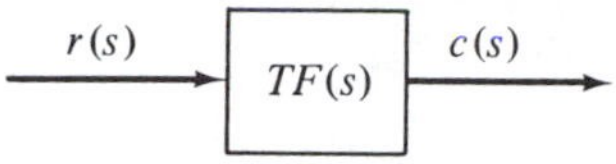

$r(s)$ → $TF(s)$ → $c(s)$

Figure 1–4–1. Block diagram of a linear element or system.

Moving in the direction of the arrow denotes multiplication; thus

$$c(s) = TF(s)r(s) \tag{1–4–1}$$

Moving against the arrows denotes division:

$$r(s) = \frac{c(s)}{TF(s)} \tag{1–4–2}$$

If two linear elements are connected in series so that the output of the first becomes the input to the second, as in Fig. 1–4–2a, the individual blocks represent the equations

$$r_1 = G_1 r \qquad c = G_2 r_1 \tag{1–4–3}$$

where G_1 and G_2 are the respective transfer functions.* These equations may be combined to give

$$c = G_1 G_2 r \tag{1–4–4}$$

This equation shows that the transfer function for the output c for an input r is $G_1 G_2$ or the product of the individual transfer functions. Therefore, the two blocks in series can be replaced by a single block as shown in Fig. 1–4–2b. Any number of blocks in series can be similarly reduced to a single block. Implicit

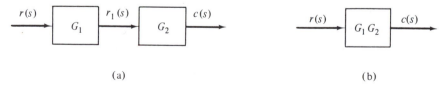

(a)   (b)

Figure 1–4–2. Two linear elements in series: (a) two blocks (b) a single block.

* Specification of the variable s will be dropped except where confusion might arise.

in this reduction and in the transfer function concept is the assumption that the transfer function of any block is not affected by any of the succeeding blocks, i.e., impedances are properly matched. Furthermore, the sequence of individual transfer functions need not be maintained; this is why the transfer function in Fig. 1–4–2b can be shown as G_1G_2 as well as G_2G_1.

The *comparator* or *summation point* is a symbol for the subtraction or addition of two or more variables. The symbol for a comparator and its use are shown in Fig. 1–4–3. A comparator may be a physical piece of hardware, such as a gyroscope or a difference amplifier, or it may be abstract, representing such things as kinematic or gravitational inputs.

With the use of the comparator, the series rule, and elementary arithmetic, block diagrams can be rearranged and reduced to a single block representing the transfer function of the complete system. The next-to-last step in the reduction process for a negative feedback control system is shown in Fig. 1–4–4a, where G is the accepted symbol for forward transfer functions and H for feedback transfer functions. At the comparator,

$$\varepsilon = r - Hc \qquad (1\text{–}4\text{–}5)$$

From the forward path,

$$c = G\varepsilon \qquad (1\text{–}4\text{–}6)$$

Combining Eqs. (1–4–5) and (1–4–6) yields the transfer function

$$\frac{c}{r} = \frac{G}{1 + GH} \qquad (1\text{–}4\text{–}7)$$

This is the transfer function of the system, often called the *closed-loop transfer function*. This final reduction is shown in Fig. 1–4–4b; it is frequently used and should be memorized.

The product of all the individual transfer functions in a loop is called the *open-loop transfer function*. In this example it is GH. If the feedback loop were opened just before the comparator, the feedback variable at that point would be the product of the input and the open-loop transfer function: $GHr(s)$. A closed-loop transfer function, therefore, can be described as the product of the transfer functions in the forward path, divided by unity plus the open-loop transfer function, where unity plus the open-loop transfer function form the characteristic equation of the system.

Block diagrams can be rearranged to give different input-output relationships.

(a)   (b)

Figure 1–4–3.   The comparator: (a) as summation point (b) as subtraction point.

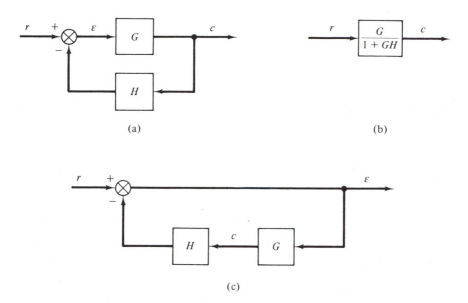

Figure 1–4–4. Common block diagrams.

For example, to find the relationship between the actuating signal ε and the input r, the block diagram of Fig. 1–4–4a can be redrawn with ε as the output, as in Fig. 1–4–4c. Since the forward path transfer function is simply unity, the transfer function for this input-output combination is

$$\frac{\varepsilon}{r} = \frac{1}{1 + GH} \tag{1–4–8}$$

This particular transfer function is of importance in determining steady-state errors.

Some typical block diagram identities are illustrated in Fig. 1–4–5. They may easily be verified by showing that the output from the two equivalent diagrams is the same. A multiple-loop reduction is shown in Fig. 1–4–6. The general procedure is to put the inner loops in the form of Fig. 1–4–4a, reduce the innermost loops first, and proceed sequentially to the outermost loop. The use of the inverse transform function simplifies the reduction of complicated multiple-loop systems but results in the same transfer function.

It can be seen that the reduction process progressively obscures the internal operation of the system and can mask the physical nature, dimensions, and magnitudes of internal variables. It is wise, therefore, to maintain a detailed block diagram of the system for reference and a check at the end of the analysis and design process.

In closing this section, it may be of interest to note that the transfer function concept with its representation of a component, subsystem, or system by a block or box has added the adjectival phrase "black box" to the engineering and operational vocabularies. We often hear such expressions as black box design, black box engineering, and black box maintenance.

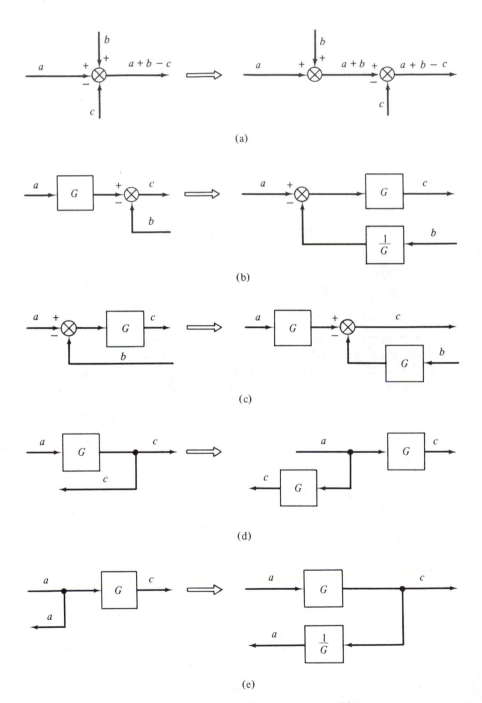

Figure 1–4–5. Some block diagram identities.

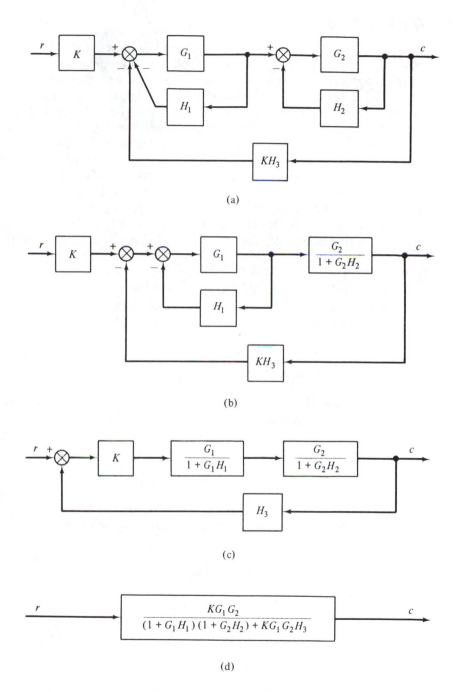

Figure 1–4–6.   Reduction of a multiple-loop block diagram to a single block.

*1–5 SIGNAL FLOW GRAPHS

An alternative to the block diagram is the signal flow graph which, as might be expected, has its advantages and disadvantages. It is convenient for multivariable systems defined by a system of linear algebraic equations in that each node represents a variable. System transfer equations can be obtained directly from the signal flow graph by means of Mason's gain formula without the need for sequential reduction of the block diagram, and the signal flow diagram can be extended with minimum effort to a state diagram that is useful in writing a state-variable model. The signal flow diagram, however, is restricted to linear systems and is constrained by more rigid mathematical relationships than the block diagram. It usually does not give as clear a picture of physical interactions as the block diagram. The block diagram will be the principal representation in this book.

Although the rules for the signal flow graph will be developed by example, there are a few basic rules and definitions that need to be stated first:

1. An input node has only outgoing branches.
2. An output node has only incoming branches.
3. Nodes are arranged in order from left to right, starting with an input node and ending with an output node. There can be multiple input and output nodes.
4. A branch can be traveled only in the direction of the arrow. Otherwise, the algebra is that of the block diagram.

Let us now look at the system (or plant) represented by the following set of linear algebraic equations, where u is the single input, x_3 is the principal output, and x_1 and x_2 are secondary outputs of interest.

$$x_1 = t_{21}x_2 + t_1u \tag{1–5–1a}$$

$$x_2 = t_{12}x_1 - t_{32}x_3 \tag{1–5–1b}$$

$$x_3 = t_{13}x_1 + t_{23}x_2 + t_3u \tag{1–5–1c}$$

The synthesis of the signal flow diagram is shown in Fig. 1–5–1 where (a) is the diagram of the first equation, (b) has the second equation added, and (c) is the complete signal flow graph. Notice in (c) that the input and output nodes have been highlighted by making them single-branch nodes connected to themselves by a unity gain branch.

Since a loop is a path that starts and ends at the same node, we see that there are three loops in Fig. 1–5–1c. Remember that a path can be traveled only in the direction of the arrow. The three loops and their loop gains are

$$L_1 = -t_{23}t_{32}$$
$$L_2 = t_{12}t_{21} \tag{1–5–2}$$
$$L_3 = -t_{13}t_{32}t_{21}$$

Mason's gain formula, which establishes the relationship between an input

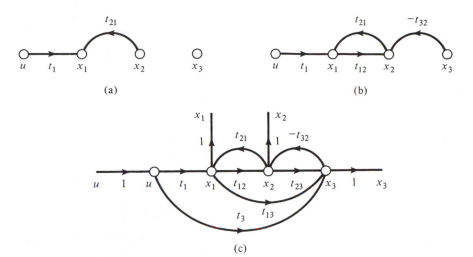

Figure 1–5–1. Signal flow graph synthesis for system of Eq. (1–5–1).

node and an output node, includes a denominator D that can be found from the expression

$D = 1 -$ (the sum of all the individual loop gains)
    $+$ (the sum of the products of the gains of all the possible combinations of two nontouching loops)
    $-$ (the sum of the products of the gains of all the possible combinations of three nontouching loops) $+ \cdots$

Since nontouching loops do not have any common nodes and all three of these loops do indeed have common nodes, they are all touching loops. Therefore, D (often called the determinant of the graph) for this system is given by

$$D = 1 + t_{23}t_{32} - t_{12}t_{21} + t_{13}t_{32}t_{21} \qquad (1\text{–}5\text{–}3)$$

The complete gain formula is given by

$$M = \frac{1}{D}\sum_k P_k \Delta_k \qquad (1\text{–}5\text{–}4)$$

where M, the system gain, is the relationship between the two nodes (variables) of interest; in this case, $M = x_3/u$. P_k is the gain of the kth forward path between u and x_3, and Δ_k is the cofactor of the kth path which is found by removing the loops that touch the kth path from the denominator (determinant) of the graph.

There are three forward paths between u and x_3, and their path gains are

$$P_1 = t_1 t_{12} t_{23}$$
$$P_2 = t_3 \qquad (1\text{–}5\text{–}5)$$
$$P_3 = t_1 t_{13}$$

All three loops touch P_1 and P_3, so that Δ_1 and Δ_3 are both equal to unity. Only L_1 and L_3 touch P_2, however, so that

$$\Delta_2 = 1 - L_2 = 1 - t_{12}t_{21}$$

With the appropriate substitutions into Eq. (1–5–4), the relationship between the input u and the output of interest x_3 is

$$\frac{x_3}{u} = \frac{t_1 t_{12} t_{23} + t_1 t_{13} + t_3 (1 - t_{12} t_{21})}{1 + t_{23} t_{32} + t_{13} t_{32} t_{21} - t_{21} t_{12}} \qquad (1\text{–}5\text{–}6)$$

The gain between u and x_1 can be found by first defining the two forward paths: $P_1 = t_1$ and $P_2 = -t_3 t_{32} t_{21}$. Since P_1 does not touch L_1, $\Delta_1 = 1 + t_{23} t_{32}$, and since P_2 touches all the loops, $\Delta_2 = 1$. Consequently,

$$\frac{x_1}{u} = \frac{t_1(1 + t_{23} t_{32}) - t_3 t_{32} t_{21}}{D} \qquad (1\text{–}5\text{–}7)$$

In applying Mason's gain rule, it is necessary to identify the loops that touch each other in order to write the determinant, and to identify the loops that touch the paths of interest in order to write the cofactors.

The gains in a signal flow graph can be functions of the Laplace variable; in fact, they can even be transfer functions. Consider the system of Fig. 1–4–6a. The corresponding signal flow graph is shown in Fig. 1–5–2. There are three loops, and their respective loop gains are

$$L_1 = -G_1 H_1$$
$$L_2 = -G_2 H_2 \qquad (1\text{–}5\text{–}8)$$
$$L_3 = -K G_1 G_2 H_3$$

L_1 and L_2 do not touch, so

$$D = 1 + G_1 H_1 + G_2 H_2 \, K G_1 G_2 H_3 + G_1 G_2 H_1 H_2 \qquad (1\text{–}5\text{–}9)$$

There is only one forward path between r and c, $K G_1 G_2$. It is touched by all three loops, and its cofactor, therefore, is unity. Consequently, the system gain can be written directly as

$$\frac{c}{r} = \frac{K G_1 G_2}{(1 + G_1 H_1)(1 + G_2 H_2) + K G_1 G_2 H_3} \qquad (1\text{–}5\text{–}10)$$

which is identical to the transfer function obtained by block diagram reduction.

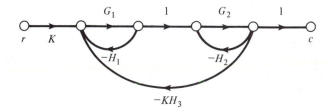

Figure 1–5–2. Signal flow graph of system of Fig. 1–4–6a.

1–6 THE CONCEPT OF STABILITY

The concept of stability is important and should be understood before proceeding further. Let us consider a system, either linear or nonlinear, that is in equilibrium or at rest. First, we shall disturb the system, displacing it from its equilibrium position. If the system eventually returns by itself to its original equilibrium position, we say it is *stable;* if it does not, we say it is *unstable.* If we apply a bounded input to a stable system, the output will pass through a transient phase and settle down to a steady-state response that will be of the same form as, or bounded by, the input. If we apply the same input to an unstable system, the output will never settle down to a steady-state phase; it will increase in an unbounded manner, usually exponentially or with oscillations of increasing amplitude. Typical examples of stable and unstable responses are shown in Fig. 1-6-1.

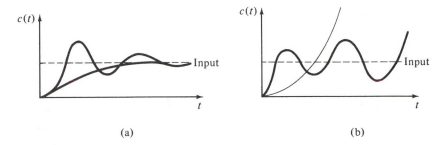

Figure 1-6-1.   Typical system responses: (a) stable systems (b) unstable systems.

A system should be stable to be useful.* Stability can be defined in various ways and is difficult to define for nonlinear systems. The stability of stationary linear systems can be defined without ambiguity in terms of the time response of the system to a unit impulse input. The unit impulse $I(t)$ is defined and shown in Fig. 1-6-2a. Although a true unit impulse cannot be generated physically, it can be satisfactorily approximated when the application time a can be kept small with respect to the characteristic times of the system and its variables. System (and plant) stability can be determined and defined by applying a unit impulse to the system in equilibrium, or at rest, and examining the output as time increases. If the perturbations of the output about the equilibrium value approach zero, as in Fig. 1-6-2b, the system is *stable.* If the output increases indefinitely with time, as in Fig. 1-6-2c, the system is *unstable.* If the output settles down to a bounded oscillation, as in Fig. 1-6-2d, it is *marginally stable.* If the output approaches a constant value other than the original equilibrium condition, as in Fig. 1-6-2e, the system is called *limitedly stable.*† It will be shown subsequently that the stability of a system can be determined from the location of the roots of the characteristic equation in the s plane.

Conditions of instability are a potential hazard in feedback systems and result from unfortunate timing between the feedback variable and the input variable, causing increasing surges in the external power supplied to the system through the control elements. Mathematically, the output of an unstable system

---

* Although the system should be stable, the plant need not necessarily be. Compare the unstable bicycle with the stable tricycle, each by itself and with riders of varying skill.

  † A *limitedly stable* plant or system is really unstable; however, it can be useful and is treated as a special stability condition.

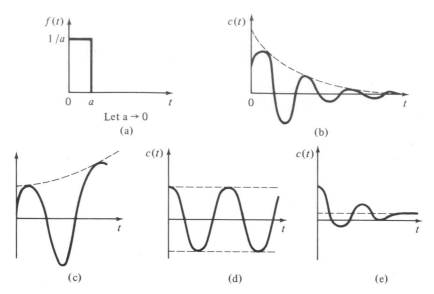

Figure 1-6-2. Response of a linear system to a unit impulse: (a) unit impulse defined (b) stable (c) unstable (d) marginally stable (e) limitedly stable.

would be unbounded. In practice, bounds are imposed by such physical constraints as limit stops, structural limitations, and amplifier and motor saturation. An open-loop system with stable components will always be stable; the same system with a negative feedback loop can become unstable under certain conditions.

1-7 PERFORMANCE AND DESIGN CRITERIA

It is important that the control system designer have a set of significant performance criteria to serve first as design objectives and then as figures of merit in evaluating the performance of the completed design. The definition and selection of the proper criteria are not necessarily simple tasks; often they may be the most significant decisions the designer makes.

The user's requirements should be the basis of such criteria and generally include such factors as accuracy, speed, cost, reliability, ease of operation and maintenance, and definition of the operating environment. The user's requirements may range from the very explicit to the very general. If the former, the designer still has the responsibility to determine the compatibility of the specifications among themselves and the effects of overly stringent specifications upon the complexity and cost of the system. If the user's specifications are general, the designer must develop a set of more definitive specifications for his or her own use.

In the classical approach, the technical criteria can usually be considered satisfied if the answers to the following three questions are all affirmative:

1. Is the system stable?
2. Does the system have acceptable accuracy?
3. Is the transient response acceptable?

The first question is straightforward; a system is either stable or not stable. The second question is relatively straightforward. Accuracy is generally specified in terms of acceptable steady-state errors, where a steady-state error is the difference between the desired output and the actual output after all the transients have died out. The types and magnitudes of inputs to be considered must be specified along with the allowable error for each. Error constants and error coefficients* are sometimes used to specify required accuracy but are directly applicable only to unity feedback systems.

The specification of an acceptable transient response is not straightforward, nor are the associated criteria precise. The three basic characteristics of a transient response are the maximum overshoot of the desired output by the actual output, the initial speed of response to an input, and the time required for the output to reach a steady-state value. These characteristics are normally expressed in the time domain in terms describing the transient response to a specified input, usually a unit step in position, or in the frequency domain in terms describing the steady-state response to a specified input, usually a sinusoid.

Some time-domain performance criteria based on the unit step response are shown in Fig. 1–7–1. The maximum or *peak overshoot* is experienced on the first oscillation and is commonly expressed in terms of a percentage of the

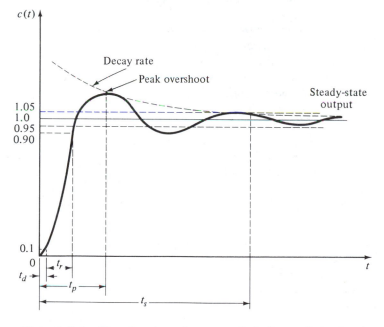

Figure 1–7–1.   Time-domain performance criteria for a unit step input.

---

* See Secs. 5–3 and 5–4.

steady-state output. The initial speed of response can be described by one or more of the following times: the *delay time t_d*, the *rise time t_r*, and the *peak time t_p*. The *settling time t_s* is defined as the time required for the output to reach and remain within a certain percentage of its steady-state value; ±5 percent will be used in this book to define the settling time. Other criteria that are sometimes used are the number of oscillations, the frequency of oscillation, and the decay rate.

Figure 1–7–2 represents a plot of the ratio of the amplitude of the output to the amplitude of the input (referred to as either the system amplitude ratio or magnification) designated by the symbol M versus the common logarithm of the frequency of the input sinusoid ω_f, where ω_f has the dimensions of radians per unit time. Specification of M_p, the *peak magnification* or peak response, primarily describes the maximum overshoot, although it also relates to the settling time. The *peak frequency ω_p* and the *bandwidth ω_b* (*BW*) are principally descriptive of the response speeds, both initial and steady state. The slope of the curve beyond the bandwidth is called the *cutoff rate* and is a measure of the filtering efficiency of a system. The gain margin *GM* and phase margin Φ_m are also used to describe and specify the transient behavior of a system; these two terms will be defined and discussed in Chap. 7.

In the time domain our normal design objectives are small peak overshoot, fast response time (small t_d, t_r, t_p, and t_s), large decay rate, and small number and low frequency of oscillations. In the frequency domain we seek small peak response M_p, small bandwidth in the proper frequency range, and high cutoff rate. Unfortunately, it is not possible to satisfy all these criteria simultaneously; for example, decreasing the bandwidth normally increases the rise time. The control system designer's skill is a measure of how well he or she resolves the conflicts among individual criteria.

These criteria, based on steady-state accuracy and acceptable transient response, are subjective and as such are not too helpful in evaluating and comparing competitive designs. Design techniques using these criteria are essentially trial and error, leading to solutions that are not unique. Finally, these criteria are difficult to apply to multivariable systems. Performance indices are an attempt to remedy these shortcomings.

A *performance index* (PI) is an expression of key variables of the system or control problem that we wish to optimize, usually by minimizing. For a spacecraft designer, minimum fuel or minimum time to rendezvous are apt to

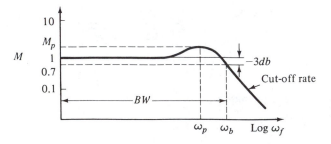

Figure 1–7–2. Frequency-domain performance criteria for a sinusoidal input.

be the basis for a performance index, rather than cost; whereas for an automobile designer, cost may be of primary importance. The performance index J is often an integral of a function of the key variable to be minimized; the function may also include constraints, such as a maximum allowable magnitude for the key variable and any additional factors such as cost.

 There are performance indices (criteria) based on the error that can be used with either the classical or modern approach and are often used in, but not limited to, the evaluation of process control systems. The choice of a particular PI is determined by the objectives of the control system. For example, the *minimum area criterion,* also referred to as the *integrated absolute error* (IAE), is applicable to systems in which deviation of the controlled variable above or below a given value for any appreciable period of time may result in damage or an off-specification product. The IAE is given by

$$J = \int |E(t)|\,dt \qquad\qquad (1\text{–}7\text{–}1)$$

where $E(t)$ is the system error for a specified input, as shown in Fig. 1–7–3.

 The *integral of the squared error* (ISE) is similar to the IAE, but with more weight given to the magnitude of the error. It is found from

$$J = \int E^2(t)\,dt \qquad\qquad (1\text{–}7\text{–}2)$$

Either of these criteria can be time-weighted, imposing penalties for errors that do not die out quickly, leading to the ITAE and ITSE performances indices. Among other PIs that might be appropriate for different systems are the *minimum-disturbance criterion* for restricting output oscillations to avoid resonant excitation of subsequent loops, and the *minimum-amplitude criterion* for avoiding damage when a large error, no matter how short its duration, might result in damage to the plant or product.

 Performance indices are more common to modern control theory and are more complicated than those described above. An optimal control system is designed directly and uniquely by minimizing a performance index. This approach may appear to be more straightforward than the apparent trial-and-error approach of classical control theory, but the designer still has the subjective task of selecting and defining the "best" and most appropriate performance index.

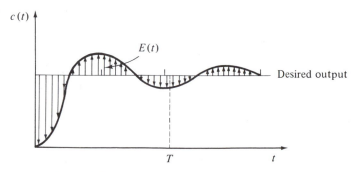

Figure 1–7–3.   The system error for a step input.

1–8 THE ROLE OF THE COMPUTER

Although seldom mentioned in this book, the computer plays an important—in some cases, essential—role in control theory and its application. The computer can be used to perform arithmetic functions (solve equations, etc.), to simulate a plant or a complete control system, or to control a plant. The computer should not be used without checks; one good check is an approximate solution obtained with the classical techniques and without the aid of the computer. Computers can be analog, digital, or hybrid; the last type is a combination of analog and digital that is becoming increasingly popular.

The analog computer is older than the digital; its proponents also consider it closer to nature and more accessible. Continuous variables are represented by voltages, integrations are performed by high-gain dc (operational) amplifiers, and parametric values are controlled by potentiometers. The analog computer can be used to solve differential equations—both linear and nonlinear—but is particularly useful in simulating physical systems or processes since parametric changes can be made by simply turning a potentiometer knob. In addition, physical objects and equipment for which adequate mathematical representations are not possible (such as a human pilot in a flight control system) can be introduced directly into the analog simulation. The speed of the simulation is controllable, as is the relationship between the system variables and the computer voltages representing them. The proper scaling of the amplitude of the voltages and the speed of the simulation can be critical for a successful simulation; the classical analyses are an aid in selecting the scale factors.

The principles of the analog computer and simulation can be illustrated with the simple mechanical system of Fig. 1–8–1a showing a mass suspended by a spring and dashpot and subjected to an input force. We are interested in

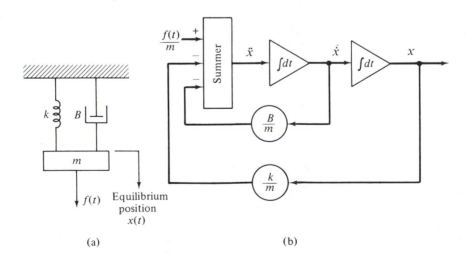

Figure 1–8–1. A simple mechanical system: (a) functional diagram (b) analog simulation.

the displacement of the mass x as measured from the equilibrium position. The linearized, lumped parameter, differential equation is

$$m\ddot{x} + B\dot{x} + kx = f(t) \tag{1-8-1}$$

This equation is rearranged to express the highest derivative as a function of the other terms and coefficients, so that

$$\ddot{x}(t) = \frac{f(t)}{m} - \frac{B}{m}\dot{x} - \frac{k}{m}x \tag{1-8-2}$$

If we know $\ddot{x}$, we can integrate once to find $\dot{x}$, and once more to find x. A schematic of the analog computer or simulation is shown in Fig. 1-8-1b; the diagram of an actual analog solution would differ in detail for practical reasons concerning the equipment. To time-scale this solution, we can select a new time variable τ and let it be equal to at. Substituting in Eq. (1-8-2) yields

$$\ddot{x}(\tau) = \frac{f(\tau/a)}{a^2 m} - \frac{B}{am}\dot{x}(\tau) - \frac{k}{a^2 m}x(\tau) \tag{1-8-3}$$

Time-scaling is accomplished by choosing a value of a and adjusting the potentiometer settings accordingly. If $a = 1$, the solution is a real-time solution*; if a is greater than 1, the solution is slowed down, and if less than 1, speeded up. Nonlinear and nonstationary simulations require the addition of function generators and function multipliers.

The digital computer is excellent for solving complex equations and for obtaining answers in large quantities or with high accuracy. It is particularly useful in finding the complete time response of a system and in optimizing a system; the digital computer is essential to modern control theory. Since the computer is characterized by numbers in and numbers out, finding the effects of changes in parameters, inputs, or operating conditions upon the output(s) is generally difficult even with trade-off and exchange ratio plots. The digital computer can be programmed to perform simulation either alone or in conjunction with the analog computer; the combination is termed a *hybrid* and hopefully exploits the advantages of each type of computer.

The analog computer can be characterized by such terms as continuous variables, small memory, limited accuracy, parallel operation, good and continuous communication with the operator, and scaling of both time and variable magnitudes. Similarly the digital computer characteristics are discrete variables, large memory, high accuracy, series operation, limited communication with the operator, and no scaling. Use the computer of your choice whenever you feel it can help you in your analysis and design. Just remember to keep it in its proper place; be sure it works for you and not the reverse.

Even though the simplest controller is one form of an analog computer, computer control implies to most people the use of a digital computer to control a plant. The computer may be devoted exclusively to the control of a particular system, in which case we speak of *on-line control* or *direct digital control* (DDC). If, on the other hand, the computer is used on a part-time basis (it may be controlling several plants and computing payrolls), we speak of *off-line control*.

* A real-time solution is necessary if humans or equipment are used in the simulation.

1–9 DESIGN PHILOSOPHY AND SEQUENCE

Designing a control system is not a precise or well-defined process; rather, it is a sequence of interrelated events. A typical sequence might be

1. Modeling of the plant
2. Linearization of the plant model
3. Dynamic analysis of the plant
4. Nonlinear simulation of the plant
5. Establishment of the control philosophy and strategy
6. Selection of the performance criteria and indices
7. Design of the controller
8. Dynamic analysis of the complete system
9. Nonlinear simulation of the complete system
10. Selection of the hardware to be used
11. Construction and test of the development system
12. Design of the production model
13. Test of the production model.

This sequence is not rigid, all-inclusive, or necessarily sequential. It is given here to establish a rationale and context for the techniques developed and discussed in the subsequent chapters of this book. In the design sequence there are many decision points and consequently many iterations which require the exercise of engineering judgment. For example, the development of a manageable mathematical model of the plant to be used for the design of a control system may easily involve many iterations among the first four steps plus verification of the model by actual tests if the plant exists and is available.

The control philosophy and strategy step involves answering such questions as whether the plant should be redesigned and whether control should be open loop or closed loop, optimal or suboptimal, linear or nonlinear. This step is obviously important and is a measure of the judgment, creativity, and ingenuity of the designer. Although the remaining steps are self-explanatory and perhaps obvious, their importance to a successful control system should not be overlooked.

Finally, the control designer should rigorously examine the completed design for the existence of a simpler or more acceptable solution before committing it to production.

PROBLEMS

1–1. Using the definitions of Sec. 1–2, describe the systems represented by each of the following differential equations:

(a) $3\dfrac{dx}{dt} - 4x + \dfrac{d^2y}{dt^2} - \dfrac{2dy}{dt} + 3y = 2z$

(b) $\dfrac{d^2y}{dt^2} + 3\left(\dfrac{dy}{dt}\right)^2 + 2ty = 4\dfrac{dx}{dt} + 3x$

(c) $\dfrac{d^2y}{dx^2} + 10\dfrac{d^2y}{dt^2} = 5$

(d) $\dfrac{d^2y}{dt^2} + 5t\dfrac{dy}{dt} + 3y = 10x$

(e) $\dfrac{d^2x}{dt^2} + 6\dfrac{dx}{dt} + 10x = 3y$

1–2. A control system has the block diagram of Fig. P1–2.

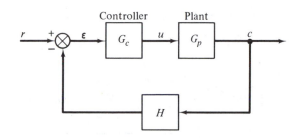

Figure P1–2

(a) Find the system transfer function c/r.

(b) Redraw the block diagram to show the control force u as the output and find the transfer function u/r.

(c) Redraw the diagram with the actuating signal ε as the output and find the transfer function ε/r.

(d) Discuss the similarities and differences among these three transfer functions.

1–3. For each of the block diagrams of Fig. P1–3, find the transfer function $c(s)/r(s)$ by block diagram reduction. Clear any fractions.

1–4. Use block diagram reduction to find the various transfer functions of the system represented by the block diagram of Fig. P1–4.

(a) With $d = 0$, find the system TF, $c(s)/r(s)$.

(b) With $d = 0$, find the control force TF, $u(s)/r(s)$.

(c) With $r = 0$, find the disturbance TF, $c(s)/d(s)$.

1–5. **(a)** Draw a block diagram for the plant represented by the signal flow graph of Fig. 1–5–1c with u as the input and x_3 as the output.

(b) Apply Cramer's rule to the set of simultaneous equations of Eq. (1–5–1) to find the TF, $x_3(s)/u(s)$. It should be identical to that of Eq. (1–5–6).

1–6. Do Prob. 1–5 with x_1 as the output of interest and compare with Eq. (1–5–7).

1–7. In Fig. P1–7 the dashed line is the input and the solid line is the output. For each input-output relationship describe the stability of the corresponding system.

1–8. Fig. P1–8 shows the time response of a particular system to a step input. Find the delay, rise, peak, and settling times, as well as the peak overshoot as a percent of the input and the number of oscillations.

1–9. Fig. P1–9 shows the frequency response of a stable system. Find the peak magnification, peak frequency, and bandwidth. How would you describe the steady-state accuracy of the system for step inputs and for sinusoidal inputs with frequencies less than 1 rad/sec?

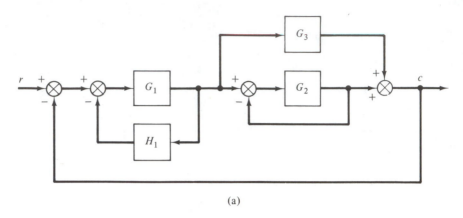

(a)

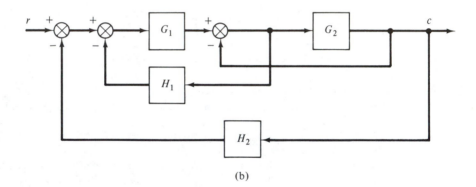

(b)

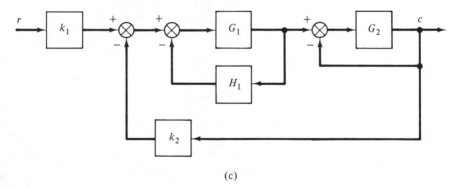

(c)

Figure P1–3

1–10. A linear system is represented by the differential equation $3_3\dddot{c} + 10\ddot{c} + 50\dot{c} + 100c = 5\dot{r} + 100r$. Draw the analog schematic for a real-time solution; remember to avoid differentiation.

1–11. Draw the analog schematic for $A_3\dddot{c} + A_2\ddot{c} + A_1\dot{c} + A_0c = A_0r$ with the response speeded up by a factor of 4; i.e., 1 sec on the computer equals 4 sec of real time.

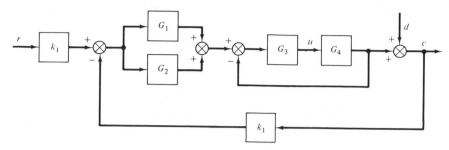

Figure P1–4

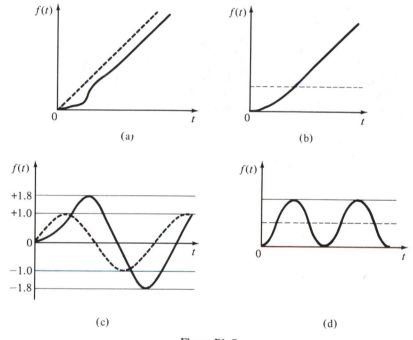

Figure P1–7

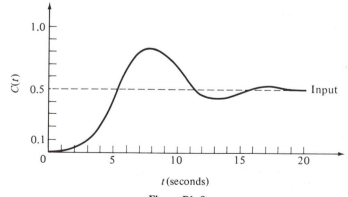

t (seconds)

Figure P1–8

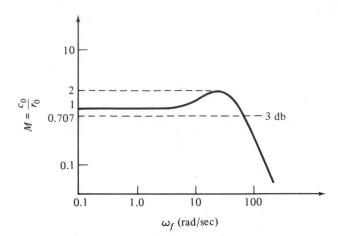

Figure P1-9

2

Linearization and the
Laplace Transformation

2–1 INTRODUCTION

The physical world is inherently nonlinear; unfortunately, general analytical methods are known only for linear systems. Consequently, it is to the designer's advantage, especially for the preliminary analysis and design, to have a linearized mathematical model. A distinction is made between those nonlinearities inherent in a plant or device and those deliberately introduced into a system to improve its performance. Nonlinear control,* although more difficult to analyze, can often be both more effective and cheaper than linear control. Inherent nonlinearities can be, and generally are, kept small by proper design, whereas those deliberately introduced are usually large.

There are several techniques for linearizing inherent nonlinearities. One is the method of first approximations; simply ignore the nonlinearity as is done with the backlash of a gear train in Fig. 2–1–1. A second technique is to restrict the operating range of a physical device so as to avoid regions of nonlinearity. This technique is illustrated in Fig. 2–1–2 for a double-acting spring which is restricted to maximum displacements of $+a$ and $-b$. A third technique, called piecewise linearization, is an extension of the second whereby a nonlinear function is approximated by a series of straight-line sections as is done for the saturating amplifier of Fig. 2–1–3 and the nonlinear spring of Fig. 2–1–4. Multivariable

---

* The on-off control systems of Sec. 10-3 are examples of nonlinear control.

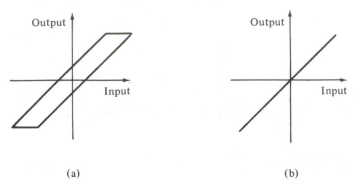

(a)   (b)

Figure 2–1–1.   A gear train: (a) with backlash (b) without backlash.

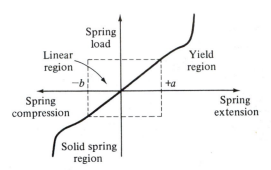

Figure 2–1–2.   A double-acting spring.

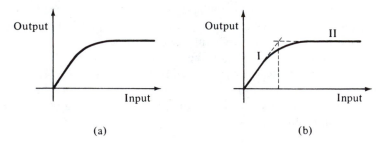

(a)   (b)

Figure 2–1–3.   A saturated amplifier: (a) nonlinear (b) two-piece linearization.

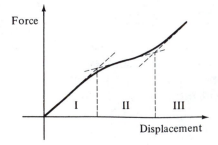

Figure 2–1–4.   Three-piece linearization of a nonlinear spring.

functions and operating data can be linearized by assuming that the perturbations of the variables with respect to a set of reference conditions are small. This useful technique is known as *small perturbation theory* and will be discussed in the next section.

The fact that relatively simple techniques are available for the analysis and design of linear systems is a strong justification in itself for the use of linearization techniques. Moreover, negative feedback tends to reduce inherent nonlinearities, and linear solutions are very useful for scaling nonlinear simulations and checking computer solutions. Finally, the performance of the linearized system is often indicative of the best performance that can be expected since inherent nonlinearities generally degrade performance.

2-2 LINEARIZATION BY SMALL PERTURBATION THEORY

A *linear system* is defined as a system for which the *principle of superposition* holds. This principle states that the output of a linear system in response to the simultaneous application of any number of known inputs is the sum of the individual outputs resulting from the application of each input alone. Consider the system of Fig. 2-2-1. With $r_2 = 0$, apply r_1 to obtain an output c_1. Now set $r_1 = 0$ and apply r_2 to obtain c_2. If r_1 and r_2 are applied simultaneously, the system output c will be the sum of c_1 and c_2 only if the system is linear.

An equation is linear if the dependent variables and their derivatives are of the first degree; that is, they do not appear in combination as products, are not raised to a power other than 1, and are not in nonlinear functions. There are no restrictions on the independent variables. Examples of linear and nonlinear equations follow in which x and y are the dependent variables and are functions of the independent variable t. Consider first the equation

$$\frac{d^2y}{dt^2} + 4x\frac{dy}{dt} + 2x^2 = 3t \tag{2-2-1}$$

The second and third terms are nonlinear, and thus the equation is nonlinear. The equation

$$\frac{d^2y}{dt^2} + 4t\frac{dy}{dt} + 4y = \sin t \tag{2-2-2}$$

is linear; the terms $t\,dy/dt$ and $\sin t$ may appear to be nonlinear, but they are not; it is the *independent* variable t that appears in suspicious circumstances.

A nonlinear system or equation can be linearized by assuming that the perturbations (changes) of the dependent variables from their values at an arbitrary steady-state equilibrium condition are sufficiently small so that products and powers of the perturbed variables and their derivatives can be neglected. The

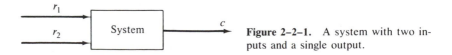

Figure 2-2-1.   A system with two inputs and a single output.

steady-state equilibrium need not be a stationary point. It can be a time-varying trajectory as is the case in Prob. 2–1d, where the equilibrium (reference) condition is a sinusoid.

Consider the simple nonlinear equation

$$\dot{Y} + 2Y - 3X^2 = 0 \tag{2-2-3}$$

where X and Y are functions of time and $\dot{Y}$ denotes differentiation with respect to time. Let us express the dependent variables as

$$X(t) = X_0 + x(t) \qquad Y(t) = Y_0 + y(t) \tag{2-2-4}$$

where X_0 and Y_0 are the values at some equilibrium condition and are constants, and $x(t)$ and $y(t)$ are the time-varying perturbations of X and Y about the equilibrium point. Since Y_0 is constant,

$$\dot{Y}(t) = \frac{d}{dt}[Y_0 + y(t)] = \dot{y}(t) \tag{2-2-5}$$

Substituting Eqs. (2–2–4) and (2–2–5) into Eq. (2–2–3) and expanding the binomial $(X_0 + x)^2$ yields

$$\dot{y} + 2Y_0 + 2y - 3X_0^2 - 6X_0x - 3x^2 = 0 \tag{2-2-6}$$

With the assumption of small perturbations, the nonlinear term containing x^2 is dropped, leaving the linear approximation

$$\dot{y} + 2Y_0 + 2y - 3X_0^2 - 6X_0x = 0 \tag{2-2-7}$$

The *steady-state equation* describing the equilibrium condition can be obtained directly from Eq. (2–2–7) by setting all the time-varying perturbations ($\dot{y}$, y, and x) equal to zero so that

$$2Y_0 - 3X_0^2 = 0 \tag{2-2-8}$$

In spite of the fact that this steady-state equation was obtained from the linearized equation, it is exact and can also be obtained from the original nonlinear equation, Eq. (2–2–3), by setting all differentiations with respect to time equal to zero.

The steady-state equation, Eq. (2–2–8), can now be subtracted from Eq. (2–2–7) to obtain the linear differential equation with constant coefficients that describes the dynamic behavior of the variables with respect to the equilibrium or steady-state condition:

$$\dot{y} + 2y - 6X_0x = 0 \tag{2-2-9}$$

This equation is known as the *dynamic equation*. The initial conditions of the perturbed variables are equal to zero, which facilitates use of the transfer function.

This linearization technique of substitution and binomial expansion can be quite tedious if the nonlinearities are numerous or involved. It may be easier to use a truncated Taylor's series expansion about the equilibrium condition. If Z is a nonlinear function of the two time-dependent variables, X and Y, then the Taylor's series expansion about the equilibrium condition Z_0 can be written as

$$Z_0 + z(t) = Z_0 + \left.\frac{\partial Z}{\partial X}\right|_0 x + \left.\frac{\partial Z}{\partial Y}\right|_0 y + \left.\frac{\partial^2 Z}{\partial X^2}\right|_0 \frac{x^2}{2!} + \left.\frac{\partial^2 Z}{\partial Y^2}\right|_0 \frac{y^2}{2!} + \left.\frac{\partial^2 Z}{\partial X \partial Y}\right|_0 xy + \cdots \tag{2-2-10}$$

where x and y are the time-varying perturbations of the dependent variables with respect to their equilibrium values. If these perturbations are assumed to be small, the series can be truncated by dropping all terms containing products and powers of the perturbed variables. Doing this and removing the equilibrium condition Z_0 from both sides reduces Eq. (2-2-10) to the dynamic equation

$$z(t) \cong \left.\frac{\partial Z}{\partial X}\right|_0 x + \left.\frac{\partial Z}{\partial Y}\right|_0 y \qquad (2-2-11)$$

Extending this procedure to a function of n variables X_n, we can write the generalized expression for a truncated Taylor's series expansion as

$$z(t) = \left.\frac{\partial Z}{\partial X_1}\right|_0 x_1 + \left.\frac{\partial Z}{\partial X_2}\right|_0 x_2 + \cdots + \left.\frac{\partial Z}{\partial X_n}\right|_0 x_n \qquad (2-2-12)$$

This dynamic equation is a linear ordinary differential equation whose coefficients are the partial derivatives of the function evaluated at the equilibrium condition; they are assumed to be constant. The steady-state equation defining the equilibrium conditions must be obtained from the original nonlinear equation.

In using this technique to linearize Eq. (2-2-3), rearrange the equation so as to express the highest derivative as a function of the other variables:

$$\dot{Y} = -2Y + 3X^2 = \dot{Y}(Y, X) \qquad (2-2-13)$$

Expanding Y in a truncated Taylor's series yields

$$\dot{y} = \left.\frac{\partial \dot{Y}}{\partial Y}\right|_0 y + \left.\frac{\partial \dot{Y}}{\partial X}\right|_0 x \qquad (2-2-14)$$

When the function is known, as is the case here, the partials can be evaluated as

$$\left.\frac{\partial \dot{Y}}{\partial Y}\right|_0 = -2|_0 = -2 \qquad (2-2-15a)$$

and

$$\left.\frac{\partial \dot{Y}}{\partial X}\right|_0 = +6X|_0 = 6X_0 \qquad (2-2-15b)$$

Substituting Eqs. (2-2-15) into Eq. (2-2-14) results in the dynamic equation

$$\dot{y} + 2y - 6X_0 x = 0 \qquad (2-2-16)$$

which is identical to Eq. (2-2-9), as is to be expected. The steady-state equation is found from Eq. (2-2-3) or (2-2-13) to be

$$2Y_0 - 3X_0^2 = 0 \qquad (2-2-17)$$

which is identical to Eq. (2-2-8).

Next let us consider the more complicated equation

$$\ddot{X} + Y\dot{X} + X^2 + XY + \frac{\dot{Y}}{Z} + Y^2 Z = 0 \qquad (2-2-18)$$

where all terms but one are nonlinear. Express $\ddot{X}$ as a function of the remaining dependent variables and their derivatives

$$\ddot{X} = -Y\dot{X} - X^2 - XY - \frac{\dot{Y}}{Z} - Y^2 Z \qquad (2\text{-}2\text{-}19)$$

The truncated Taylor's series expansion is

$$\ddot{x} = \frac{\partial \ddot{X}}{\partial \dot{X}}\bigg|_0 \dot{x} + \frac{\partial \ddot{X}}{\partial X}\bigg|_0 x + \frac{\partial \ddot{X}}{\partial \dot{Y}}\bigg|_0 \dot{y} + \frac{\partial \ddot{X}}{\partial Y}\bigg|_0 y + \frac{\partial \ddot{X}}{\partial Z}\bigg|_0 z \qquad (2\text{-}2\text{-}20)$$

Expressions for the partial derivatives can be obtained from Eq. (2–2–19) and can be evaluated at the equilibrium conditions as

$$\frac{\partial \ddot{X}}{\partial \dot{X}}\bigg|_0 = -Y\big|_0 = -Y_0 \qquad (2\text{-}2\text{-}21\text{a})$$

$$\frac{\partial \ddot{X}}{\partial X}\bigg|_0 = -2X_0 - Y_0 \qquad (2\text{-}2\text{-}21\text{b})$$

$$\frac{\partial \ddot{X}}{\partial \dot{Y}}\bigg|_0 = -\frac{1}{Z_0} \; (Z_0 \text{ cannot be zero}) \qquad (2\text{-}2\text{-}21\text{c})$$

$$\frac{\partial \ddot{X}}{\partial Y}\bigg|_0 = -X_0 - 2Y_0 Z_0 \text{ since } \dot{X}_0 = 0 \qquad (2\text{-}2\text{-}21\text{d})$$

$$\frac{\partial \ddot{X}}{\partial Z}\bigg|_0 = -Y_0^2 \text{ since } \dot{Y}_0 = 0 \qquad (2\text{-}2\text{-}21\text{e})$$

Substituting Eq. (2–2–21) back into Eq. (2–2–20) results in the linearized dynamic equation

$$\ddot{x} + Y_0\dot{x} + (2X_0 + Y_0)x + \frac{\dot{y}}{Z_0} + (X_0 + 2Y_0 Z_0)y + Y_0^2 z = 0 \quad (2\text{-}2\text{-}22)$$

The steady-state equation is obtained directly from Eq. (2–2–18) and is

$$X_0^2 + X_0 Y_0 + Y_0^2 Z_0 = 0 \qquad (2\text{-}2\text{-}23)$$

Notice that the magnitudes of the coefficients* of the linearized terms vary with the equilibrium conditions. This is a manifestation of the nonlinearity of the equation. In physical situations the expressions for the coefficients can often be simplified by relationships developed from the steady-state equation(s).

Implicit in and essential to the technique of linearization about an equilibrium condition is the existence of an equilibrium condition and of all the partial derivatives at the equilibrium condition. The partials may be zero but may not be infinite, thus the restriction in Eq. (2–2–21c) that Z_0 cannot be zero. Furthermore, the equations should be in the form of ordinary differential equations before linearization; this means that partial differentiation with respect to any of the dependent variables must be carried out before linearization to preclude losing any essential terms. Finally, how large a small perturbation can be before it introduces unacceptable errors is difficult if not impossible to determine prior

---

* These coefficients are referred to at times as *stability derivatives* or *influence coefficients*.

to linearization; the errors are dependent upon the equations themselves and upon the specific equilibrium conditions of interest.

When equations or functional relationships are impossible or difficult to obtain, it may be possible to obtain linearized equations from operating data. The operating curves for an engine shown in Fig. 2–2–2 indicate that the output shaft speed N is a nonlinear function of both the fuel flow rate Q and the load torque T; i.e., $N = N(Q, T)$. Expanding this generalized function in a truncated Taylor's series expansion about the operating point 0 yields the linearized dynamic equation

$$n(t) = \frac{\partial N}{\partial Q}\bigg|_0 q(t) + \frac{\partial N}{\partial T}\bigg|_0 \hat{t}(t) \tag{2–2–24}$$

where n, q, and $\hat{t}$ are the perturbations of N, Q, and T with respect to the operating point. The partial derivatives can be evaluated by determining the incremental change in N for incremental changes in Q and T alone:

$$\frac{\partial N}{\partial Q}\bigg|_0 = \frac{\Delta N}{\Delta Q}\bigg|_{T=T_0} = C_Q \tag{2–2–25a}$$

and

$$\frac{\partial N}{\partial T}\bigg|_0 = \frac{\Delta N}{\Delta T}\bigg|_{Q=Q_0} = C_T \tag{2–2–25b}$$

Thus, the dynamic behavior of the engine may be approximated by the linear differential equation with constant coefficients

$$n(t) = C_Q q(t) + C_T \hat{t}(t) \tag{2–2–26}$$

C_Q and C_T should be reevaluated if the operating point 0 changes significantly. There is no explicit steady-state equation; values of N_0, Q_0, and T_0 must be obtained directly from the operating data. If the engine must operate in a highly nonlinear region, such as at 0_1, consideration should be given to selecting another engine with less nonlinearity in the desired operating region.

The transport lag approximations of Sec. 9–7 are interesting examples of linearization using a power series expansion in combination with the small perturbation assumption.

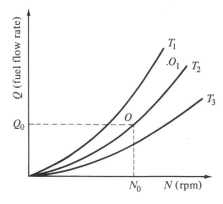

Figure 2–2–2. Engine operating curves.

2–3 THE LAPLACE TRANSFORMATION

The Laplace transformation comes from the area of operational mathematics and is extremely useful in the analysis and design of linear systems. Ordinary differential equations with constant coefficients transform into algebraic equations that can be used to implement the transfer function concept. Furthermore, the Laplace domain is a nice place in which to work. Transfer functions may be easily manipulated, modified, and analyzed. The designer quickly becomes adept in relating changes in the Laplace domain to behavior in the time domain without actually having to solve the system equations. When time domain solutions are required, the Laplace transform method is straightforward. The solution is complete, including both the homogeneous (transient) and particular (steady-state) solutions, and initial conditions are automatically included. Finally, it is easy to move from the Laplace domain into the frequency domain.

The Laplace transform is an evolution from the unilateral Fourier integral and is defined as

$$L[f(t)] = \int_0^\infty f(t)e^{-st}\, dt = f(s) \qquad (2\text{–}3\text{–}1)$$

where $f(s)$ is the Laplace transform of $f(t)$. Conversely, $f(t)$ is the inverse transform of $f(s)$ and can be represented by the relationship

$$L^{-1}[f(s)] = f(t) \qquad (2\text{–}3\text{–}2)$$

The symbol s denotes the Laplace variable and is a complex variable ($\sigma + j\omega$); consequently, s is sometimes referred to as a *complex frequency* and the Laplace domain called the *complex frequency domain*.

Since the definite integral of Eq. (2–3–1) is improper, not all functions are Laplace-transformable; fortunately, the functions of interest to a control system designer usually are. The conditions for existence, proofs of theorems, and other uses of the Laplace transformation can be found in standard works on operational mathematics.

The definition of Eq. (2–3–1) can be used to find the Laplace transform of the functions we are most likely to encounter or use. For example, the Laplace transform of the unit step function of Fig. 2–3–1 is

$$L[u(t)] = \int_0^\infty e^{-st}dt = \left. \frac{-e^{-st}}{s} \right|_0^\infty = \frac{1}{s} \qquad (2\text{–}3\text{–}3)$$

Conversely, the inverse transform of the function $1/s$ is the unit step function

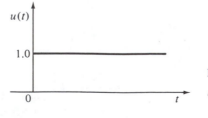

Figure 2–3–1. The unit step function
$u(t)$; $u(t)\begin{cases} 0, t < 0 \\ 1, t > 0 \end{cases}$

$u(t)$, as shown by the relationship

$$L^{-1}\left[\frac{1}{s}\right] = u(t) \tag{2–3–4}$$

Other functions of interest can be transformed, either directly or by use of appropriate theorems, and used to construct a table of transform pairs such as Table 2–3–1. Such tables simplify transformation into and out of the Laplace domain. More elaborate tables exist, but this one is adequate for our needs.

There are certain theorems and properties of the Laplace transformation that are either essential or helpful. The first of these, all without proof, is linearity:

$$L[cf(t)] = cL[f(t)] = cf(s) \tag{2–3–5}$$

TABLE 2–3–1. LAPLACE TRANSFORM PAIRS

$f(t)$	$f(s)$
1. Unit impulse: $I(t)$	1. 1
2. Unit step: $u(t)$	2. $1/s$
3. Ramp: t	3. $1/s^2$
4. $t^n/n!$	4. $\dfrac{1}{s^{n+1}}$
5. e^{-at}	5. $\dfrac{1}{s+a}$
6. $\dfrac{t^n e^{-at}}{n!}$	6. $\dfrac{1}{(s+a)^{n+1}}$
7. $\sin \omega t$	7. $\dfrac{\omega}{s^2+\omega^2}$
8. $\cos \omega t$	8. $\dfrac{s}{s^2+\omega^2}$
9. $e^{-at} \sin \omega t$	9. $\dfrac{\omega}{(s+a)^2+\omega^2}$
10. $e^{-at} \cos \omega t$	10. $\dfrac{(s+a)}{(s+a)^2+\omega^2}$
11. $\dfrac{K}{\omega} e^{-\alpha t} \sin(\omega t + \psi)$	11. $\dfrac{(s+a_0)}{(s+\alpha)^2+\omega^2}$

$$\psi = \tan^{-1}\frac{\omega}{a_0 - \alpha}$$

$$K = [(a_0 - \alpha)^2 + \omega^2]^{1/2}$$

$f(t)$	$f(s)$
12. $\dfrac{d}{dt}[f(t)]$	12. $sf(s) - f(0)$
13. $\dfrac{d^2}{dt^2}[f(t)]$	13. $s^2 f(s) - sf(0) - \dfrac{d}{dt}f(0)$
14. $\displaystyle\int_0^t f(t)\,dt$	14. $\dfrac{f(s)}{s} + \dfrac{1}{s}\displaystyle\int f(t)\,dt\,\big|_0$
15. $f(t-a)u(t-a)$	15. $e^{-as} f(s)$

where c is a constant. The second is superposition:

$$L[c_1 f_1(t) + c_2 f_2(t) + \cdots] = c_1 f_1(s) + c_2 f_2(s) + \cdots \qquad (2\text{–}3\text{–}6)$$

which is helpful in direct and inverse transformations.

The next theorems deal with differentiation and integration. The Laplace transformation of derivatives with respect to time can be shown to be

$$L\left[\frac{d}{dt} f(t)\right] = sf(s) - f(0) \qquad (2\text{–}3\text{–}7a)$$

$$L\left[\frac{d^2}{dt^2} f(t)\right] = s^2 f(s) - sf(0) - \frac{df}{dt}(0) \qquad (2\text{–}3\text{–}7b)$$

$$L\left[\frac{d^n}{dt^n} f(t)\right] = s^n f(s) - s^{n-1} f(0) - \cdots - \frac{d^{n-1}}{dt^{n-1}} f(0) \qquad (2\text{–}3\text{–}7c)$$

where $f(0)$, $df(0)/dt$, etc., are the initial conditions. If the initial conditions are zero, as is generally the case for control system analysis and design, then Eq. (2–3–7c) reduces to

$$L\left[\frac{d^n}{dt^n} f(t)\right] = s^n f(s) \qquad (2\text{–}3\text{–}8)$$

This equation shows that differentiation in the time domain is equivalent to multiplying by s in the Laplace domain. The Laplace transform of an integral is

$$L\left[\int_0^t f(t)\, dt\right] = \frac{1}{s} f(s) + \frac{1}{s} \int f(t)\, dt \bigg|_{t=0} \qquad (2\text{–}3\text{–}9)$$

which with zero initial conditions reduces to

$$L\int_0^t f(t)\, dt = \frac{1}{s} f(s) \qquad (2\text{–}3\text{–}10)$$

Thus, integration in the time domain is equivalent to division by s in the Laplace domain.

The initial value and final value theorems relate the Laplace domain to the time domain at two specific points in time. The *initial value* theorem* states that

$$\lim_{t \to 0+} f(t) = f(0+) = \lim_{s \to \infty} sf(s) \qquad (2\text{–}3\text{–}11)$$

and can be useful at times in the inverse transformation, particularly when the initial conditions are known to be zero. The *final value theorem* states that

$$\lim_{t \to \infty} f(t) = f_{ss} = \lim_{s \to 0} sf(s) \qquad (2\text{–}3\text{–}12)$$

where f_{ss} is the steady-state value of $f(t)$. This theorem is very useful in determining the steady-state accuracy and static sensitivities of a system. The theorem, however, is valid only when the function $sf(s)$ is analytic on the imaginary axis

---

* The distinction between initial values and initial conditions will be discussed in Sec. 2–4.

and in the right half of the s plane, which means that it is valid only for stable systems. Consequently, *never* use the final value theorem to determine the stability of a system or plant.

The two shifting theorems express the translational properties of the Laplace transformation. The first shifting theorem states that

$$L[e^{-at}f(t)] = L[f(t)]_{s \to s+a} = f(s + a) \tag{2-3-13}$$

or conversely that

$$L^{-1}[f(s + a)] = e^{-at}f(t) \tag{2-3-14}$$

Equation (2–3–14) indicates that translation through a units in the Laplace domain results in multiplication by e^{-at} in the time domain. This theorem can be useful in returning to the time domain from the Laplace domain. In Table 2–3–1 transform pairs 5, 6, 9, 10, and 11 illustrate the use of this theorem.

The second shifting theorem states that translation through a units in the time domain is equivalent to multiplication by e^{-as} in the Laplace domain. This theorem can be expressed in many forms, two of which are

$$L[f(t - a)u(t - a)] = e^{-as}L[f(t)] = e^{-as}f(s) \tag{2-3-15}$$

and

$$L[f(t)u(t - a)] = e^{-as}L[f(t + a)] = e^{-as}f(s + a) \tag{2-3-16}$$

This theorem is useful in transforming delayed inputs and signals such as transport lags and piecewise continuous inputs that are represented by analytic functions. As an illustration of the latter use, consider the pulse of Fig. 2–3–2a. It can be regarded as a magnitude $1/a$ multiplied by a function of unity magnitude that exists only when $0 < t < a$. This multiplying function is known as the filter function and, as shown in Figs. 2–3–2b and c, can be represented by the sum

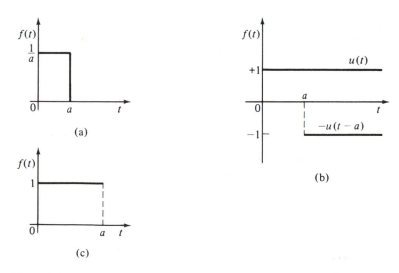

Figure 2–3–2.   (a) A pulse of unity strength (b) two unit steps of the filter function (c) the filter function itself.

of two unit step functions: $u(t) - u(t - a)$. The pulse can now be expressed as

$$f(t) = \frac{1}{a}[u(t) - u(t - a)] \tag{2-3-17}$$

Its Laplace transform is

$$f(s) = \frac{1}{a}\{L[u(t)] - L[u(t - a)]\} = \frac{1}{as}(1 - e^{-as}) \tag{2-3-18}$$

By definition a unit impulse $I(t)$ is the limiting case of this particular pulse as the time of application approaches zero. If we let a approach zero in Eq. (2-3-17), $f(s)$ becomes equal to 0/0 or indeterminate. Using l'Hôpital's rule the transform of a unit impulse becomes

$$L[I(t)] = \lim_{a \to 0} \frac{(d/da)(1 - e^{-as})}{(d/da)(as)} = 1 \tag{2-3-19}$$

The final theorem to be mentioned equates the product of two Laplace transforms to the Laplace transform of the *convolution integral**:

$$L[f(t)]\cdot L[g(t)] = L\left[\int_0^t f(\tau)g(t - \tau)\,d\tau\right] = L\left[\int_0^t g(\tau)f(t - \tau)\,d\tau\right] \tag{2-3-20}$$

where τ is a dummy time variable. One use of this theorem is in the determination of the time response of a linear system to an input that is not Laplace-transformable. Such an input is usually an experimentally obtained input. Suppose the system of Fig. 2-3-3a with a transfer function $W(s)$ is subjected to the arbitrary input $r(t)$ shown in Fig. 2-3-3b. From the definition of the transfer function and Eq. (2-3-20),

$$c(s) = W(s)r(s) = L\left[\int_0^t r(\tau)w(t - \tau)\,d\tau\right] \tag{2-3-21}$$

and the time response is

$$c(t) = L^{-1}[c(s)] = \int_0^t r(\tau)w(t - \tau)\,d\tau \tag{2-3-22}$$

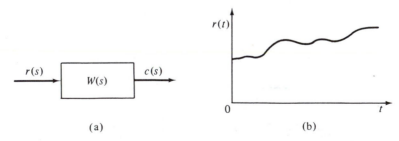

(a)   (b)

Figure 2–3–3. (a) Linear system with known transfer function $W(s)$ (b) arbitrary input $r(t)$.

---

* Notice that $L^{-1}[f(s)\cdot g(s)]$ is *not* equal to $f(t)\cdot g(t)$.

The function $w(t)$ is called the *weighting function;* it is the inverse transform of the transfer function or the experimentally determined response of the system to a unit impulse output. With $w(t)$ known, $c(t)$ can be found from Eq. (2–3–22), either graphically or by representing $r(t)$ as a series of pulses.

2–4 THE LAPLACE AND TIME DOMAINS

Let us now apply the Laplace transformation to ordinary linear differential equations with constant coefficients and zero initial conditions to obtain transfer functions and then solutions in the time domain; the latter requiring inverse transformation techniques. As the first example, consider the equation

$$\frac{d^2y}{dt^2} + 7\frac{dy}{dt} + 12y = 6\frac{dx}{dt} + 12x \qquad (2\text{–}4\text{–}1)$$

where $y(t)$ and $x(t)$ are the output and input, respectively. Using the theorems and the table of transform pairs of the preceding section, the transformed equation with zero initial conditions is

$$s^2y(s) + 7sy(s) + 12y(s) = 6sx(s) + 12x(s) \qquad (2\text{–}4\text{–}2)$$

This is now an algebraic equation and can be written as

$$(s^2 + 7s + 12)y(s) = 6(s + 2)x(s) \qquad (2\text{–}4\text{–}3)$$

Solving for the ratio of the transformed output to the transformed input yields the transfer function of the system:

$$W(s) = \frac{y(s)}{x(s)} = \frac{6(s + 2)}{s^2 + 7s + 12} = \frac{6(s + 2)}{(s + 3)(s + 4)} \qquad (2\text{–}4\text{–}4)$$

The system of Eq. (2–4–1) can now be represented by the block diagram of Fig. 2–4–1. If we wish to know the time response of the system $y(t)$, we must first specify the input. A common input is the unit step in position; e.g., $x(t) = u(t)$ with a Laplace transform $x(s) = 1/s$. Since $y(s) = W(s)x(s)$, Eq. (2–4–4) and the expression for $x(s)$ result in

$$y(s) = \frac{6(s + 2)}{s(s + 3)(s + 4)} \qquad (2\text{–}4\text{–}5)$$

The equation for the inverse transformation of a function in the Laplace domain is the complex integral

$$f(t) = \frac{1}{2\pi j} \int_{\sigma-j\infty}^{\sigma+j\infty} f(s)e^{st}\, dt \qquad (2\text{–}4\text{–}6)$$

Rather than perform this integration, it is simpler to express the right-hand side of Eq. (2–4–5) as the sum of partial fractions

$$y(s) = \frac{6(s + 2)}{s(s + 3)(s + 4)} = \frac{C_1}{s} + \frac{C_2}{s + 3} + \frac{C_3}{s + 4} \qquad (2\text{–}4\text{–}7)$$

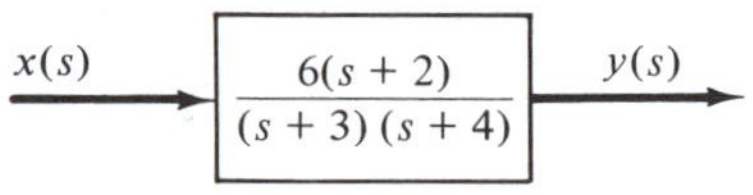

$x(s) \longrightarrow \boxed{\dfrac{6(s + 2)}{(s + 3)\,(s + 4)}} \longrightarrow y(s)$

Figure 2–4–1. Block diagram of the system of Eq. (2–4–1).

and then use a table of transform pairs to invert each fraction individually. From Table 2–3–1 we can invert the expression for $y(s)$ to obtain

$$y(t) = C_1 + C_2 e^{-3t} + C_3 e^{-4t} \qquad (2\text{–}4\text{–}8)$$

The constant coefficients can be evaluated in several ways. The first is the method of undetermined coefficients. Multiply both sides of Eq. (2–4–7) by the lowest common denominator and collect terms to obtain the identity

$$6s + 12 = (C_1 + C_2 + C_3)s^2 + (7C_1 + 4C_2 + 3C_3)s + 12C_1 \quad (2\text{–}4\text{–}9)$$

For this identity to be true, the coefficient of each term on one side must be identical to the corresponding coefficients on the other side. Equating coefficients results in a set of three simultaneous equations:

$$C_1 + C_2 + C_3 = 0$$
$$7C_1 + 4C_2 + 3C_3 = 6 \qquad (2\text{–}4\text{–}10)$$
$$12C_1 = 12$$

Solving this set, we find $C_1 = 1$, $C_2 = +2$, and $C_3 = -3$. Substituting these values into Eq. (2–4–8) produces the time response

$$y(t) = 1 + 2e^{-3t} - 3e^{-4t} \qquad (2\text{–}4\text{–}11)$$

which is sketched in Fig. 2–4–2.

At this point let us explore the difference between *initial conditions* (ICs) and *initial values* (IVs). Initial conditions exist prior to the application of the input at $t = 0$, i.e., at $t = 0-$. Initial values, on the other hand, appear immediately after the application of the input, i.e., at $t = 0+$. The initial conditions and initial values are not necessarily identical, as can be seen by an examination of the example we have just worked. In the statement of the problem, the initial conditions were set equal to zero, which means that $y(0-) = \dot{y}(0-) = 0$. The time solution for a unit step input is given by Eq. (2–4–11). If we set $t = 0+$, i.e., let t approach zero from positive time, we find that the first initial value $y(0+)$ is equal to zero and does indeed match the first initial condition that $y(0-)$ is zero. If, however, we differentiate Eq. (2–4–11) with respect to time to obtain an expression for $y(t)$ and then let $t = 0+$, we find the second initial value $y(0+)$ to be equal to $+6$. This second initial value obviously does not match the second initial condition, indicating a discontinuity in the derivative at $t = 0$.

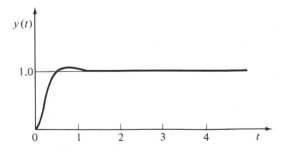

Figure 2–4–2. Time response of the system of Fig. 2–4–1 to a unit step input.

The degree of match between *zero initial conditions* and the corresponding initial values can be determined by an examination of the fraction representing the transformed variable of interest. If the order of the denominator is n and that of the numerator is m, then $(n - m - 1)$ initial values will be equal to zero. For example, in Eq. (2–4–5), which is the Laplace transform of y, the order of the denominator is three and that of the numerator is one. Since $(n - m - 1)$ is unity, we should expect only the first initial value to be zero, as is the case. This matching criterion is valid only when all the initial conditions are zero.

In Eq. (2–4–7), the constants can also be evaluated by application of the Heaviside expansion theorems. This is a simpler technique that can also be used for repeated factors. To find C_1, multiply Eq. (2–4–7) by the denominator of C_1, s in this case, to obtain

$$sy(s) = \frac{6(s + 2)}{(s + 3)(s + 4)} = C_1 + \frac{sC_2}{s + 3} + \frac{sC_3}{s + 4} \qquad (2\text{–}4\text{–}12)$$

Now set the multiplying factor, s in this case, equal to zero. The C_2 and C_3 terms become equal to zero, so that

$$C_1 = [sy(s)]_{s=0} = \left[\frac{6(s + 2)}{(s + 3)(s + 4)}\right]_{s=0} = 1 \qquad (2\text{–}4\text{–}13)$$

To find C_2, multiply each term by the denominator of C_2 and then set the denominator $(s + 3)$ to zero, which in turn makes the C_1 and C_3 terms zero, leaving

$$C_2 = [(s + 3)y(s)]_{s=-3} = \left[\frac{6(s + 2)}{s(s + 4)}\right]_{s=-3} = 2 \qquad (2\text{–}4\text{–}14)$$

Similarly,

$$C_3 = [(s + 4)y(s)]_{s=-4} = \left[\frac{6(s + 2)}{s(s + 3)}\right]_{s=-4} = -3 \qquad (2\text{–}4\text{–}15)$$

A Heaviside expansion theorem involving differentiation can be used for repeated factors. As an example, let

$$y(s) = \frac{s + 4}{s(s + 2)^2} \qquad (2\text{–}4\text{–}16)$$

When expanding repeated factors, include a partial fraction for each included power of the repeated factor so that

$$y(s) = \frac{s + 4}{s(s + 2)^2} = \frac{C_1}{s} + \frac{C_2}{s + 2} + \frac{C_3}{(s + 2)^2} \qquad (2\text{–}4\text{–}17)$$

The constants* C_1 and C_3 can be evaluated as in the previous example as

$$C_1 = [sy(s)]_{s=0} = \left[\frac{s + 4}{(s + 2)^2}\right]_{s=0} = 1 \qquad (2\text{–}4\text{–}18)$$

---

* These constants are sometimes called *residues*.

and

$$C_3 = [(s + 2)^2 y(s)]_{s=-2} = \left[\frac{s + 4}{s}\right]_{s=-2} = -1 \qquad (2\text{-}4\text{-}19)$$

To find C_2, however, it is necessary to differentiate the expression $(s + 2)^2 y(s)$ with respect to s and then set s equal to -2. Thus

$$C_2 = \frac{1}{1!}\frac{d}{ds}\left[\frac{s + 4}{s}\right]_{s=-2} = \frac{1}{1!}\left[\frac{1}{s} - \frac{s + 4}{s^2}\right]_{s=-2} = -1 \qquad (2\text{-}4\text{-}20)$$

With the constants evaluated, Eq. (2-4-17) becomes

$$y(s) = \frac{1}{s} - \frac{1}{s + 2} - \frac{1}{(s + 2)^2} \qquad (2\text{-}4\text{-}21)$$

and

$$y(t) = 1 - e^{-2t}(1 + t) \qquad (2\text{-}4\text{-}22)$$

Notice that repeated factors introduce t explicitly into the coefficients of the time response. If the repeated factor were a triple rather than a double factor, a second differentiation would be required to find the added constant, and t^2 would appear in one or more of the time response coefficients. Repeated factors to the nth power require n partial fractions and $(n - 1)$ differentiations* in order to evaluate all the coefficients.

Repeated roots do not have to be real, as in this example, and do not commonly exceed two. The method of undetermined coefficients can also be used to find the constants of the inverse transformation with repeated factors.

Since we have restricted ourselves to differential equations with real coefficients, complex factors will always appear in conjugate pairs. Although the first-order complex factors can be treated in the same manner as the real factors of the preceding example, it may be simpler to leave conjugate pairs in quadratic form when expanding in partial fractions, as in the following example.

$$y(s) = \frac{13}{s(s^2 + 4s + 13)} = \frac{C_1}{s} + \frac{C_2 s + C_3}{s^2 + 4s + 13} \qquad (2\text{-}4\text{-}23)$$

Do not forget to include an s term in the numerator of the quadratic fraction. The constant C_1 is easily evaluated as

$$C_1 = \left[\frac{13}{s^2 + 4s + 13}\right]_{s=0} = 1 \qquad (2\text{-}4\text{-}24)$$

and Eq. (2-4-23) can be written as

$$y(s) = \frac{13}{s(s^2 + 4s + 13)} = \frac{1}{s} + \frac{C_2 s + C_3}{s^2 + 4s + 13} \qquad (2\text{-}4\text{-}25)$$

The constants C_2 and C_3 can now be found from the method of undetermined coefficients to be equal to -1 and -4, respectively. Substituting these values into Eq. (2-4-25) and expressing the quadratic factor as the sum of two squares,

---

* The jth differentiation must be divided by the j factorial, as in Eq. (2-4-20).

by completing the square of the first two terms, yields

$$y(s) = \frac{1}{s} - \frac{s + 4}{(s + 2)^2 + 3^2} \tag{2–4–26}$$

Both fractions appear in Table 2–3–1 in pairs 1 and 11, so that

$$y(t) = 1 - 1.2e^{-2t} \sin(3t + \tan^{-1} 1.5) \tag{2–4–27}$$

The constants C_2 and C_3 in Eq. (2–4–25) can also be evaluated by subtracting any known fractions from the complete expression and then dividing the resulting fraction by the already evaluated factors, s in this case. Equation (2–4–25) becomes

$$\frac{C_2 s + C_3}{(s^2 + 4s + 13)} = \frac{13}{s(s^2 + 4s + 13)} - \frac{1}{s}$$

$$= \frac{-s^2 - 4s}{s(s^2 + 4s + 13)} = \frac{-(s + 4)}{s^2 + 4s + 13} \tag{2–4–28}$$

The initial value theorem sometimes can provide a shortcut to determining one or more of the constants. For example, the inverse of Eq. (2–4–17) can be written as

$$y(t) = C_1 + C_2 e^{-2t} + C_3 t e^{-2t} \tag{2–4–29}$$

The coefficients C_1 and C_3 are evaluated as $+1$ and -1, respectively, without differentiation, so that

$$y(t) = 1 + C_2 e^{-2t} - t e^{-2t} \tag{2–4–30}$$

Using the initial value theorem we find that y at $t = 0$ is zero, and Eq. (2–4–30) becomes

$$0 = 1 + C_2 \tag{2–4–31}$$

which is solved for $C_2 = -1$.

Finally, there is a graphical technique based on the representation of complex functions by vectors. It is described in App. A and can be useful in estimating the relative magnitudes of the transient modes.

In each of the preceding examples $y(s)$ is a proper fraction with the power of s in the numerator less than that in the denominator. This is the case for functions representing physical systems; mathematical manipulations, however, may result in improper fractions. In this event divide the numerator by the denominator to obtain a series of terms in s that will transform directly,* plus a proper fraction that can be expanded and evaluated by the methods of this section.

2–5 NONZERO INITIAL CONDITIONS

In the preceding section, the initial conditions were assumed equal to zero. Zero initial conditions imply that the differential equations represent the dynamic behavior of the system with respect to an equilibrium condition and that the

---

* The inverse transform of s is a unit doublet, and of s^2 a unit triplet. These singularity functions are without physical significance.

system is in equilibrium at time t equal to zero when an input is applied. Nonzero initial conditions indicate that the system is not in equilibrium at $t = 0$. Nonzero initial conditions may be deliberately introduced to provide a specific input or forcing function, or they may be the result of an input or disturbance introduced prior to t equal to zero.

Consider the system of Eq. (2–4–1), but this time with nonzero initial conditions, i.e., $\dot{y}(0) \neq 0$ and $y(0) \neq 0$. Now the Laplace transform of this equation becomes

$$s^2 y(s) - sy(0) - \dot{y}(0) + 7sy(s) - 7y(0) + 12y(s) = 6(s + 2)x(s) \quad (2\text{–}5\text{–}1)$$

Rearranging this equation as

$$(s^2 + 7s + 12)y(s) = 6(s + 2)x(s) + (s + 7)y(0) + \dot{y}(0) \quad (2\text{–}5\text{–}2)$$

and solving for $y(s)$ yields

$$y(s) = \frac{6(s + 2)x(s)}{s^2 + 7s + 12} + \frac{(s + 7)y(0) + \dot{y}(0)}{s^2 + 7s + 12} \quad (2\text{–}5\text{–}3)$$

We recognize the coefficient of $x(s)$ as the transfer function of the system; the additional term on the right is sometimes called the *initial-conditions operator*.

Since the system is linear, we can use the principle of superposition to obtain the total system response to a specified input plus nonzero initial conditions. In the preceding section we found the time response to a unit step input with zero initial conditions to be

$$y(t) = 1 + 2e^{-3t} - 3e^{-4t} \quad (2\text{–}5\text{–}4)$$

If we let $\dot{y}(0) = 1$ and $y(0) = 1$ and assume $x(t) = 0$, then $y(s)$ due to the initial conditions alone is

$$y(s) = \frac{s + 8}{s^2 + 7s + 12} = \frac{C_1}{s + 3} + \frac{C_2}{s + 4} \quad (2\text{–}5\text{–}5)$$

Since

$$C_1 = \left[\frac{s + 8}{s + 4}\right]_{s=-3} = 5 \quad (2\text{–}5\text{–}6a)$$

and

$$C_2 = \left[\frac{s + 8}{s + 3}\right]_{s=-4} = -4 \quad (2\text{–}5\text{–}6b)$$

$$y(t) = 5e^{-3t} - 4e^{-4t} \quad (2\text{–}5\text{–}7)$$

Equation (2–5–7) is the response due to the initial conditions alone, and Eq. (2–5–4) is the response due to the unit step input alone. The complete response is the sum of the two, or

$$y(t) = 1 + 7e^{-3t} - 7e^{-4t} \quad (2\text{–}5\text{–}8)$$

The complete response can also be obtained by substituting values for $x(s)$ and the initial conditions directly into Eq. (2–5–2) and solving:

$$y(s) = \frac{s^2 + 14s + 12}{s(s + 3)(s + 4)} = \frac{C_1}{s} + \frac{C_2}{s + 3} + \frac{C_3}{s + 4} \quad (2\text{–}5\text{–}9)$$

The constants are evaluated as $C_1 = 1$, $C_2 = +7$, and $C_3 = -7$, so that

$$y(t) = 1 + 7e^{-3t} - 7e^{-4t} \qquad (2\text{-}5\text{-}10)$$

Comparing Eq. (2-5-8) with Eq. (2-5-4), we see that the presence of initial conditions does not increase or decrease the number of time-dependent terms (the transient modes) or affect the constant term (the steady-state response). Consequently, it is convenient and customary to assume that initial conditions are zero, introducing nonzero initial conditions only when necessary and then treating them as additional inputs.

*2-6 MULTIVARIABLE (MIMO) PLANTS AND SYSTEMS

Although the examples of the preceding sections were single variable (SISO) systems, the techniques discussed can be extended to multivariable systems. Consider a plant described by the following set of differential equations:

$$\dot{y}_1 + 2y_1 + \dot{y}_2 + y_2 = u_1 \qquad (2\text{-}6\text{-}1a)$$

$$5y_1 + \ddot{y}_2 + 9\dot{y}_2 + 8y_2 = 2u_1 + u_2 \qquad (2\text{-}6\text{-}1b)$$

where y_1 and y_2 are the two output (and state) variables and u_1 and u_2 are the two input (control) variables. With two inputs and two outputs, there will be four transfer functions, all with the same characteristic function (denominator). With zero initial conditions the transformed equations are

$$(s + 2)y_1(s) + (s + 1)y_2(s) = u_1(s) \qquad (2\text{-}6\text{-}2a)$$

$$5y_1(s) + (s^2 + 9s + 8)y_2(s) = 2u_1(s) + u_2(s) \qquad (2\text{-}6\text{-}2b)$$

The coefficient matrix for this set of equations is

$$\mathbf{F} = \begin{vmatrix} (s + 2) & (s + 1) \\ 5 & (s + 1)(s + 8) \end{vmatrix} \qquad (2\text{-}6\text{-}3)$$

Since the determinant of the coefficient matrix $\mathbf{F}$ is the characteristic function

$$D(s) = |\mathbf{F}| = (s + 1)(s + 1.26)(s + 8.74) \qquad (2\text{-}6\text{-}4)$$

and since the characteristic equation is $D(s) = 0$, we see that there are three negative real roots, at $s = -1$, -1.26, and -7.48. By applying Cramer's rule,* the four transfer functions are found to be

$$G_{11} = \frac{y_1}{u_1} = \frac{(s + 1)(s + 6)}{(s + 1)(s + 1.26)(s + 8.74)} \qquad (2\text{-}6\text{-}4a)$$

$$G_{12} = \frac{y_1}{u_2} = \frac{-(s + 1)}{(s + 1)(s + 1.26)(s + 8.74)} \qquad (2\text{-}6\text{-}4b)$$

$$G_{21} = \frac{y_2}{u_1} = \frac{2s - 1}{(s + 1)(s + 1.26)(s + 8.74)} \qquad (2\text{-}6\text{-}4c)$$

$$G_{22} = \frac{y_2}{u_2} = \frac{s + 2}{(s + 1)(s + 1.26)(s + 8.74)} \qquad (2\text{-}6\text{-}4d)$$

* See App. C for a description of Cramer's rule and the matrix operations to follow.

Notice that all four transfer functions have the same denominator, as is to be expected, and that in the numerator of the first two transfer functions there is a common factor $(s + 1)$ that can be used to cancel the $(s + 1)$ factor in the characteristic function (the denominator). If this were done, the resulting transfer functions would give the impression of being second-order with only two roots, whereas the characteristic function and the other two transfer functions show that there indeed is a third root. So, be careful when canceling and do not forget that the canceled root is still there, even though suppressed in two of the transfer functions.

It might be interesting to write the differential equations represented by each of the transfer functions shown above. With cancellation of the common factor and the knowledge that multiplying by s in the Laplace domain is equivalent to taking a derivative in the time domain, the four equations are

$$\ddot{y}_1 + 10\dot{y}_1 + 11y_1 = \dot{u}_1 + 6u_1 \tag{2-6-5a}$$

$$\ddot{y}_1 + 10\dot{y}_1 + 11y_1 = -u_2 \tag{2-6-5b}$$

$$\ddot{y}_2 + 11\ddot{y}_2 + 21\dot{y}_2 + 11y_2 = 2\dot{u}_1 - u_1 \tag{2-6-5c}$$

$$\ddot{y}_2 + 11\ddot{y}_2 + 21\dot{y}_2 + 11y_2 = \dot{u}_2 + 2u_2 \tag{2-6-5d}$$

We see that we have apparently* *decoupled* the two output variables and the two input variables and now have four SISO relationships in the form of the four transfer functions. These transfer function relationships are shown in Fig. 2–6–1a.

It may be convenient to represent this set of transfer functions by a vector matrix equation:

$$\mathbf{y}(s) = \mathbf{G}_p(s)\,\mathbf{u}(s) \tag{2-6-6}$$

where $\mathbf{y}(s)$ and $\mathbf{u}(s)$ are 2×1 vectors defined as

$$\mathbf{y}(s) = \begin{vmatrix} y_1 \\ y_2 \end{vmatrix} \qquad \mathbf{u}(s) = \begin{vmatrix} u_1 \\ u_2 \end{vmatrix} \tag{2-6-7a}$$

and $\mathbf{G}_p(s)$ is the 2×2 matrix

$$\mathbf{G}_p(s) = \begin{vmatrix} G_{11} & G_{12} \\ G_{21} & G_{22} \end{vmatrix} \tag{2-6-7b}$$

The transfer function matrix relationship for this plant is shown in Fig. 2–6–1b, where now any number of inputs and outputs could be represented.

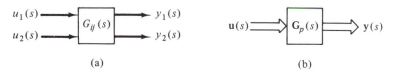

(a)   (b)

Figure 2–6–1. A multivariable plant: (a) transfer function representation (b) transfer matrix representation.

* Actually, there is still dynamic coupling.

It is also possible to represent the original set of two differential equations by the single vector matrix equation

$$\mathbf{F}\mathbf{y}(s) = \mathbf{B}\mathbf{u}(s) \qquad (2\text{–}6\text{–}8a)$$

where **F** and **B** are the 2 × 2 matrices

$$\mathbf{F} = \begin{vmatrix} (s + 2) & (s + 1) \\ 5 & (s + 1)(s + 8) \end{vmatrix} \qquad (2\text{–}6\text{–}8b)$$

$$\mathbf{B} = \begin{vmatrix} 1 & 0 \\ 2 & 1 \end{vmatrix} \qquad (2\text{–}6\text{–}8c)$$

If both sides of Eq. (2–6–8a) are multiplied by the inverse of the matrix **F** (**F**$^{-1}$), where **F**$^{-1}$ = adj **F**/$D(s)$ [where $D(s)$ is the determinant of **F**], the equation becomes

$$\mathbf{y}(s) = \mathbf{F}^{-1}\mathbf{B}\mathbf{u}(s) \qquad (2\text{–}6\text{–}9)$$

Comparison with Eq. (2–6–6) shows that

$$\mathbf{F}^{-1}\mathbf{B} = G_p(s) \qquad (2\text{–}6\text{–}10)$$

and that Eqs. (2–6–6) and (2–6–9) are equivalent and that Eq. (2–6–10) represents a method other than Cramer's rule (but requiring more effort) for finding the transfer functions of multivariable plants and systems.

Transfer function matrices can be manipulated and combined in the same manner as transfer functions, as, for example, is done in Fig. 2–6–2. Other than as a shorthand representation of more than one transfer function, transfer function matrices are not particularly useful, however. When the number of variables becomes sufficiently large so that the transfer functions become unwieldy, it would probably be better to switch to the state-variable approach and bypass the transfer function matrices, or to restrict their use to aiding the transition between state-variable representation and transfer functions.

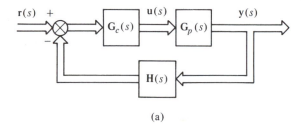

(a)

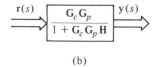

(b)

Figure 2–6–2.   A multivariable system: (a) general block diagram (b) single block representation.

PROBLEMS

2–1. Identify the nonlinear terms in the following differential equations:

(a) $\dot{X} + X^2 + 2XY + Y^2 + 3Y = 0$

(b) $\dot{X} + \dfrac{X}{Z} + \dfrac{Y^2}{Z} - Z = 0$

(c) $\ddot{X} + \dot{X}\cos X + XY + \dot{Y} = 0$

(d) $4\dot{Y} - \dfrac{XZ}{Y} + Z = \cos t$

(e) $\ddot{X} + 3\dot{X}^2 + 3X + Y^3 = 0$

2–2. Linearize the equations of Prob. 2–1 with respect to an arbitrary equilibrium condition by direct substitution and then by application of the small-perturbation assumption. Separate and identify the steady-state and dynamic equations.

2–3. Linearize the equations of Prob. 2–1 by expansion in a truncated Taylor's series expansion. Write the steady-state and dynamic equations in each case.

2–4. The volume of a spherical sector is given by the expression $V = \frac{2}{3}\pi R^2 h$.

(a) Linearize this expression with respect to an arbitrary R_0, h_0, and V_0 and write the reference expression and the linearized expression.

(b) If $R_0 = 10$ in. (0.25 m) and $h_0 = 4$ in. (0.1 m), find the reference volume V_0.

(c) If R and h are each increased by 10 percent, find the new volume using both the linearized and nonlinear expressions and determine the percentage error incurred by using the former.

(d) Do (c) for a 5 percent increase in the two dimensions.

2–5. One of the equations describing a singly excited magnetic field transducer contains the nonlinear term

$$\frac{1}{2}I^2(t)\frac{\partial L(t)}{\partial X(t)}$$

where

$$L(t) = L_0\left(\frac{d + X_0}{d + X(t)}\right)$$

where X_0 and L_0 are the values of $X(t)$ and $L(t)$ at an arbitrary equilibrium condition and d is a constant.

(a) Linearize this term correctly by first finding $\partial L/\partial X$.

(b) Now linearize this term by linearizing $L(t)$ prior to performing the partial differentiation and compare the results with those obtained in (a).

2–6. Find the transfer functions of the plants and systems represented by the following differential equations:

(a) $\dddot{c} + 15\ddot{c} + 50\dot{c} + 500c = \dot{r} + 2r$

(b) $5\ddot{c} + 25\dot{c} = 0.5u$

(c) $\ddot{c} + 25c = 0.5r$

(d) $\ddot{c} + 3\dot{c} + 6c + 4\int c\,dt = 4u$

2–7. Use Cramer's rule to find the two TFs, $x(s)/r(s)$ and $y(s)/r(s)$ for each of the following pairs of differential equations:

(a) $7\dot{x} + 2x - 7\dot{y} - 0.035y = r$

$\quad 0.03\dot{x} + 0.5x + 0.5\ddot{y} + 0.2\dot{y} = 26$

(b) $14\dot{x} + 2x + y = 0$

$\quad 2x - 10\ddot{y} - 2y = \dot{r} + 2r$

2–8. For each of the following transformed responses (i) expand in partial fractions, (ii) write an expression for $y(t)$ with the coefficients unspecified, (iii) find the coefficients and rewrite $y(t)$, and (iv) find the steady-state value of $y(t)$:

(a) $y(s) = \dfrac{50(s + 0.6)}{s(s + 10)(s^2 + 4s + 13)}$

(b) $y(s) = \dfrac{5(s^2 + 3s + 5)}{s^2(s + 5)^2}$

(c) $y(s) = \dfrac{2(s + 1)}{s^3(s + 3)}$

(d) $y(s) = \dfrac{247.5(s + 1)}{s(s + 4.5)(s + 5.5)(s + 10)}$

2–9. For each of the expressions in Prob. 2–8, find the initial value $y(0+)$ by application of the initial value theorem and compare with the value obtained by substituting $t = 0+$ into the expression for $y(t)$.

2–10. Since $L[\dot{y}(t)] = sy(s)$, then $\dot{y}(0+) = \lim\limits_{s \to \infty} s^2 y(s)$. Use the initial value theorem to determine $\dot{y}(0+)$ for each of the expressions in Prob. 2–8.

2–11. The systems whose responses are given in Prob. 2–8 are all stable. Apply the final value theorem to determine the steady-state response and compare with the value obtained by letting t go to ∞ in the expression for $y(t)$.

2–12. A stable system has the transfer function

$$\frac{c(s)}{r(s)} = \frac{20(s + 1)}{(s + 2)(s + 10)}$$

(a) Write the corresponding differential equation.
(b) For zero initial conditions, find the time response for a unit ramp input, i.e., $r(t) = t$. Also find the initial and final values of c.
(c) For the initial conditions $\dot{c}(0-) = +1$ and $c(0-) = +2$, do (b) above.

2–13. Same as Prob. 2–12 but with the transfer function

$$\frac{c(s)}{r(s)} = \frac{50}{(s + 5)(s^2 + 4s + 13)}$$

2–14. The use of the convolution integral in Eq. (2–3–20) implies that $L^{-1}[W(s) \cdot r(s)] \neq W(t) \cdot c(t)$. Show this nonidentity to be true for $W(s) = 10/(s + 10)$ and $r(s) = r_0/s$.

3

Modeling and
Transfer Functions

3-1 INTRODUCTION

Analytical techniques require mathematical models. The transfer function, which has been discussed in Sec. 1–3, is a convenient model form for the analysis and design of stationary linear systems with lumped parameters and with a limited number of continuous inputs and outputs. In this chapter, we shall find the transfer functions of certain systems directly from the differential equations and by block diagram algebra, with emphasis on the latter technique. Experimental determination of transfer functions will be discussed briefly in a subsequent chapter.

Formulation of the differential or integro-differential equations describing the behavior of a particular plant, process, or component and the experimental determination of transfer functions both require the specialized knowledge of a particular discipline or specialty and as such are not essential to the objectives of this book. Consequently, the components and systems of the following sections are kept simple and serve to demonstrate techniques and the analogous forms of transfer functions in different disciplines.

3-2 SIMPLE MECHANICAL SYSTEMS

The basic elements of mechanical systems with translational motion are mass, the compression-tension spring, and the dashpot or viscous damper. For each of these elements, Fig. 3–2–1 shows its symbol, its differential equation relating

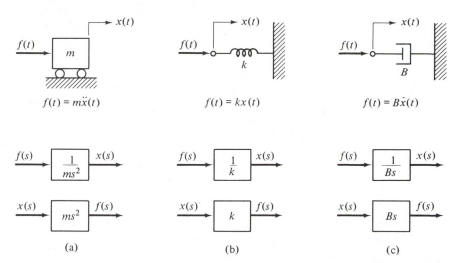

Figure 3–2–1. The basic mechanical elements: (a) the mass (b) the spring (c) the dashpot.

force and displacement, and two block diagrams. One diagram has force as the input and displacement as the output; the other has the input and output reversed. Remember that these elements are idealized in that they are considered to be both linear and lumped parameter elements.

Let us combine these basic elements in the one-degree-of-freedom system of Fig. 3–2–2a where the displacement of the mass may be measured from either the unstretched spring position, denoted $X(t)$, or from the static equilibrium position, denoted $x(t)$. Assume the former and write the Newtonian equation of motion as

$$f(t) - B\dot{X}(t) - kX(t) + mg = m\ddot{X}(t) \qquad (3\text{–}2\text{–}1)$$

If $X(t)$ is written as $X_0 + x(t)$, where X_0 is the static equilibrium position and $x(t)$ is the time-varying displacement with respect to X_0, then Eq. (3–2–1) with $f(0) = 0$ becomes

$$f(t) - B\dot{x}(t) - kX_0 - kx(t) + mg = m\ddot{x}(t) \qquad (3\text{–}2\text{–}2)$$

Letting t go to zero yields the steady-state or equilibrium equation

$$-kX_0 + mg = 0 \qquad (3\text{–}2\text{–}3)$$

which subtracted from Eq. (3–2–2) produces the dynamic equation of motion

$$f(t) - B\dot{x}(t) - kx(t) = m\ddot{x}(t) \qquad (3\text{–}2\text{–}4)$$

Notice that initial conditions can be set equal to zero by using the equilibrium position as the reference. Rearranging the terms in Eq. (3–2–4) and using the Laplace transform results in

$$(ms^2 + Bs + k)x(s) = f(s) \qquad (3\text{–}2\text{–}5)$$

which can be solved for the transfer function

$$W(s) = \frac{x(s)}{f(s)} = \frac{1}{ms^2 + Bs + k} \qquad (3\text{–}2\text{–}6)$$

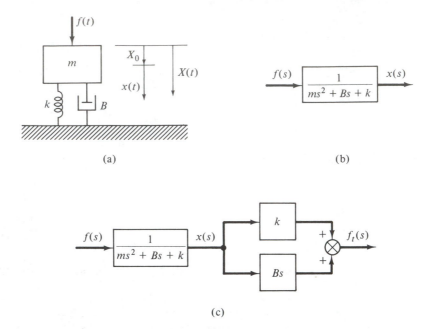

Figure 3–2–2. A simple mechanical system: (a) functional diagram (b) transfer function for displacement as output (c) block diagram for transmitted force as output.

The system can now be represented by the single block and transfer function of Fig. 3–2–2b.

The transfer function can also be obtained by constructing a block diagram of the system, as shown in Fig. 3–2–3a, by using the block diagrams and transfer functions of the individual elements. Note the inherent negative feedback loops and the role of the mass as a force comparator. This block diagram is successively reduced in Fig. 3–2–3b and c to the single block and transfer function of Fig. 3–2–3d by use of block diagram algebra and the identities of Fig. 1–4–5.

When f_t, the sum of the forces transmitted through the spring and the dashpot to the floor, is of interest (as in shock-mounting a piece of rotating machinery), the transfer function relating it to f, the input force, can be found from the block diagram of Fig. 3–2–2c to be

$$\frac{f_t(s)}{f(s)} = \frac{Bs + k}{ms^2 + Bs + k} \tag{3–2–7}$$

In the frequency domain, the magnitude of this transfer function is often referred to as the *transmissibility* or *transmission* factor.

Let us now combine the basic mechanical elements into the system of Fig. 3–2–4a, which could conceivably represent the suspension system for one wheel of an automobile. The equilibrium condition would comprise a smooth road with the spring compressed so as to balance its proportionate share of the weight of the body. The input $y(t)$ is the displacement of the wheel due to bumps and potholes in the road, and the output $x(t)$ is the displacement of the body with

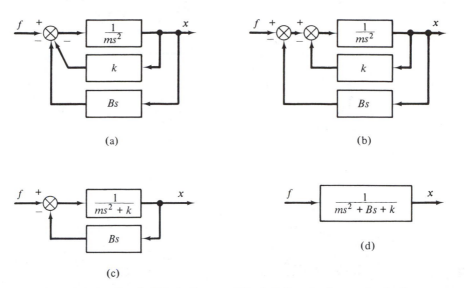

(a)   (b)

(c)

(d)

Figure 3-2-3.   Detailed block diagram of Fig. 3-2-2a and subsequent reduction.

respect to its equilibrium position. The displacement of the wheel will transmit forces through the spring and shock absorber to the mass, and in turn the forces will be modified by the motion of the mass. The equation of motion can be written as

$$k(y - x) + B(\dot{y} - \dot{x}) = m\ddot{x} \qquad (3\text{-}2\text{-}8)$$

Transforming and solving for the transfer function yields

$$\frac{x}{y}(s) = \frac{Bs + k}{ms^2 + Bs + k} \qquad (3\text{-}2\text{-}9)$$

as shown in Fig. 3-2-4b.

This system could also have been represented by the block diagram of Fig. 3-2-5a, with a single loop feeding back the body displacement x, or by the block diagram of Fig. 3-2-6, with two loops feeding back force rather than displacement. Figure 3-2-5a is reduced to obtain the transfer function of Eq. (3-2-9).

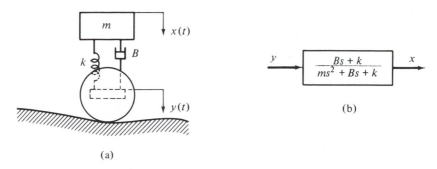

(a)

(b)

Figure 3-2-4.   (a) Automobile suspension system (b) single block diagram.

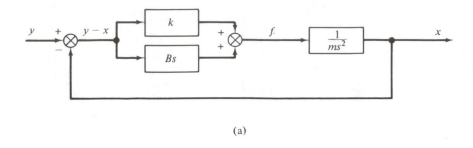

(a)

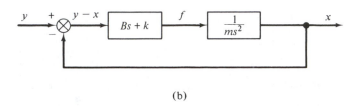

(b)

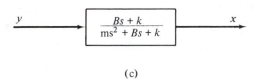

(c)

Figure 3–2–5. A detailed block diagram of the automobile suspension system of Fig. 3–2–4a.

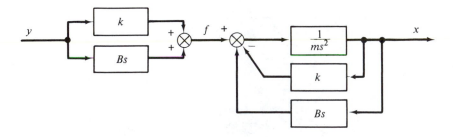

Figure 3–2–6. Another block diagram of the automobile suspension system of Fig. 3–2–4a.

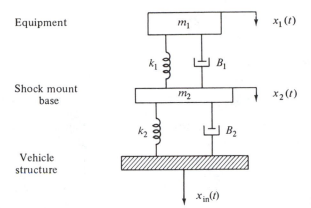

Equipment

Shock mount
base

Vehicle
structure

Figure 3–2–7. A shock mount.

The two-degree-of-freedom system of Fig. 3–2–7 could represent a shock mount for a radio transceiver with a mass m_1 that we wish to isolate from the vertical motion $x_{in}(t)$ of a vehicle. The shock mount base m_2 is connected to the equipment and the vehicle by springs and shock absorbers that are lumped together as k_1, k_2, B_1, and B_2. The displacements x_1 and x_2 are measured with respect to a static equilibrium position where x_{in} is zero. The detailed block diagram of this shock mount system is constructed in Fig. 3–2–8a showing the various feedback loops involving both displacement and force. This block diagram

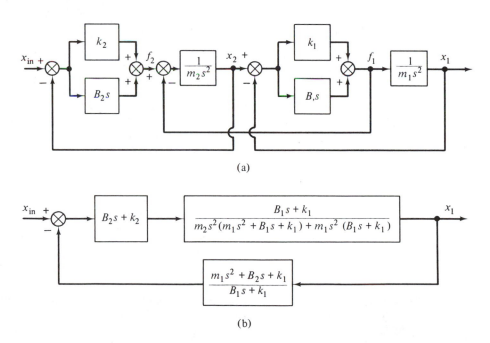

(a)

(b)

Figure 3–2–8. Block diagram of Fig. 3–2–7; (a) detailed (b) next to last step.

can be successively reduced to the form shown in Fig. 3–2–8b, and the transfer function can be found from the relationship $G/(1 + GH)$ to be

$$\frac{x_1}{x_{in}} = \frac{B_1 B_2 s^2 + (B_1 k_2 + B_2 k_1)s + k_1 k_2}{m_1 m_2 s^4 + (m_1 B_2 + m_2 B_1 + m_1 B_1)s^3}$$

$$+ (B_1 B_2 + m_1 k_2 + m_2 k_1 + m_1 k_1)s^2 + (B_2 k_1 + B_1 k_2)s + k_1 k_2 \quad (3\text{–}2\text{–}10)$$

If we want the displacement of the base x_2, any of the block diagrams in which x_2 can still be identified can be redrawn with x_2 as the output and everything beyond becoming part of the feedback path.

The transfer function of Eq. (3–2–10) could also have been obtained by applying Newton's law to each of the masses to obtain the two differential equations

$$k_2(x_{in} - x_2) + B_2(\dot{x}_{in} - \dot{x}_2) - k_1(x_2 - x_1) - B_1(\dot{x}_2 - \dot{x}_1) = m_2 \ddot{x}_2 \quad (3\text{–}2\text{–}11a)$$

$$k_1(x_2 - x_1) + B_1(\dot{x}_2 - \dot{x}_1) = m_1 \ddot{x}_1 \quad (3\text{–}2\text{–}11b)$$

These equations are then transformed with zero initial conditions, rearranged with $x_{in}(s)$ as the forcing function, and solved for x_1 in terms of x_{in} by using Cramer's rule* to obtain the transfer function of Eq. (3–2–10).

The treatment of rotational mechanical systems is similar to that of translational systems. A simple rotational system and its block diagram are shown in Fig. 3–2–9, where θ is measured from the equilibrium position. The similarity to Fig. 3–2–3a is quite apparent.

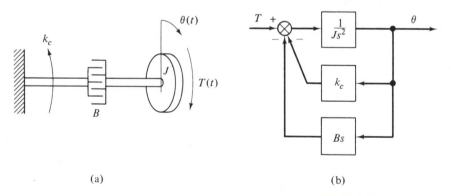

(a)   (b)

Figure 3–2–9.   (a) A simple rotational mechanical system (b) its block diagram.

In developing analogies, which will be discussed in the next section, it is often convenient to speak of parallel and series mechanical elements. Elements having the same displacement are said to be in series, such as the spring and dashpot in Fig. 3–2–10a. Elements whose displacements are additive (the force across each element being the same) are said to be in parallel as shown in Fig. 3–2–10b.

* See App. C.

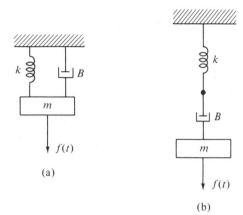

(a)

(b)

Figure 3–2–10. Mechanical elements:
(a) in series (b) in parallel.

3–3 PASSIVE ELECTRICAL NETWORKS

The basic passive electrical elements are the resistor, the capacitor, and the inductor. Figure 3–3–1 shows the symbol, differential equation, and transfer function for each element in terms of the voltage drop across and the current through each element.

The transfer function for a system comprising any combination of these elements (a passive network) can be obtained, as for the mechanical systems, from either the block diagram or the differential equations. We will use block diagram reduction exclusively in this section to increase our familiarity with that process.

For elements in series, the current is the same through each element and the total voltage drop is the sum of the individual voltage drop across each element. For elements in parallel, the converse is true; the voltage drop across each element is the same, and the total current is the sum of the individual currents through each element. Consider the simple *RLC* series network of Fig. 3–3–2a for which we wish to find the total current *I* resulting from any applied

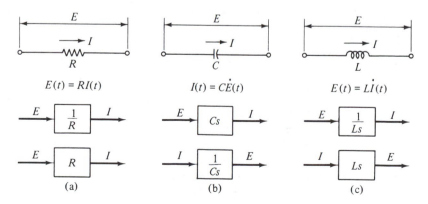

Figure 3–3–1. Passive electrical elements: (a) resistor (b) capacitor (c) inductor.

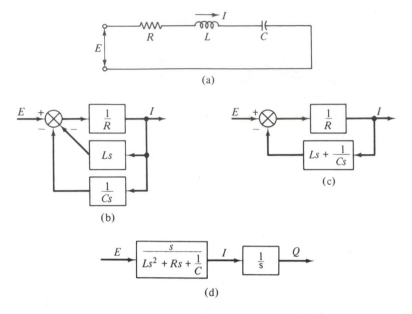

Figure 3–3–2. A simple *RLC* passive network.

voltage *E*. The block diagram is shown in Fig. 3–3–2b, reduced to one loop in Fig. 3–3–2c and to a single block and transfer function in Fig. 3–3–2d. Since current is defined as the time rate of change of the charge *Q*, i.e., $I(t) = \dot{Q}(t)$ and $I(s) = sQ(s)$, the transfer function for voltage in and charge out can be written as

$$\frac{Q}{E}(s) = \frac{1}{Ls^2 + Rs + 1/C} \tag{3–3–1}$$

Comparing the transfer function of Eq. (3–3–1) with that of Eq. (3–2–6) for the mechanical system of Fig. 3–2–2a, we see that the forms are identical. It is therefore possible to represent the mechanical system by the electrical system, or vice-versa, or to deduce the dynamic behavior of one of these systems from the dynamic behavior of the other. Two physical systems that have this type of similarity are called *analogies*, each being an *analog** of the other. Based on the similarity between Eqs. (3–3–1) and (3–2–6), an electrical analog of the translational mechanical system can be constructed by replacing mechanical elements in parallel by electrical elements in parallel, and by replacing series mechanical elements by series electrical elements with voltage representing force, inductance mass, resistance viscous damping, and so forth.

Passive networks are frequently used in control systems for stability compensation, which is modification of the system response to meet the design criteria. A simple compensation network is the *lead network* of Fig. 3–3–3a

---

* An analog is not the same thing as an analog computer.

with the block diagram of Fig. 3–3–3b. This block diagram is reduced to obtain a transfer function

$$\frac{E_{\text{out}}}{E_{\text{in}}} = \frac{s + 1/\alpha\tau}{s + 1/\tau} \tag{3–3–2}$$

where $\tau = R_1 R_2 C/(R_1 + R_2)$ and $\alpha = (R_1 + R_2)/R_2$.

Another compensation network is the *lag network* of Fig. 3–3–4a. Its block diagram is given and reduced in the same figure to obtain the transfer function

$$\frac{E_{\text{out}}}{E_{\text{in}}} = \frac{1}{\alpha}\frac{s + 1/\tau}{s + 1/\alpha\tau} \tag{3–3–3}$$

where $\tau = R_2 C$ and $\alpha = (R_1 + R_2)/R_2$. The application of these two networks will be illustrated in the chapter on system compensation.

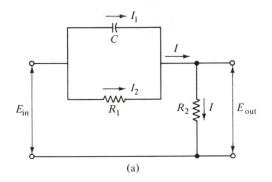

(a)

(b)

(c)

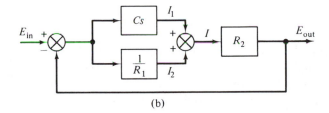

$$\tau = \frac{R_1 R_2 C}{R_1 + R_2}$$

$$\alpha = \frac{R_1 + R_2}{R_2}$$

Figure 3–3–3. A simple passive lead network.

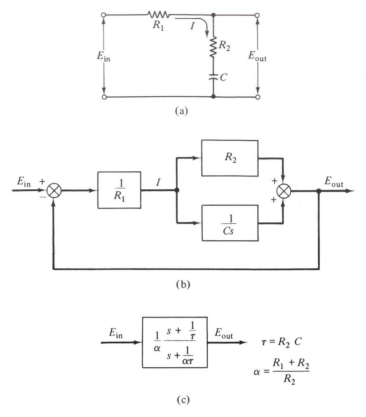

(a)

(b)

(c)

Figure 3–3–4. A simple passive lag network.

3–4 THE AMPLIFIER

An amplifier is an active element and is present in one form or another in most feedback control systems. Amplification per se allows low-power signals to be used for information and control; furthermore, varying the gain (amplification factor) of the amplifier is a simple method of modifying both the accuracy and the dynamic response of a system.

Amplifiers may be electronic, either vacuum tube or solid state; magnetic; or fluid, either with moving parts or purely fluidic. With a few exceptions, the transfer function of an open-loop amplifier can be represented by an amplification factor μ and a first-order time lag. For example, the transfer function of an open-loop electronic voltage amplifier would be

$$\frac{E_{out}}{E_{in}} = \frac{-\mu}{\tau s + 1} \qquad (3\text{--}4\text{--}1)$$

The minus sign in Eq. (3–4–1) appears because the polarity of the input is reversed in passing through the amplifier. A well-designed amplifier will have a time constant τ that is small with respect to the other constants in the system and to the response times. Consequently, τ is generally assumed equal to zero and the transfer function taken to be simply the negative of the amplification factor or gain of the amplifying element. Equation (3–4–1) becomes

$$\frac{E_{out}}{E_{in}} = -\mu \tag{3–4–2}$$

and the open-loop amplifier can be represented by the block diagram of Fig. 3–4–1.

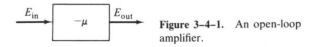

Figure 3–4–1. An open-loop amplifier.

Open-loop amplifiers do have inherent deficiencies in that the gain is seldom linear over a range of input values and tends to drift or change with time (to be nonstationary). This tendency to drift is particularly evident in dc amplifiers. The gain can be linearized to some extent and also can be stabilized by feeding back a portion β of the output, which is added to the input as shown in Fig. 3–4–2a. This is not a positive feedback system since the minus sign associated with the gain μ has the effect of reversing the sign at the comparator. The transfer function for the feedback amplifier can easily be found from simple algebra to be

$$\frac{E_{out}}{E_{in}} = \frac{-\mu}{1 + \beta\mu} \tag{3–4–3}$$

and the closed-loop amplifier can be represented by the single block of Fig. 3–4–2b.

This example of a feedback amplifier demonstrates two characteristics common to all negative feedback systems. The first is the reduction of the gain.

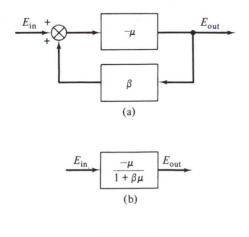

(a)

(b)

(c)

Figure 3–4–2. A feedback amplifier: (a) detailed block diagram (b) single block diagram (c) customary representation.

Compare Eq. (3–4–2) with Eq. (3–4–3). In the former the gain is μ, but in the latter it has been reduced by the factor $1/(1 + \beta\mu)$. If μ, for example, is 100 and β is 0.1, then the overall gain has been reduced from 100 to approximately 9.1. This is an obvious disadvantage; if an overall gain of 100 is required, additional amplifier stage(s) must be added. Now for the good news: the feedback loop reduces the error in the output arising from changes in the value of μ. In the open-loop case, a 10 percent change in μ results in a 10 percent error in the output—a one-to-one correspondence—whereas with feedback and $\beta\mu$ equal to 10, the error in the output is less than 1 percent. *To generalize, negative feedback reduces the overall gain but minimizes the effects of changes in forward-path parameters upon the input-output relationship.* It should be noted, however, that effects of parameter changes in the feedback path are not similarly reduced (see Sec. 5–8). For example, a 10 percent increase in β decreases the overall gain by a little more than 8 percent, almost a one-to-one relationship.

If $\beta\mu$ is much greater than unity, which is the case for a well-designed amplifier, then Eq. (3–4–3) can be simplified and the transfer function approximated by

$$\frac{E_{out}}{E_{in}} = \frac{+1}{\beta} = +A \qquad (3\text{–}4\text{–}4)$$

The sign of the amplifier gain has been changed from negative to positive as a mathematical convenience. The actual negative sign or phase reversal is accounted for in physical systems by reversing the polarity of the input to one of the other components in the system, such as a motor. Henceforth, amplifiers will be assumed to be of the feedback type and will be represented by the single block of Fig. 3–4–2c.

Feedback amplifiers used as comparators are referred to as *difference amplifiers*. A special type of feedback amplifier is the *operational amplifier*. It is a dc amplifier with extremely high gain (10^4 to 10^8 amplification) and high input impedance. When used with an RC network as shown in Fig. 3–4–3, it can

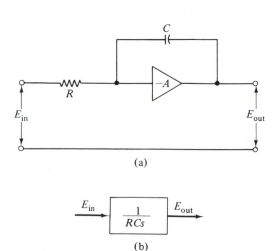

(a)

(b)

Figure 3–4–3. An operational amplifier used as an integrator: (a) network diagram (b) block diagram.

function as an integrator. Such operational amplifiers are often used in active controllers and are the basic element of analog computers.

3-5 THE ELECTRIC MOTOR

The electric motor or its equivalent is also a common control system component. It drives the output variable to a desired position often at a desired rate or acceleration. Motors come in many types, both ac and dc, but a generalized transfer function can be developed that will represent to a first approximation any type of motor.

The torque developed by a motor can be related to an input or control voltage (or current) by the simple proportionality

$$T_m = k_m e_c \tag{3-5-1}$$

If we consider first a pure inertia load on the output shaft, the transformed equation of motion is

$$T_m = k_m e_c = Js^2\theta_{out} \tag{3-5-2}$$

where J is the combined moments of inertia of the motor and the load and k_m is the torque constant of the motor. The block diagrams for this motor-load combination are shown in Fig. 3-5-1. From Eq. (3-5-2) we can see that a constant control voltage results in a constantly accelerating output shaft angle. The motor acts as a double integrator and is known as a runaway motor; it cannot be used in a control system without external damping.

Alternating-current motors and most dc motors have internal damping that is viscous in nature and arises from the induced currents or back emf generated by the rotation of the armature in the presence of magnetic fields. This damping is represented in Fig. 3-5-2a as a negative feedback voltage proportional to the angular velocity of the output shaft ($s\theta_{out}$ in the Laplace domain). Turn the shaft

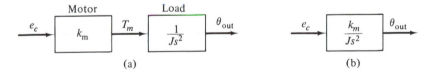

(a)   (b)

Figure 3-5-1.   An inertially loaded motor with no feedback.

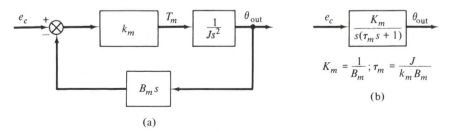

(a)

Figure 3-5-2.   An inertially loaded motor with internal damping.

of a small electric motor, such as an electric drill, by hand; the resistance you feel is the internal damping. The block diagram of Fig. 3–5–2a is reduced to a single block in Fig. 3–5–2b with the transfer function

$$\frac{\theta_{\text{out}}}{e_c} = \frac{K_m}{s(\tau_m s + 1)} \tag{3-5-3}$$

where the effective motor gain K_m and the time constant τ_m are defined in the figure. Equation (3–5–3) shows that the motor now acts as an integrator with a first-order time lag. Since the motor speed is $\dot{\theta}(t)$ or $s\theta(s)$ in the Laplace domain, Eq. (3–5–3) can be written as

$$\frac{\dot{\theta}}{e_c}(s) = \frac{K_m}{\tau_m s + 1} \tag{3-5-4}$$

If a constant control voltage e_c is applied, the motor will settle down to a constant speed after a time period of the order of three times the motor time constant τ_m.

The expression for τ_m in Fig. 3–5–2b shows τ_m to be inversely proportional to B_m, the measure of internal damping. When the internal damping is too small to provide the desired τ_m, it can be augmented by using a tachometer to feed back a voltage proportional to the motor speed. This type of damping feedback is referred to as *tachometer* or *rate feedback*.* With B_t denoting the tachometer feedback, the block diagram is shown in Fig. 3–5–3a and reduced in Fig. 3–5–

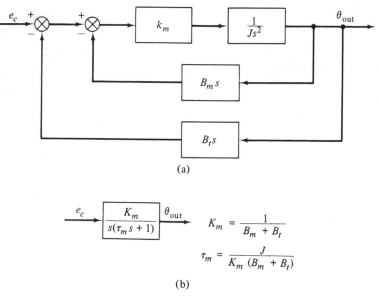

(a)

(b)

$$K_m = \frac{1}{B_m + B_t}$$

$$\tau_m = \frac{J}{K_m (B_m + B_t)}$$

Figure 3–5–3. An inertially loaded motor with internal damping and tachometer feedback.

* Tachometer feedback used for the control of a motor-generator set is called a *Ward-Leonard system*.

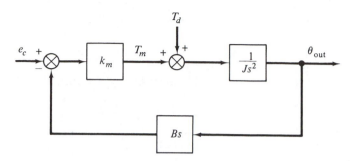

Figure 3–5–4. Changes in load torque as an additional input to a motor.

3b with K_m and τ_m redefined. The form of the transfer function remains unchanged by the addition of tachometer feedback.

If the load torque or moment of inertia changes from the nominal value or if there are disturbance torques, such as those arising from the wind loading of a large parabolic antenna, they are treated as additional inputs. Care must be taken to introduce them properly into the block diagram. Denoting these disturbances as T_d, we can introduce them as shown in Fig. 3–5–4.

If the output shaft is elastically restrained in any way, the load torque T_l can be represented by the transformed expression

$$T_l = (Js^2 + k_l)\theta_{\text{out}} \tag{3–5–5}$$

where k_l is the spring constant. This motor-load combination can be represented by the block diagrams of Fig. 3–5–5. If a constant control voltage e_c is applied to this motor-load combination, the output shaft will pass through a transient phase and come to rest with an angular displacement from the equilibrium position

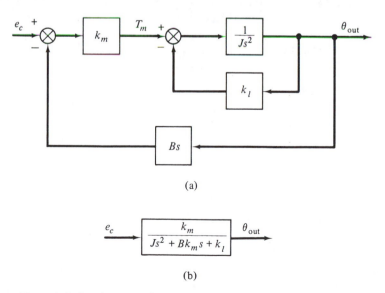

(a)

(b)

Figure 3–5–5. A motor with an elastic restraint on the output shaft.

equal to k_m/k_l radians. Synchro motors and repeater systems have this characteristic.

Other time lags within a well-designed motor, such as those resulting from inductance in the field and armature, are small with respect to τ_m and are neglected. The effects of stiction and coulomb friction from brushes and commutators are minimized by the presence of the amplifier in the controller, which raises the level of the motor control voltage and thus increases the torque; stiction and friction are, therefore, neglected. A motor, like the engine of Sec. 2–2, can be considered linear only over a limited range of operating conditions and therefore must be matched to the expected range of operating conditions.

*3–6 OTHER SYSTEMS

Transfer functions and analogies can be developed in a similar manner for other systems in other disciplines. The general procedure is to develop a system of differential equations or input-output relationships that will describe the system with the precision required by the performance and design criteria. These equations and relationships will generally be nonlinear; linearization must be accomplished with awareness of the importance of the expected operating conditions. In establishing analogies, do not be hasty; it is usually wiser to develop the equations or transfer functions first, and then the analogies.

No attempt will be made to develop transfer functions illustrative of a wide variety of systems. We will look briefly at a few examples of hydraulic components and subsystems to show the similarity to some of the electrical components previously examined.

Let us consider first a simple hydraulic actuator as shown in Fig. 3–6–1a, comprising a valve and a cylinder-piston assembly. If the load is small compared to the force on the piston, then the displacement of the piston from its equilibrium position can be considered to result from the volume flow rate alone. The flow rate q through the valve is proportional to the valve opening x_v so that

$$q = k_v x_v \qquad (3\text{–}6\text{–}1)$$

where k_v is assumed to be constant. If the hydraulic fluid is assumed to be incompressible, then within the cylinder the flow rate must equal the rate of change of the volume behind the piston, which in turn is the product of the piston area A and the velocity of the piston $\dot{x}_{out}$. We now have another expression for the volume flow rate:

$$q = A\dot{x}_{out} \qquad (3\text{–}6\text{–}2)$$

The block diagram with component transfer functions shown in Fig. 3–6–1b is reduced to obtain the transfer function

$$\frac{x_{out}}{x_v}(s) = \frac{k_v}{As} \qquad (3\text{–}6\text{–}3)$$

This is the transfer function of an integrator; a constant displacement x_v of the valve will cause the piston to move at a constant velocity. To use this actuator as a hydraulic lift, the assumption that the load be small compared to the force

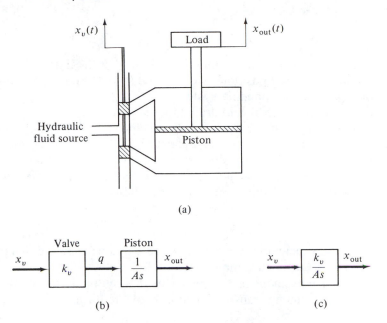

Figure 3-6-1.   An open-loop hydraulic actuator.

on the piston implies a large piston area A (if the fluid pressure is to be kept economically low) and thus a low output velocity.

If the displacement of the piston x_{out} is subtracted from the input displacement x_{in}, say by a mechanical linkage, so that the valve displacement is proportional to their difference, we now have a closed-loop position servomechanism. The block diagram is shown and reduced in Fig. 3-6-2, and the transfer function is

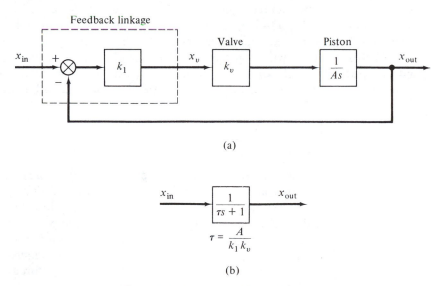

Figure 3-6-2.   A hydraulic position servo.

$$\frac{x_{\text{out}}}{x_{\text{in}}} = \frac{1}{\tau s + 1} \tag{3-6-4}$$

where the time constant τ, which is a measure of the time required for the piston to attain the desired position, is equal to $A/k_1 k_v$. The large piston area condition implies a slowly responding system.

In reference to the hydraulic actuator of Fig. 3–6–1, if the load on the piston is large or if we want a larger velocity with a smaller-diameter piston, the analysis becomes more complicated. The velocity of the piston and the pressure drop across the piston are now coupled in a nonlinear manner involving fluid compressibility, leakage, and any viscous damping. After linearization, the transfer function of the open-loop hydraulic actuator can be written as

$$\frac{x_{\text{out}}}{x_v} = \frac{K_m}{s(\tau_m s + 1)} \tag{3-6-5}$$

The hydraulic actuator now acts as a linear motor; compare this transfer function with that of the electric motor in Eq. (3–5–3).

If the movement of the piston of this hydraulic motor is restrained by a spring, the resulting transfer function is of the form

$$\frac{x_{\text{out}}}{x_v}(s) = \frac{K_a}{\tau_a s + 1} \tag{3-6-6}$$

where K_a can be made larger than unity. We now have a hydraulic amplifier with a first-order time lag.

The hydraulic valve may well be a closed-loop servomechanism itself and when driven by an electrical signal from an amplifier is commonly referred to as an *electrohydraulic servovalve* (EHV). It is important that the dynamics of the EHV (or of any other component) be considered in the analysis and design of any system in which it is being used. Sometimes it is possible to ignore the dynamics of a component or sensor completely, as in the case of the slowly moving actuator of Eq. (3–6–3), or to represent higher-order dynamics by a first-order time lag, as in the case of the actuator represented by Eq. (3–6–5).

At this point we will stop our development of transfer functions and state that the successful application of control theory, which is universal, does require a specialized knowledge and understanding of the problems to be solved.

3–7 CLOSING REMARKS

As has been mentioned several times, no attempt has been made to cover the various components that are in use today. Time and space do not permit such coverage, nor is it necessary to an understanding of the principles of control system analysis and design. Knowledge and selection of the proper components and sensors are of great importance in a specific design problem but are related to the area of expertise of the individual control system designer. There are times when a system may require the design of a component tailored specifically for that one application.

Gear trains, transformers, and other coupling and impedance-matching devices have been explicitly and deliberately omitted, as have power sources. Such components are often required and must be given careful consideration in that they can be sources of nonlinearities and undesirable loading of other components.

In this book, electrical signals are taken to be dc only. There are situations where ac components or systems have distinct advantages over their dc counterparts. Systems can be all dc, all ac, or hybrid, as illustrated in Fig. 3–7–1. The analyses of ac and dc systems are similar, and to a first approximation the transfer functions are identical. The principal difference lies in the design of compensation networks to be used in the controller. Care must be exercised in ac compensation to modify the signal but not the carrier. Since compensation is easier with dc signals, an ac system may have the ac actuating signal reduced to a dc signal, compensated, and then returned to its ac state, as illustrated in Fig. 3–7–2.

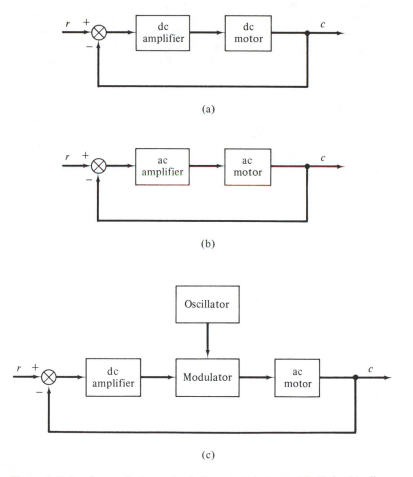

Figure 3–7–1. Some electromechanical control systems: (a) all dc (b) all ac (c) hybrid.

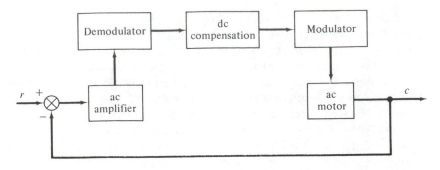

Figure 3-7-2.   Direct-current compensation of an ac control system.

PROBLEMS

3-1. Use Cramer's rule to solve Eqs. (3-2-11a) and (3-2-11b) and find the transfer functions x_1/x_{in} and x_2/x_{in}. Compare x_1/x_{in} with x_2/x_{in} and with Eq. (3-2-10).

3-2. Draw a block diagram for each of the electrical networks (a, b, and c) of Fig. P3-2; assume zero initial conditions. Then reduce the block diagram to obtain the transfer function e_0/e_{in}.

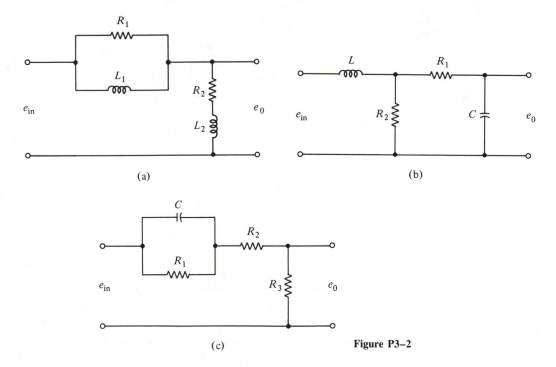

Figure P3-2

3-3. (a) Draw a block diagram for each of the mechanical networks shown in Fig. P3-3; assume zero initial conditions. Hint for Fig. P3-3a: Insert a fictitious mass m_3 between k_2 and B; after reduction, set $m_3 = 0$.

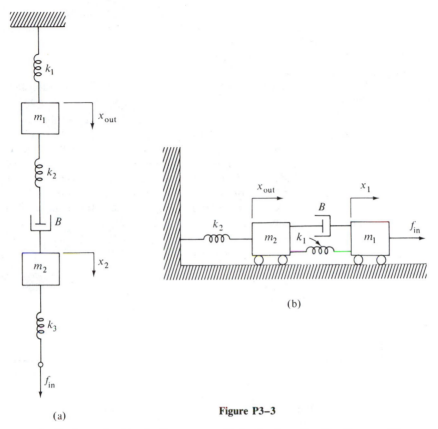

Figure P3-3

(a)

(b) Reduce the block diagram to obtain the transfer function x_{out}/f_{in}.

(c) Write the differential equations of motion and use them to obtain the transfer function x_{out}/f_{in}.

3-4. Figure P3-4 shows a typical speedometer configuration in which a viscous drag in a fluid coupler deflects a spring-restrained indicator needle through an angular displacement θ_{out}, which is to be proportional to an input angular velocity $\omega_{in}(t)$. J denotes a moment of inertia about the rotational axis, k an elastic spring constant,

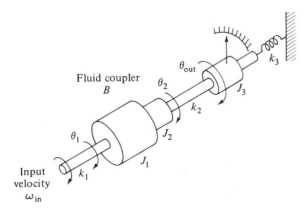

Figure P3-4

and B the viscous damping coefficient. Assume that k_3 is much less than k_2 so that the reaction shift can be considered a rigid shaft, i.e., $\theta_2 = \theta_{out}$.
(a) Write the equations of motion.
(b) Obtain the transfer function $\theta_{out}(s)/\omega_{in}(s)$.
(c) Draw a block diagram.
(d) Assuming the speedometer to be stable, find the steady-state relationship between ω_{in} and θ_{out} for a step in $\omega_{in}(t)$.

3–5. Figure P3–5 is the block diagram of a control system used in an inertial navigation system to maintain a platform in a desired orientation. ω_{IP} is the angular velocity of the platform with respect to inertial space, ω_{IV} is the angular velocity of the vehicle carrying the platform with respect to inertial space, and ω_{VP} is the angular velocity of the platform with respect to the vehicle (motor speed). One design objective of the stabilization loop is to isolate the platform from any angular motion of the vehicle ω_{IV}, i.e., keep $\omega_{IPss} = \phi_{IPss} = 0$, where ϕ_{IP}, the angular displacement of the platform with respect to inertial space, is equal to $\int \omega_{IP}\, dt$.

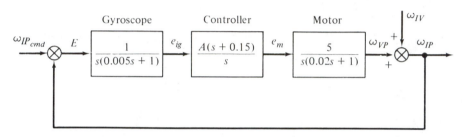

Figure P3–5

(a) Set $\omega_{iPcmd} = 0$ and find the TFs: ω_{IP}/ω_{IV}, ϕ_{IP}/ω_{IV}, and e_m/ω_{IV}.
(b) Assuming the system is stable, find the final values of ω_{IP}, ϕ_{IP}, and e_m for a constant ω_{IV}, i.e., a step.
(c) Does this system isolate the platform from any angular motion of the vehicle?

3–6. Another design objective of the stabilization loop of Fig. P3–5 and Prob. 3–5 is to move the platform at an angular velocity ω_{IP} that is equal to a desired angular velocity ω_{IPcmd}.
(a) Set $\omega_{IV} = 0$ and find the TFs: $\omega_{IP}/\omega_{IPcmd}$ and e_{ig}/ω_{IPcmd}.
(b) Assuming that the system is stable, find the initial and final values of ω_{IP} and e_{ig} for a constant ω_{IPcmd}.
(c) Will the platform move at the desired angular velocity?

3–7. Figure P3–7 is the schematic of a displacement autopilot for a subsonic jet transport where θ is the pitch angle. The component transfer functions are

$$\text{Vertical gyro:} \frac{e_g}{\theta_{in}} = K_g \text{ volts/degree}$$

$$\text{Amplifier:} \frac{e_0}{e_g} = A \text{ volts}$$

$$\text{Elevator servo:} \frac{\delta_e}{e_{in}} = \frac{1}{(0.1s + 1)} \text{ degrees/volt}$$

$$\text{Aircraft dynamics:} \frac{\theta_{A/C}}{\delta e} = \frac{0.8(s + 0.5)}{s(s^2 + 1.06s + 1.2)} \text{ degrees/degree}$$

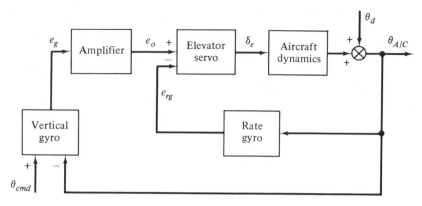

Figure P3–7

Rate gyro: $\dfrac{e_{rg}}{\theta_{A/C}} = 0.05\,s$ volt-seconds/degree

Set $\theta_d = 0$ and

(a) Draw the block diagram using the component TFs.

(b) Find the TFs: $\theta_{A/C}/\theta_{cmd}$ and δ_e/θ_{cmd}.

(c) Assuming the autopilot is stable, find the final values of $\theta_{A/C}$ and δ_e for a step in θ_{cmd}. Is $[\theta_{A/C}/\theta_{cmd}]_{ss} = 1$?

3–8. The autopilot of Prob. 3–7 and Fig. P3–7, in addition to responding to a θ_{cmd}, should not respond to any disturbances represented by θ_d; that is, $\theta_{A/C}$ should return to zero for a step in θ_d when $\theta_{cmd} = 0$. This is the regulator problem.

    (a) Set $\theta_{cmd} = 0$ and draw the block diagram of the autopilt with θ_d as the input and $\theta_{A/C}$ as the output.

    (b) Find the TF: $\theta_{A/C}/\theta_d$.

    (c) Assuming the autopilot is stable, find the final value of $\theta_{A/C}$ for a step in θ_d. Is $[\theta_{A/C}/\theta_d]_{ss} = 0$?

3–9. The hydraulic relay of Fig. P3–9 is designed to produce a displacement x_0 that is

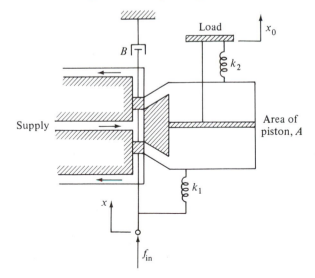

Figure P3–9

proportional to an input force f_{in}. The area of the piston is sufficiently large so that all masses in the relay can be neglected.

(a) Draw the block diagram.

(b) Find the transfer function x_0/f_{in}. Assuming the system is stable, find the final values of x_0 for a constant input force, i.e., a step in $f_{in}(t)$.

(c) Do (b) for the transfer function x/f_{in}.

4

Stability and
the Time Response

4–1 INTRODUCTION

An acceptable system must at minimum satisfy the three basic criteria of stability, accuracy, and a satisfactory transient response. These three criteria are implied in the statement that an acceptable system must have a satisfactory time response to specified inputs and disturbances. So, although, we work in the Laplace and frequency domains for convenience, we must be able to relate these two domains, at least qualitatively, to the time domain.

The time response of a stable system comprises a transient response that disappears with time and a steady-state response that is a measure of the system accuracy. An unstable system has only a transient response. The Laplace domain provides information about the transient response of both stable and unstable systems and about the steady-state response of stable systems. The frequency domain, on the other hand, deals explicitly with the steady-state response of stable systems to sinusoidal inputs, although transient behavior may be implied with varying degrees of accuracy.

This chapter is concerned with relating the Laplace domain to the time response with emphasis on the transient response, and with establishing specific criteria in the Laplace domain for system stability. Accuracy will be treated in the next chapter, and the frequency response in subsequent chapters.

4–2 THE CHARACTERISTIC EQUATION

The time response of a system to any input can be expressed as

$$c(t) = L^{-1} c(s) = c_{ss}(t) + c_{tr}(t) \qquad (4\text{--}2\text{--}1)$$

where $c_{ss}(t)$ is the steady-state response and $c_{tr}(t)$ the transient response. If the system is unstable, there will be no steady-state response, only a transient response.

Without transport lag, the transfer function of a system can be expressed as a ratio of polynomials in the complex Laplace variable s:

$$W(s) = \frac{c}{r}(s) = \frac{N(s)}{D(s)} \tag{4-2-2}$$

The characteristic equation is formed by setting the denominator polynomial equal to zero and can be written in factored form as

$$D(s) = \prod_{i=1}^{n} (s - r_i) = 0 \tag{4-2-3}$$

where r_i denotes the *roots* * *of the characteristic equation*—the values of s that make $D(s)$ equal to zero. These roots may be real, complex, or equal to zero; if complex, they will always occur in conjugate pairs since the coefficients of the differential equations are real.

It can be shown that the transient response for n distinct roots in the Laplace domain is

$$c_{tr}(s) = \frac{C_1}{s - r_1} + \frac{C_2}{s - r_2} + \frac{C_3}{s - r_3} + \cdots + \frac{C_n}{s - r_n} \tag{4-2-4}$$

and in the time domain is

$$c_{tr}(t) = C_1 e^{+r_1 t} + C_2 e^{+r_2 t} + C_3 e^{+r_3 t} + \cdots + C_n e^{+r_n t} \tag{4-2-5}$$

Each term in Eq. (4–2–5) is called a *transient mode*. There is a transient mode for each root with a shape determined solely by the location of the root in the s plane. The sign and size of the coefficients of the modes are determined, however, by both the input and the numerator polynomial $N(s)$.

It is possible to define a stable system as one whose transient response goes to zero as time increases. Therefore, if

$$\lim_{t \to \infty} c_{tr}(t) = 0 \tag{4-2-6}$$

then the system is stable. This criterion for stability is consistent with that of Sec. 1–5, since a unit impulse response can be thought of as a transient response only. To meet the stability criterion of Eq. (4–2–6), each and every transient mode of Eq. (4–2–5) must disappear with time, which means that each mode must contain a negative exponential.

Similarly we can state that a system is unstable if

$$\lim_{t \to \infty} c_{tr}(t) = \infty \tag{4-2-7}$$

This would be the case if one or more of the transient modes increases with time. If any of the transient modes approaches a constant in the limit, that mode is limitedly stable. If any approaches a sinusoid, that mode is marginally stable.

The transient mode shapes associated with different locations of the roots

---

* The roots are sometimes called the *closed-loop poles* and are the *eigenvalues* of state-variable methods.

in the complex s plane are sketched in Fig. 4–2–1 along with a description of the mode's stability. Note the use of the symbol $\square$ to denote the roots of the characteristic equation. The only two stable modes are shown in Fig. 4–2–1a and b, and they occur when the roots are in the left half of the s plane, i.e., when the roots are negative or have negative real parts. If these roots were

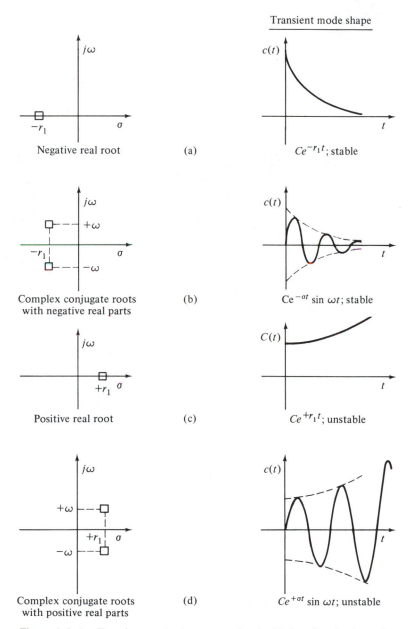

Figure 4–2–1 (a) Negative real root — $Ce^{-r_1 t}$; stable

(b) Complex conjugate roots with negative real parts — $Ce^{-\sigma t} \sin \omega t$; stable

(c) Positive real root — $Ce^{+r_1 t}$; unstable

(d) Complex conjugate roots with positive real parts — $Ce^{+\sigma t} \sin \omega t$; unstable

Figure 4–2–1. Transient mode shapes associated with locations in the s plane of the roots of the characteristic equation.

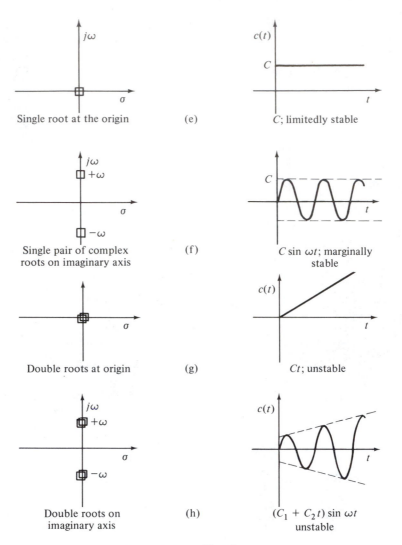

Figure 4–2–1. (Cont.).

multiple (double, triple, etc.), their transient modes would have coefficients that increase with time but the modes would still be stable since the decaying exponential would dominate.

We can now define the stability of a system in terms of the location in the *s* plane of the roots of the characteristic equation:

A system is *stable* if *all* the roots lie in the left-half plane.

A system is *unstable* if *any* root lies in the right-half plane or if *multiple* (or repeated) roots lie on the imaginary axis or at the origin.

A system is limitedly stable if a *single* root lies at the origin and *all other* roots lie in the left-half plane.

A system is *marginally* stable if only *single* pairs of complex conjugate roots lie on the imaginary axis and *all other* roots lie in the left-half plane.

4–3 THE ROUTH-HURWITZ CRITERION

If the characteristic equation of a system is in factored form, we know by inspection the location of the roots in the s plane and thus the stability of the system. Unfortunately, this is seldom the case, and we are faced with the task of factoring a high-order polynomial. The Routh-Hurwitz criterion is a simpler method that will tell us how many roots, if any, are located in the right-half s plane or on the imaginary axis. With this information we know whether the system is stable, unstable, marginally stable, or limitedly stable.

With the characteristic equation in the form

$$D(s) = a_n s^n + a_{n-1} s^{n-1} + \cdots + a_1 s + a_0 = 0 \qquad (4\text{--}3\text{--}1)$$

we can make a quick check to see if the system might be unstable or marginally stable. It is a necessary but not sufficient condition for stability that all coefficients be nonzero and positive.* If any coefficient is zero or negative, there are roots either in the right-half plane or on the imaginary axis, and the system is not stable.

To determine the number of roots in the right-half plane, form the Routhian array as follows:

$$
\begin{array}{cccccc}
s^n: & a_n & a_{n-2} & a_{n-4} & \cdots & 0 \\
s^{n-1}: & a_{n-1} & a_{n-3} & a_{n-5} & \cdots & 0 \\
s^{n-2}: & b_1 & b_2 & b_3 & \cdots & 0 \\
s^{n-3}: & c_1 & c_2 & c_3 & \cdots & 0 \\
 & \cdot & \cdot & \cdot & & \\
 & \cdot & \cdot & \cdot & & \\
 & \cdot & \cdot & \cdot & & \\
s^1: & i & 0 & & & \\
s^0: & j & 0 & & &
\end{array}
\qquad (4\text{--}3\text{--}2)
$$

The first and second rows of the array are formed from the coefficients of the characteristic equation. The elements of succeeding rows are found by cross-multiplying as follows:

$$b_1 = \frac{a_{n-1} a_{n-2} - a_n a_{n-3}}{a_{n-1}}$$

$$b_2 = \frac{a_{n-1} a_{n-4} - a_n a_{n-5}}{a_{n-1}}$$

$$\cdot$$
$$\cdot$$
$$\cdot$$

$$(4\text{--}3\text{--}3)$$

* If all the coefficients are negative, divide through by -1.

until b becomes and remains zero. Then

$$c_1 = \frac{b_1 a_{n-3} - b_2 a_{n-1}}{b_1}$$

$$\quad (4\text{-}3\text{-}4)$$

$$c_2 = \frac{b_1 a_{n-5} - b_3 a_{n-1}}{b_1}$$

until c becomes zero, and so on. There will be only one element in each of the last two rows. Any row in the array can be multiplied or divided by a nonzero positive number. When the array is complete, the *number of roots in the right-half plane will be equal to the number of sign changes in the first column.*

Let us apply the Routh-Hurwitz criterion to a system with the transfer function

$$W(s) = \frac{c}{r}(s) = \frac{2(s^2 + 2s + 5)}{s^5 + s^4 + 3s^3 + 9s^2 + 16s + 10} \quad (4\text{-}3\text{-}5)$$

The characteristic equation is

$$s^5 + s^4 + 3s^3 + 9s^2 + 16s + 10 = 0 \quad (4\text{-}3\text{-}6)$$

which passes the quick test since all the coefficients are present and positive. The Routhian array is

$$
\begin{array}{llllll}
s^5: & +1 & 3 & 16 & 0 \\
s^4: & +1 & 9 & 10 & 0 \\
s^3: & -1 & +1 & 0 \\
s^2: & +1 & +1 & 0 \\
s^1: & +2 & 0 \\
s^0: & +1 & 0
\end{array}
\qquad (4\text{-}3\text{-}7)
$$

Common factors have been removed from the s^3 and s^2 rows. There are two sign changes in the first column: one from $+1$ in the s^4 row to -1 in the s^3 row, and another when the sign changes back to plus in the s^2 row. Two sign changes indicate the presence of two roots in the right-half plane and an unstable system. As soon as a minus sign appears anywhere in the Routhian array we know that the system is not stable.

There are two special cases to be considered. The first of these is a zero in the first column. For the characteristic equation

$$s^5 + 3s^4 + 4s^3 + 12s^2 + 10s + 15 = 0 \quad (4\text{-}3\text{-}8)$$

the Routhian array is

$$
\begin{array}{llll}
s^5: & 1 & 4 & 10 & 0 \\
s^4: & 1 & 4 & 5 & 0 \\
s^3: & 0 & 5 & 0 \\
s^2: & \infty
\end{array}
\qquad (4\text{-}3\text{-}9)
$$

The array blows up at the s^2 row. There are two techniques that may be used to get around this singularity. The first is to substitute $1/x$ for s in the characteristic equation, multiply by x^n, and form a new array. Equation (4-3-8) becomes

$$\frac{1}{x^5} + \frac{3}{x^4} + \frac{4}{x^3} + \frac{12}{x^2} + \frac{10}{x} + 15 = 0 \qquad (4\text{–}3\text{–}10)$$

Multiplying by x^5 and rearranging yields

$$15x^5 + 10x^4 + 12x^3 + 4x^2 + 3x + 1 = 0 \qquad (4\text{–}3\text{–}11)$$

The new Routhian array is

$$
\begin{array}{lllll}
x^5: & +\ 5 & 4 & 1 & 0 \\
x^4: & +10 & 4 & 1 & 0 \\
x^3: & +\ 4 & 1 & 0 \\
x^2: & +\tfrac{3}{2} & 1 & 0 \\
x^1: & -\tfrac{5}{3} & 0 \\
x^0: & +\ 1 & 0 \\
\end{array}
\qquad (4\text{–}3\text{–}12)
$$

There are two sign changes in the first column signifying two roots in the right-half plane and an unstable system.

The second technique for bypassing the zero is to replace it by a very small positive number ϵ, complete the array, and then evaluate the signs in the first column by letting ϵ go to zero. The array in Eq. (4–3–9) becomes

$$
\begin{array}{lccc}
s^5: + & 1 & 4 & 10 \quad 0 \\
s^4: + & 1 & 4 & 5 \quad 0 \\
s^3: + & \epsilon & 5 & 0 \\
s^2: - & \dfrac{4\epsilon - 5}{\epsilon} & 5 & 0 \\
s^1: + & \dfrac{-(\epsilon^2 - 4\epsilon + 5)}{(4\epsilon - 5)} & 0 \\
s^0: + & 5 \\
\end{array}
\qquad (4\text{–}3\text{–}13)
$$

As ϵ goes to zero, the sign of the first element in the s^2 row becomes negative, and that of the s row positive. There are two sign changes, and the system is unstable. When using the first technique, another zero will occasionally appear in the first column. If this happens, the second technique with ϵ must be used.

The second special case is the occurrence of an *entire row of zeros* (not just a zero in the first column), indicating pairs of equal roots with opposite signs located either on the real axis or on the imaginary axis. Consider the characteristic equation

$$D(s) = s^5 + 3s^4 + 10s^3 + 16s^2 + 24s + 16 = 0 \qquad (4\text{–}3\text{–}14)$$

The corresponding Routhian array is

$$
\begin{array}{lcccc}
s^5: & 1 & 10 & 24 & 0 \\
s^4: & 3 & 16 & 16 & 0 \\
s^3: & 1 & 4 & 0 \\
s^2: & 1 & 4 & 0 \\
s^1: & 0 & 0 \\
s^0: & & \\
\end{array}
\qquad (4\text{–}3\text{–}15)
$$

The s row contains all zeros; in the array in Eq. (4–3–9) only the first element of the s^3 row is zero. To proceed with the array in the presence of an entire row of zeros, form an auxiliary equation using the coefficients from the row above, being careful to alternate the powers of s. In this example, the auxiliary equation is

$$A(s) = s^2 + 4 = 0 \qquad (4–3–16)$$

To obtain the elements of the all-zero row, differentiate the auxiliary equation with respect to s to obtain

$$2s + 0 = 0 \qquad (4–3–17)$$

The coefficients of Eq. (4–3–17) become the elements of the s row. The array of Eq. (4–3–15) can now be completed and is

s^5:	1	10	24	0
s^4:	3	16	16	0
s^3:	1	4	0	
s^2:	1	4	0	
s^1:	2	0		
s^0:	4	0		

$$(4–3–18)$$

There are no sign changes in the first column and no roots in the right-half plane. The system is *not* stable, however, because of the row of zeros in the original array. It so happens that the roots of the auxiliary equation are also roots of the characteristic equation. Solving Eq. (4–3–16) we find a pair of complex conjugate roots lying on the imaginary axis at $s = \pm j2$. This system, therefore, is marginally stable.

For a system to be stable, there must be no sign changes in the first column (to ensure that there are no roots in the right-half plane) and no rows of zeros (to ensure that there are no pairs of roots on the imaginary axis). One sign change in the first column and a row of zeros would imply one real root in the right-half plane and a real root of the same magnitude in the left-half plane.

The second special case, involving an entire row of zeros, can be used to establish conditions for the stable operation of a system in terms of any parameter or grouping of parameters. As an example, consider the servomechanism of Fig. 4–3–1. For a specified motor, load, and amplifier ($\tau_m = 0.2$ sec, $\tau_a = 0.1$ sec, $K_m = 5$) the characteristic equation will be

$$D(s) = s^3 + 15s^2 + 50s + 250A = 0 \qquad (4–3–19)$$

We should like to know what values of A will produce a stable system. We construct the Routhian array

s^3:	1	50	0
s^2:	3	50A	0
s^1:	$\dfrac{150 - 50A}{3}$	0	
s^0:	50A	0	

$$(4–3–20)$$

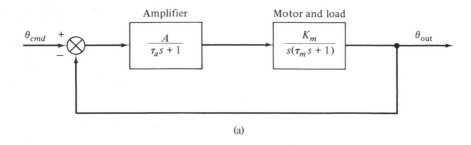

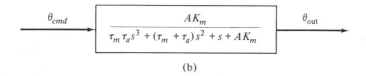

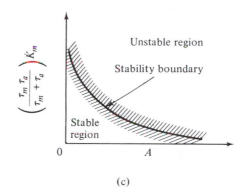

Figure 4–3–1. A third-order servomechanism: (a) block diagram (b) transfer function (c) stability relationships among system parameters.

For the system to be stable, all elements in the first column must be positive. From the s and s^0 rows we have the set of conditions that

$$150 - 50A > 0 \qquad\qquad\qquad (4\text{–}3\text{–}21a)$$

$$\text{or} \quad A < 3 \qquad\qquad\qquad (4\text{–}3\text{–}21b)$$

and

$$50A > 0 \qquad\qquad\qquad (4\text{–}3\text{–}22a)$$

$$\text{or} \quad A > 0 \qquad\qquad\qquad (4\text{–}3\text{–}22b)$$

Consequently, the amplifier gain A must be greater than 0 but less than 3. Such a system is called a *conditionally stable system* since the system is stable only when the parameter A meets certain conditions.

In this example, when A is equal to 3, the s row is an all-zero row. The auxiliary equation is

$$3s^2 + 150 = 0 \qquad (4\text{-}3\text{-}23)$$

with roots equal to $\pm j7.07$ lying on the imaginary axis. The system is marginally stable for this value of A, which is called the *critical value* of A and is denoted by A_c. When A is equal to zero, the s^0 row becomes the all-zero row. The corresponding auxiliary equation is

$$50s = 0 \qquad (4\text{-}3\text{-}24)$$

with a single root at the origin, and the system is now limitedly stable. If A is greater than 3, the sign of the first element in the s row is negative. All other signs in the first column are positive, so there are two sign changes, there are two roots in the right-half plane, and the system is unstable. If A is less than zero (negative), the first element in the s^0 row will be negative. There will be one sign change, there will be one root in the right-half plane, and the system will be ustable.

The Routh-Hurwitz criterion can also be used to determine relationships among system parameters necessary for stability. In parametric form the char acteristic equation of the system of Fig. 4-3-1 is

$$D(s) = \tau_m \tau_a s^3 + (\tau_m + \tau_a)s^2 + s + AK_m = 0 \qquad (4\text{-}3\text{-}25)$$

with the array

$$
\begin{array}{llll}
s^3\colon & \tau_m\tau_a & 1 & 0 \\[2mm]
s^2\colon & (\tau_m + \tau_a) & AK_m & 0 \\[2mm]
s^1\colon & \dfrac{\tau_m + \tau_a - \tau_m\tau_a AK_m}{\tau_m + \tau_a} & 0 & \\[3mm]
s^0\colon & AK_m & 0 &
\end{array}
\qquad (4\text{-}3\text{-}26)
$$

The two conditions for stability are that

$$\tau_m + \tau_a > \tau_m\tau_a AK_m \qquad (4\text{-}3\text{-}27)$$

and

$$AK_m > 0 \qquad (4\text{-}3\text{-}28)$$

The stability requirement of Eq. (4-3-28) shows that with K_m positive, the usual convention, A must also be positive. Then the inequality of Eq. (4-3-27) can be rewritten as

$$\frac{\tau_m\tau_a}{\tau_m + \tau_a}K_m < \frac{1}{A} \qquad (4\text{-}3\text{-}29)$$

so that the stability conditions with respect to the amplifier gain A are that

$$0 < A < \frac{\tau_m + \tau_a}{K_m\tau_a\tau_m} \qquad (4\text{-}3\text{-}30)$$

The inequality of Eq. (4-3-29) is sketched in Fig. 4-3-1c, revealing a stability boundary separating regions of stable and unstable operation. If we hold τ_m and K_m constant (i.e., specify a particular motor), increasing τ_a reduces the maximum allowable amplifier gain for system stability. For example, with $\tau_a = 0.1$ sec, the maximum allowable gain was found to be 3. Increasing τ_a to 0.2 sec reduces the maximum allowable value of A to 2 and decreasing τ_a to 0.05 sec increases the maximum value of A to 5.

The major disadvantage of the Routh-Hurwitz criterion is that it does not specifically locate the roots of the characteristic equation. Consequently, we know only the absolute stability of the system and nothing about the degree of stability, that is, the nature of the transient response. The exact location of the roots may be found by computer solutions or by various factoring techniques, some of which are described in App. B. The Routh-Hurwitz criterion can be used in conjunction with Horner's method* to determine specific root locations; it is a tedious process and is not recommended.

4–4 FIRST-ORDER SYSTEMS

With the transfer function in the form

$$W(s) = \frac{c}{r}(s) = \frac{N(s)}{D(s)} \tag{4–4–1}$$

the order of the system is defined as the order of the characteristic function $D(s)$; i.e., the highest power of s appearing in $D(s)$ establishes the order of the system. The order of the numerator polynomial $N(s)$ is generally less than the order of the system. If the order of the system is n and that of $N(s)$ zero—no s terms—the system is called a *simple nth order system*.

We have defined the values of s that cause $D(s)$ to be equal to zero as the roots of the characteristic equation. The values of s that make $N(s)$ equal to zero are customarily called the *closed-loop zeros*. The roots determine the number and shape of the transient modes and thus the stability of a system. The closed-loop zeros affect the amplitude and sign of the transient modes and thus influence the shape of the time response.

The hydraulic position servo of Fig. 3–6–2 is a simple first-order system with the transfer function

$$\frac{x_{out}}{x_{in}} = \frac{1}{\tau s + 1} \tag{4–4–2}$$

The single root of the characteristic equation is equal to $-1/\tau$. Since the time constant τ is made up of positive quantities and is positive, the root lies in the left-half s plane and the servo is stable.

The performance criteria of Sec. 1–6 are based on the characteristics of the time response to a unit step input. With $x(t) = u(t)$ and $x(s) = 1/s$, the transformed unit step response can be written as

$$x_{out}(s) = \frac{1/\tau}{s(s + 1/\tau)} = \frac{1}{s} - \frac{1}{s + 1/\tau} \tag{4–4–3}$$

* By progressively shifting the imaginary axis to the left and searching for roots in this expanded right-half plane, the location of the roots can be determined.

The unit step time response, therefore, is

$$x_{out}(t) = 1 - e^{-t/\tau} \tag{4-4-4}$$

The constant term is the steady-state response, and the decaying exponential the transient response. The sketch of the unit step response in Fig. 4–4–1 shows that the response is determined solely by the value of the time constant τ; the smaller the time constant, the more rapidly the output approaches the input value. Since the output neither overshoots the steady-state value nor oscillates, the settling time* t_s is sufficient to define the speed of response. Setting $x_{out}(t) = 0.95 x_{in}(t)$ in Eq. (4–4–4) and solving for the settling time, we find the settling time to be exactly equal to 3τ.

Let us now examine the time response to a unit ramp input (a unit step in velocity) with $x_{in}(t) = t$ and $x_{in}(s) = 1/s^2$.

$$x_{out}(s) = \frac{1/\tau}{s^2(s + 1/\tau)} = \frac{-\tau}{s} + \frac{1}{s^2} + \frac{\tau}{s + 1/\tau} \tag{4-4-5}$$

and

$$x_{out}(t) = (t - \tau) + \tau e^{-t/\tau} \tag{4-4-6}$$

The first two terms comprise the steady-state response, and the exponential the transient response. The total time response is sketched in Fig. 4–4–2. It can be shown that the settling time for this type of input is not precisely 3τ; however, taking t_s equal to 3τ does provide a useful indication as to the order of magnitude of the settling time.

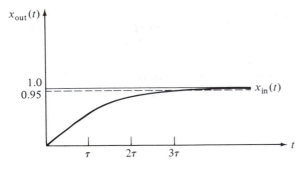

Figure 4–4–1. Time response of a stable simple first-order system to a unit step input.

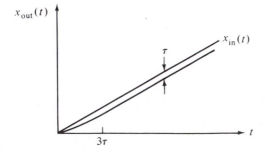

Figure 4–4–2. Time response of a stable simple first-order system to a unit ramp input.

* For first-order systems the settling time is often referred to as the *response time*.

Although the system is stable, the steady-state value of the output $(t - \tau)$ will become infinite as t goes to infinity. It is important to realize that an infinite steady-state output does not necessarily mean a system is unstable. Do *not* use the steady-state value of the output as a criterion for determining stability.

Comparison of Figs. 4–4–1 and 4–4–2 and of Eqs. (4–4–4) and (4–4–6) shows the effect of the type of input upon the accuracy of a system. For the unit step input, the steady-state error (the difference between the input and output) is zero, but for a unit ramp input it is constant and equal to the time constant τ.

It may be of interest to note that the time response to the unit step as given in Eq. (4–4–4) is the time derivative of the response of the system to a unit ramp input as given in Eq. (4–4–6). This should be expected inasmuch as a unit step is the time derivative of a unit ramp.

If our hydraulic servo is built with positive feedback, as in Fig. 4–4–3, the transfer function becomes

$$\frac{x_{\text{out}}}{x_{\text{in}}} = \frac{1}{\tau s - 1} \tag{4–4–7}$$

The single root of the characteristic equation is $+ 1/\tau$, and this system should be unstable. The transformed unit step response is

$$x_{\text{out}}(s) = \frac{1/\tau}{s(s - 1/\tau)} = \frac{-1}{s} + \frac{1}{s - 1/\tau} \tag{4–4–8}$$

and

$$x_{\text{out}}(t) = -1 + e^{+t/\tau} \tag{4–4–9}$$

The divergent and unbounded time response typical of an unstable system is shown in Fig. 4–4–4.

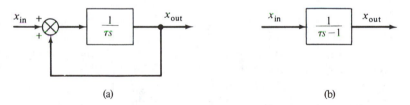

(a)   (b)

Figure 4–4–3. A hydraulic servo with positive feedback: (a) block diagram (b) transfer function.

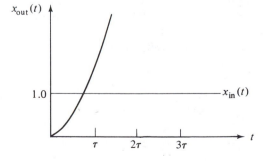

Figure 4–4–4. Time response of an unstable simple first-order system to a unit step input.

The pure integrator with the transfer function

$$\frac{x_{out}}{x_{in}} = \frac{1}{s} \tag{4-4-10}$$

may be considered a special case of a simple first-order system with an infinite time constant. The single root ($s = 0$) of the characteristic equation is located at the origin and the system is limitedly stable. The unit step time response is simply $x_{out}(t) = t$ and is sketched in Fig. 4-4-5.

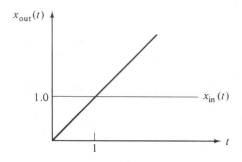

Figure 4-4-5. Time response of a limitedly stable simple first-order system to a unit step input.

4-5 SECOND-ORDER SYSTEMS

The transfer function of a simple second-order system can be written in parametric form as

$$\frac{c}{r}(s) = \frac{\omega_n^2}{s^2 + 2\zeta\omega_n s + \omega_n^2} \tag{4-5-1}$$

where ζ and ω_n are called the *damping ratio* and the *undamped natural frequency* of the system, respectively; ζ is dimensionless and the dimension of ω_n are radians per unit time.

Equation (4-5-1) could be the transfer function of the electromechanical servomechanism of Fig. 4-5-1, which might be used to position a tracking radar antenna, repeat an angular position or velocity, move a robotic arm, or rotate a piece of stock being milled. The mechanical analog of a simple second-order system is the spring-mass-dashpot system of Fig. 3-2-2. Visualize the time response to a unit step input in terms of the behavior of the mass after it has been pulled down from its equilibrium position and released. A large ω_n implies a large spring constant and a large ζ a viscous damping coefficient that is large with respect to the spring constant. In the servomechanism example, think of the amplifier gain as a spring constant and the reciprocal of the motor time constant as a measure of the viscous damping.

The characteristic equation of the transfer function is

$$s^2 + 2\zeta\omega_n s + \omega_n^2 = 0 \tag{4-5-2}$$

and has two roots

$$s = -\zeta\omega_n \pm \omega_n\sqrt{\zeta^2 - 1} \tag{4-5-3}$$

The nature of the roots is dependent upon the magnitude of the damping ratio ζ. If ζ is greater than unity, the roots are real and unequal; equal to unity, the

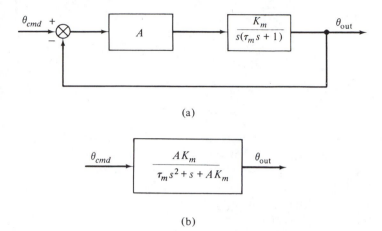

(a)

(b)

Figure 4–5–1. A simple second-order electromechanical servomechanism: (a) block diagram (b) transfer function.

roots are real and equal; and less than unity, the roots are complex conjugates. These relationships are illustrated in Fig. 4–5–2a. Both ζ and ω_n are assumed to be positive, which is the case for physical systems.

When ζ is less than unity, the system is called *underdamped*, and the expression for the roots is written as

$$s = -\zeta\omega_n \pm j\omega \qquad (4\text{–}5\text{–}4)$$

In Eq. (4–5–4) ω is called the *damped frequency* of response and is equal to $\omega_n\sqrt{1 - \zeta^2}$.* Geometric relationships among the parameters are illustrated in Fig. 4–5–2b.

The transformed response of a simple second-order system to a unit step input is

$$c(s) = \frac{\omega_n^2}{s(s^2 + 2\zeta\omega_n s + \omega_n^2)} = \frac{1}{s} - \frac{s + 2\zeta\omega_n}{(s + \zeta\omega_n)^2 + \omega^2} \qquad (4\text{–}5\text{–}5)$$

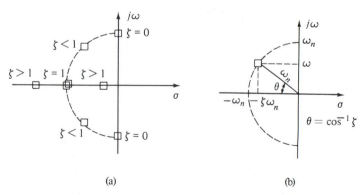

(a)   (b)

Figure 4–5–2. Roots of a simple second-order system: (a) effect of varying ζ (b) geometric relationships of ζ, ω_n, and ω.

* The *damped period P* is $2\pi/\omega$.

and the time response is

$$c(t) = 1 - \frac{1}{\sqrt{1 - \zeta^2}} e^{-\zeta \omega_n t} \sin(\omega t + \psi) \qquad (4\text{--}5\text{--}6)$$

where

$$\psi = \tan^{-1} \frac{\sqrt{1 - \zeta^2}}{\zeta} = \cos^{-1} \zeta \qquad (4\text{--}5\text{--}7)$$

When this unit step response is sketched against the dimensionless time $\omega_n t$, as in Fig. 4–5–3, we see that the shape is determined by the damping ratio ζ and that the time scaling is determined by the undamped natural frequency ω_n.

When ζ is equal to zero, the system is *undamped* and the response is a bounded sinusoidal oscillation about the input; this is the response of a marginally stable system.* As ζ is increased, the peak overshoot and the number of oscillations are reduced and the output will eventually settle down to a steady-state value equal to the input. When $0.707 \leqslant \zeta < 1$, there is only one overshoot and no undershoots. When $\zeta = 1$, the system is *critically damped* with no overshoot whatsoever. When $\zeta > 1$, the system is *overdamped* with a response that resembles that of a simple first-order system.

The *percent overshoot* (PO) is a criterion commonly used in determining the suitability of the transient response. It is defined as

$$PO = 100 \times \left(\frac{\text{peak value} - \text{steady-state value}}{\text{steady-state value}} \right) \qquad (4\text{--}5\text{--}8)$$

The PO is normalized and thus independent of the magnitude of the step input. An analytic expression can be obtained by setting the time derivative of Eq. (4–

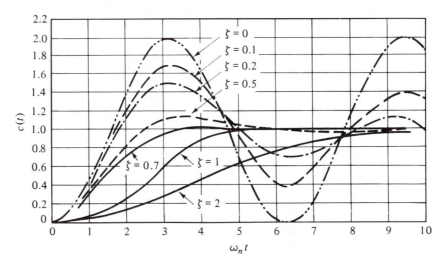

Figure 4–5–3. Time response of a stable simple second-order system to a unit step input.

* A system deliberately designed to be marginally stable is called an *oscillator*.

5–6) equal to zero and solving for the times when the output reaches maximum and minimum values. Thus

$$t_i = \frac{i\pi}{\omega_n\sqrt{1 - \zeta^2}} \qquad i = 1, 2 \dots \qquad (4\text{–}5\text{–}9)$$

The peak overshoot occurs when $i = 1$, and the peak time is

$$t_p = \frac{\pi}{\omega_n\sqrt{1 - \zeta^2}} = \frac{\pi}{\omega} \qquad (4\text{–}5\text{–}10)$$

Substituting Eq. (4–5–10) into Eq. (4–5–6) and substituting that result into Eq. (4–5–8) produces an expression for the percent overshoot that is solely a function of the damping ratio:

$$PO = 100 \exp\left(\frac{-\zeta\pi}{\sqrt{1 - \zeta^2}}\right) \qquad (4\text{–}5\text{–}11)$$

This relationship is plotted in Fig. 4–5–4; notice that a little damping does a lot to reduce the peak overshoot.

It can be shown from Eq. (4–5–6) that the output will be within 5 percent of its steady-state value when t becomes equal to $3/\zeta\omega_n$. Thus, it is customary to describe the settling time as

$$t_s = \frac{3}{\zeta\omega_n} \qquad (4\text{–}5\text{–}12)$$

The percent overshoot is a function of the damping ratio alone, but the peak and settling times are functions of both ζ and ω_n. Increasing the damping ratio will reduce the percent overshoot, the settling time, and the number of oscillations—all desirable objectives. Unfortunately, however, increasing ζ also increases the peak time, as well as the delay and rise times with the result that the system will become slower in responding to an input—more sluggish. The effect of an increasing ζ upon these times can be observed in Fig. 4–5–3. This figure also shows that increasing the undamped natural frequency reduces these times and speeds up the system response.

Analytic expressions cannot be obtained for the delay and rise times; if

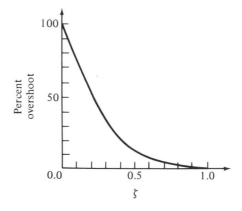

Figure 4–5–4. The percent overshoot as a function of the damping ratio ζ.

they must be known, they can be measured on a time response plot for a specific ζ and a specific ω_n, or empirical approximations can be developed. The peak and settling times are usually sufficient to indicate the speed of response of a system.

A desirable transient response would appear to call for a relatively large ζ to reduce the percent overshoot, the number of oscillations, and the settling time; a large ω_n is needed to speed up the response by reducing the delay and rise times, as well as the settling time. By drawing lines of constant ζ, ω_n, and t_s in the s plane, as in Fig. 4–5–5, we can see how these parameters are related to each other and to the location of the roots of the characteristic equation. As an illustration, the simple second-order system with roots at 1 has a larger PO, a longer t_s, and smaller delay, rise, and peak times than the system with roots at 2. Of the three parameters in Fig. 4–5–5, only two can be independently selected. Furthermore, ζ and ω_n are often adversely coupled. In the system of Fig. 4–5–1, for example, ζ and ω_n are so coupled that any increase in ω_n reduces ζ. If that were not bad enough, increasing the undamped natural frequency increases the bandwidth of the system, leading to potential problems with disturbance inputs.

The overdamped simple second-order system, $\zeta > 1$, is of less interest than the underdamped system since it is usually too sluggish to serve as a useful control system. The roots are real, and the system behaves very much like a first-order system, as can be seen from the overdamped responses in Fig. 4–5–3. In fact, we use time constants to describe the overdamped system, writing the transfer function as

$$\frac{c}{r} = \frac{1}{(\tau_1 s + 1)(\tau_2 s + 1)} \tag{4–5–13}$$

The time constants, of course, are functions of ζ and ω_n. There is no peak overshoot; the settling time is assumed to be three times the larger of the time constants, even though the actual settling time is generally somewhat longer.

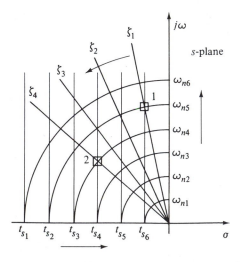

Figure 4–5–5. Constant parameter lines in the s plane.

When ζ is equal to unity, the roots are both real and equal, and the system is *critically damped*.

If we add a closed-loop zero to a simple second-order system, the system is obviously no longer simple, nor is it possible to define the transient response completely by specifying ζ and ω_n as before. The addition of a closed-loop zero does not change the nature of the transient modes, since the characteristic equation is unaltered, but it does affect the transient response. The peak overshoot is increased, and the delay, rise, and peak times are reduced; the settling time and number of oscillations remain essentially unchanged, as shown in Fig. 4–5–6. Analytic expressions can be developed to determine the effects of the closed-loop zero in terms of its location with respect to both the roots of the characteristic equation and the origin of the s plane. In essence, the influence of the zero increases as it approaches the origin, that is, as its time constant τ_z increases.

As an illustration, consider a second-order system with the transfer function

$$\frac{c}{r}(s) = \frac{16(\tau_z s + 1)}{s^2 + 4s + 16} \qquad (4\text{--}5\text{--}14)$$

This is an underdamped system with $\zeta = 0.5$ and $\omega_n = 4$ rad/sec. When $\tau_z = 0$, the closed-loop zero is located at minus infinity and Eq. (4–5–14) is the transfer function of a simple second-order system. The unit step time response is characterized by a peak overshoot of approximately 16 percent, a peak time of approximately 0.91 sec, and a settling time on the order of 1.5 sec. Setting τ_z equal to 0.1 sec puts the closed-loop zero at -10 in the s plane. The peak overshoot increases to approximately 19 percent, and the peak time decreases to 0.74 sec. Moving the zero to -2 results in a peak overshoot of approximately 70 percent and a peak time of approximately 0.45 sec. Moving the zero even closer to the origin, to -1, increases the peak overshoot to approximately 170 percent and reduces the peak time to approximately 0.40 sec.

A closed-loop zero will also increase the speed of response of an overdamped system, as might be expected. When the time constant of the zero is sufficiently large so as to place the zero between the roots and the origin, as in Fig. 4–5–7a, the output will overshoot its steady-state value, as shown in Fig. 4–5–7b. The peak overshoot is 52 percent and the peak time is approximately 0.32 sec, but the settling time is still on the order of 1.7 sec, the approximate response

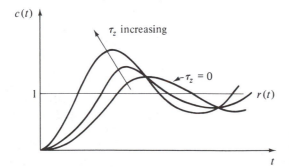

Figure 4–5–6. Effect of closed-loop zero upon unit step response of an underdamped second-order system.

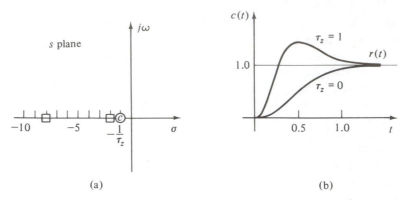

Figure 4–5–7. An overdamped second-order system with a closed-loop zero and

$$TF = \frac{16(\tau_z s + 1)}{s^2 + 10s + 16}$$

(a) s-plane locations (b) unit step response.

time of the original overdamped system. The addition of this closed-loop zero has made the first part of the step response resemble that of an underdamped system with $\zeta = 0.2$.

In summary, the unit step time response of a simple second-order system is specified if ζ and ω_n are known; ζ and ω_n, in turn, are known if the location of the roots of the characteristic equation are known. If a closed-loop zero is added to the system, ζ and ω_n no longer define completely the time response, but in conjunction with the location of the closed-loop zero, they do give a qualitative indication as to the response. The closed-loop zero may be considered a destabilizing influence in that it decreases the rise time and increases the percent overshoot, as would a reduction in the damping ratio ζ. The closer the closed-loop zero is to the origin of the s plane, the stronger its influence is.

4–6 HIGHER-ORDER SYSTEMS

As the order of a system increases, so does the difficulty of defining the transient response in terms of the root locations. Generalizations are possible, however, so that the analysis and design processes may proceed without the necessity of returning to the time domain every time a change is made.

Since a third-order system must have either three real roots or one real root plus a complex conjugate pair, we may regard it as a second-order system, either underdamped or overdamped, to which we have added a real root. With a stable system, this real root will be negative and will add a transient mode that is a decaying exponential. The effect of this additional root upon the time response of the system is a function of the location of the added root with respect to the other roots and to the origin. In general, the added root can be considered a stabilizing influence in that it tends to make the system slower to respond and tends to reduce both the percent overshoot and the number of oscillations; in fact, an underdamped system can be made to perform like an overdamped system, as shown in Fig. 4–6–1. In this figure the added root is some distance away from the complex pair. When the root is close to the pair, the time response may exhibit the characteristics of both an overdamped and underdamped system, e.g., oscillations about a slowly increasing output as in Fig. 4–6–2.

The presence of closed-loop zeros further complicates the response. As with the added root, the degree of complication is dependent upon the specific and relative locations of the zeros and roots. If a real closed-loop zero coincides with that of a real root,* one cancels the other and the third-order system effectively becomes second-order. Similarly, a pair of complex closed-loop zeros coincident with a complex pair of roots results in a first-order system. Ignoring such cancellations, we can generalize by saying that closed-loop zeros destabilize the response of a third-order system in the same way they destabilize second-order systems.

When a linear system has m inputs affecting n outputs, there are mn time responses to be considered. The mn transfer functions will have individual

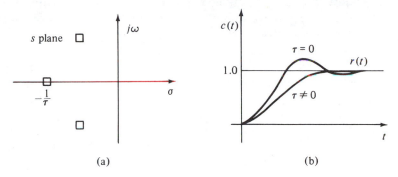

(a)   (b)

Figure 4–6–1. Real root added to underdamped second-order system: (a) root location (b) effect on unit step response.

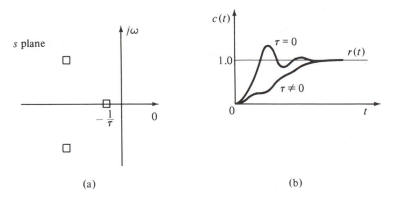

(a)   (b)

Figure 4–6–2. Real root added to underdamped second-order system: (a) root location (b) effect on unit step response.

* If the root and zero are close but not coincident, the pair is called a *dipole*.

numerator functions but identical characteristic (denominator) functions, as illustrated in Fig. 4–6–3. If the system is third-order, the transformed unit step
response can be written as

$$c_i(s) = \frac{N_i(s)}{s\left(s + \dfrac{1}{\tau}\right)(s^2 + 2\zeta\omega_n s + \omega_n^2)} \qquad i = 1, 2, \ldots \qquad (4\text{–}6\text{–}1)$$

where $N_i(s)$ represents the numerator function appropriate to the input-output
pair under consideration and the s in the denominator represents the transformed
input. After expanding Eq. (4–6–1) in partial fractions, we can write the generalized
unit step response in the form

$$c_i(t) = A_i + B_i e^{-t/\tau} + D_i e^{-\zeta\omega_n t} \sin(\omega t + \psi_i) \qquad (4\text{–}6\text{–}2)$$

The coefficients A_i, B_i, and D_i, along with ψ_i,* determine the shape of the time
response and in turn are determined by the location of the roots and of the
closed-loop zeros associated with $N_i(s)$. Since we are interested only in the
response of stable systems, A_i can easily be found with the final value theorem.
If we know B_i and D_i, or obtain their signs and relative magnitude by the graphical
method of App. A, we know which, if any, mode(s) is dominant, and thus we
know the overall shape of the system response.

If we do not wish to concern ourselves with B_i and D_i, we can still draw
some conclusions as to the general shape of the response. Assuming no cancellation
of roots by closed-loop zeros, we know the shape and approximate settling time
of each transient mode. If the real root in a third-order system is far to the left
of the complex pair, its associated mode can probably be neglected; if it lies
between the complex pair and the imaginary axis, it should not be neglected.
In this latter situation, the complex pair can probably be neglected if it is not
close to the real root. We examined these possibilities for a simple third-order
system in Fig. 4–6–1; closed-loop zeros would reduce the apparent damping of
these responses, thus increasing the speed of response and overshooting.

For systems of order higher than two or three, it is common practice to
consider only the *dominant roots* in the design process. Normally, the dominant
roots are taken to be those closest to the imaginary axis on the premise that
they will have the longest settling times and, if complex, the smallest damping
ratios. Often the dominant roots are a single complex conjugate pair leading to

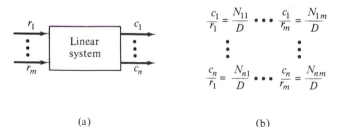

(a)   (b)

Figure 4–6–3. Multivariable system: (a) system representation (b) transfer functions.

---

* ψ is equal to $\cos^{-1} \zeta$ only for simple second-order systems.

an approximation of the system by a simple second-order system. This approximation must be made with care and with the realization that its validity may be doubtful. Some cases where the second-order approximation would be suspect are illustrated in Fig. 4–6–4. In Fig. 4–6–4a the two pairs of complex roots may be sufficiently close together to interact. A similar situation exists in Fig. 4–6–4b for a real root in conjunction with a complex pair. The situation in Fig. 4–6–4c is different in that the two pairs are separated but the frequency of the closer pair is so small, both in an absolute sense and in relation to that of the other pair, that the other pair may not be neglected and may even be the dominant pair. This last arrangement of roots is encountered in the longitudinal dynamic equations of many modern aircraft where the closer pair represents the phugoid mode and the other pair the short period mode.

In conclusion, we can visualize the unit step response of a higher-order system in terms of the response of a simple second-order system as modified by the addition of roots and closed-loop zeros. The roots, particularly the real roots, tend to stabilize, whereas the closed-loop zeros tend to destabilize. The influence of the roots and closed-loop zeros increases as they approach the imaginary axis. The second-order (dominant root) approximation must be used with caution. In any event, the design of a control system should not be considered complete until the complete time response is calculated, plotted, and thoroughly examined for acceptability.

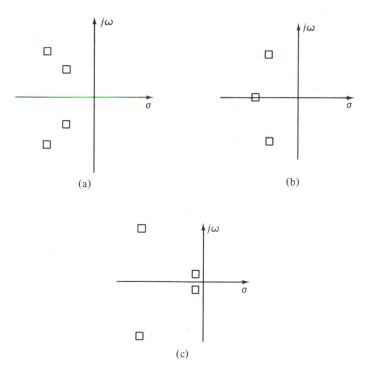

Figure 4–6–4.   Some root locations where a second-order approximation would be suspect.

PROBLEMS

4–1. For each of the following characteristic equations, describe (with justification) the stability of the system and write the expression for the transient response with the coefficients unspecified:

(a) $(s^2 - 1)(s + 5) = 0$
(b) $(s + 0.1)(s + 2)^2 = 0$
(c) $s(s - 1)(s^2 + 2s + 2) = 0$
(d) $s(s + 1)(s^2 + 2s + 2) = 0$
(e) $s^2(s + 1)(s^2 + 2s + 2) = 0$
(f) $(s^2 + 1)(s + 5) = 0$
(g) $(s^2 + 1)(s - 5) = 0$
(h) $(s + 3)(s^2 - 1)^2 = 0$
(i) $(s + 5)(s^2 - 4s + 4) = 0$
(j) $(s + 5)(s^2 + 4s + 4) = 0$

4–2. For each of the following characteristic equations, use the Routh-Hurwitz criterion to find the number of roots in the right half of the s plane or on the imaginary axis and describe the stability of the system:

(a) $s^5 + 7s^4 + 3s^3 + 30s^2 + 21s + 42 = 0$
(b) $2s^5 + 8s^4 + 5s^2 + 7s + 11 = 0$
(c) $s^4 - 8s^3 + 16s^2 - 32s + 80 = 0$
(d) $s^4 + 4s^3 + 16s^2 + 60s + 15 = 0$
(e) $s^4 - 5s^3 + 10s^2 - 20s + 24 = 0$
(f) $s^5 + 5s^4 + 2s^3 + 10s^2 + s + 5 = 0$
(g) $s^5 + 2s^4 + s^3 - 2s^2 + 2s - 4 = 0$

4–3. For each of the following characteristic equations, determine the conditions for system stability in terms of the unspecified parameter(s). Verify the conditions by substituting appropriate numerical values.

(a) $s^4 + 3s^3 + 3s^2 + 4Ks + K = 0$
(b) $s^4 + Ks^3 + 2s^2 + (5 + K)s + 5 = 0$
(c) $1 + \dfrac{K(s + 20)}{s(s + 5)(s + 10)} = 0$
(d) $s^4 + 5s^3 + 15s^2 + Ks + 3K = 0$
(e) $s^3 + 10s^2 + (K - 9)s + (5K - 90) = 0$
(f) $As^4 + 5s^3 + 10s^2 + 2Bs + B = 0$

4–4. A simple first-order system has the transfer function

$$\frac{c}{r} = \frac{N(s)}{s + 5}$$

(a) Find the time constant τ of the system. Describe the transient mode(s) with respect to stability, shape, response time, etc.
(b) With $N(s) = 5$, zero initial conditions, and a unit step input, write expressions for $c(t)$ and $dc(t)/dt$. Find the initial and final (steady-state) values of each and sufficient values of $c(t)$ to sketch the approximate time response. Find the approximate values of any peak overshoots and peak times.
(c) With $N(s) = 4$, do (b) and place the sketch on same plot with (b).
(d) Discuss any effects of $N(s)$ upon the system response.

4–5. A second-order system has the transfer function

$$\frac{c}{r} = \frac{N(s)}{s^2 + 5s + 4}$$

(a) Find ζ, ω_n, t_s, ω, and P (the damped period) and describe the stability and qualitative transient response of the system.

(b) With $N(s) = 4$, zero initial conditions, and a unit step input, write the expressions for $c(t)$ and $dc(t)/dt$. Find the initial and final values of each and sufficient values of $c(t)$ to sketch the approximate time response. Find the approximate values of any peak overshoots and peak times.

(c) Do (b) for $N(s) = 4(s + 1)$.

(d) Do (b) for $N(s) = 8(s + 0.5)$.

(e) Discuss the effects of the closed-loop zero and its location upon the stability and response of the system.

4–6. Do Prob. 4–5 for

$$\frac{c}{r} = \frac{N(s)}{s^2 + 2.4s + 4}$$

4–7. For the second-order system of Prob. 4–6 with $N(s) = 4$,

(a) Sketch the unit step time response for this simple second-order system.

(b) Add a single real root $(\tau s + 1)$ to make the system a simple third-order system. On the same figure with the sketch of (a) sketch the unit step response for $\tau = 0.1$ sec and $\tau = 10$ sec.

(c) Discuss the effects of adding these first-order time lags to the system.

4–8. For each of the following systems, write the expression for the unit step time response without specifying any coefficients or phase angles. Locate the roots and closed-loop zeros in the s plane. Indicate which root(s), if any, that you think might be dominant and roughly sketch the corresponding step response.

(a)
$$\frac{c}{r} = \frac{200}{(s + 2)(s + 100)}$$

(b)
$$\frac{c}{r} = \frac{200}{s(s + 2)(s + 100)}$$

(c)
$$\frac{c}{r} = \frac{10(s + 3)}{(s + 1)(s + 3)(s + 10)}$$

(d)
$$\frac{c}{r} = \frac{0.005(s + 0.5)(s + 5)}{(s^2 + 0.005s + 0.005)(s^2 + 0.9s + 2.5)}$$

(e)
$$\frac{c}{r} = \frac{8(s + 6)}{(s^2 + 9)(s + 1)(s^2 + 0.6s + 4)}$$

(f)
$$\frac{c}{r} = \frac{100(0.1s + 1)(0.01s + 1)}{(0.005s + 1)(0.02s + 1)(0.5s + 1)(s^2 + 6s + 100)}$$

(g)
$$\frac{c}{r} = \frac{1080(s + 10)}{(s + 100)(s + 3)(s^2 + 6s + 36)}$$

4–9. The system of Fig. P4–9 has a plant transfer function

$$\frac{c}{u} = \frac{3}{s(s - 1)(s + 5)}$$

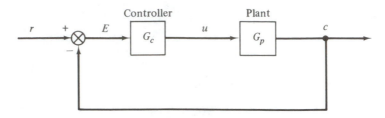

Figure P4–9

(a) Is the plant stable? Justify.

(b) If the controller is a one-mode proportional controller with a $G_c = A$, can the system be made stable? If so, what are the conditions for stability?

(c) If the controller is a proportional-derivative (PD) controller with $G_c = A(s + 1)$ can the system be made stable? If so, what are the conditions for stability?

4–10. If the plant transfer function of Fig. P4–9 is

$$\frac{c}{u} = \frac{20}{s(s + 5)(s + 10)}$$

(a) Is the plant stable? Justify.

(b) If $G_c = A$, is the system stable for all positive values of A? For $A = 20$? For $A = 50$? Find the final value of $c(t)$ for a unit step for $A = 20$. For $A = 50$.

(c) If the controller is a proportional-integral (PI) controller with $G_c = A(s + 1)/s$, what are the conditions for system stability? Find the final value of $c(t)$ for a unit ramp input, i.e., $r(t) = t$. Is this value a function of A?

4–11. For system of Fig. P4-11,

(a) If the feedback sign at comparator is negative, is the system stable for positive values of A? For negative values of A?

(b) If the feedback is positive, is the system stable for positive values of A? For negative values of A?

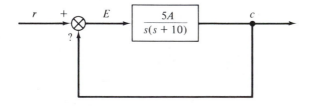

Figure P4–11

5

The Steady State

5–1 THE STEADY-STATE EQUATIONS

A control system, as generalized in Fig. 5–1–1, is designed to control the dynamic behavior (the time response) of a plant subjected to commands or disturbances. When the plant either exists or has already been designed, the principal task of the control system designer is the design of a suitable controller; in so doing, the dynamic equations are the primary tools. The designer should be fully aware, however, of the role of the steady-state equations* in the overall process, as well as their influence on the dynamic behavior of the plant.

The steady-state equations are used for the initial design of the plant. They establish the nominal operating conditions of the plant in terms of the desired output(s); these conditions in turn determine plant parameters and nominal inputs. The steady-state relationships among the plant variables and parameters are the basis for the optimization of the steady-state performance. Although the steady-state equations are not explicitly used in the dynamic analysis of the plant and in the subsequent control system design, the choice of plant parameters and operating conditions affects the coefficients of the dynamic equations and thus the dynamic behavior. The results of the dynamic analysis might well result in a modification of the plant parameters and operations, if this is possible, and thus simplify the control problem.

A simple example illustrating the interplay between the steady-state and dynamic analyses of a plant is the design of the idealized automobile suspension

---

* See Sec. 2–2 for the distinction between the steady-state and dynamic equations.

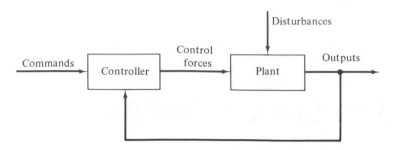

Figure 5–1–1. The generalized control system.

system of Fig. 3–2–4. This figure is redrawn in Fig. 5–1–2a to show the relationship between the undeflected position of the mass and the steady-state equilibrium position. The steady-state equation is

$$mg = kX_0 \tag{5-1-1}$$

and the dynamic equation is

$$m\ddot{x} + B\dot{x} + kx = B\dot{y} + ky \tag{5-1-2}$$

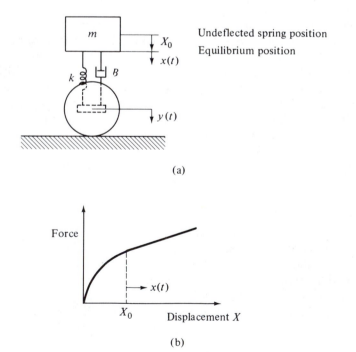

(a)

(b)

Figure 5–1–2. (a) An automobile suspension; (b) a nonlinear spring.

The transfer function is

$$\frac{x}{y}(s) = \frac{\dfrac{B}{m}\left(s + \dfrac{k}{B}\right)}{s^2 + \dfrac{B}{m}s + \dfrac{k}{m}} \tag{5–1–3}$$

Equation (5–1–3) is the transfer function of a stable second-order system with a single closed-loop zero.

   At this point a look at the design objectives of the suspension system is in order. The displacement of the wheel $y(t)$ is a disturbance input, and the purpose of the suspension system is to minimize the resulting displacement of the body $x(t)$, ideally to decouple the motion of the automobile body from that of the wheel. The shape of the input will probably be a pulse, caused by a pothole or bump, or sinusoidal in nature because of a rough road. The pulse input calls for a relatively high damping ratio to slow down the response and preclude uncomfortable overshoots and oscillations. There is obviously an upper limit on the damping ratio; if it were infinite, the body would be rigidly coupled to the wheel and would faithfully follow its motion. The rough road input calls for a small bandwidth, implying a low undamped natural frequency.

   Let us suppose that the nominal weight of our automobile is 4000 lb and that Fig. 5–1–2a represents one wheel and its associated suspension system supporting one-fourth of the weight, 1000 lb (31.1 slugs). Let us further suppose that for styling reasons X_0 is to be 1 ft (0.305 m). With the assumption of a linear spring, the steady-state equation, Eq. (5–1–1), sizes the spring; k is to be 1000 lb/ft (14,580 N/m). Since

$$\omega_n = \sqrt{\frac{k}{m}} = \sqrt{\frac{1000}{31.1}} \tag{5–1–4}$$

it is now specified as 5.674 rad/sec. The only remaining parameter to be selected is the viscous damping coefficient B. Let us assume that we will accept one overshoot. Keeping in mind the destabilizing influence of the closed-loop zero, let us select a ζ equal to 0.8. Since

$$\zeta = \frac{B}{2\sqrt{km}} \tag{5–1–5}$$

we need a shock absorber with B equal to 284 lb-sec/ft (4443.6 N/m-sec). The settling time will be approximately $3/\zeta\omega_n$ or 0.66 sec.

   An undamped natural frequency of 5.67 rad/sec is probably too high for a smooth ride over a rough road. If we specify ω_n equal to 1 rad/sec, we must use a weaker spring with k equal to 31.1 lb/ft (453.75 N/m). The damping ratio can be held at 0.8 by using a weaker shock with B approximately 50 lb-sec/ft (729 N/m-sec). The settling time has increased to approximately 3.75 sec. Returning to the steady-state equation, we find that the equilibrium deflection for a linear spring is now 32.2 ft (9.82 m)—a somewhat unrealistic figure. One solution might be a nonlinear spring, one that has a very large k for small deflections and the desired k over the expected range of operation as indicated in Fig. 5–1–2b. To complete our analysis and design, we should, as a minimum, calculate the time response for a typical input, particularly since the closed-loop zero is very close to the roots and somewhat closer to the imaginary axis.

The general design process for an aircraft is a good illustration of the interplay between the steady-state and dynamic equations. The aircraft designer is given a set of performance and operational specifications to meet; these specifications include such items as maximum range, payload, and rate of climb. The designer uses the steady-state equations to size the aircraft and to determine the type and size of the engines, etc., usually in terms of such relationships as the lift-to-drag ratio, wing loading, mass ratio, etc., as well as the flight conditions necessary to achieve these goals. The designer next considers and satisfies the static stability requirements. He or she is now ready to perform a dynamic analysis of the plant (the aircraft) using the linearized dynamic equations whose coefficients* are functions of both the aircraft parameters and its flight regime. If the time response to the pilot's commands or to disturbances, such as turbulence, is unsatisfactory, the designer must recommend changes in the parameters of the aircraft that will not degrade its performance. If that is not possible, he or she must design a controller, called a *stability augmentation system* that will produce an acceptable response. The latter approach is necessary for most modern high-speed aircraft. For example, many such aircraft have unsatisfactory Dutch roll characteristics, particularly on landing. Rather than degrade the high-speed performance, a yaw damper feedback system is used to increase the damping ratio.

5–2 THE STEADY-STATE ERROR

The accuracy of a system is a measure of how well it follows commands. It is an important performance criterion; a guidance system that cannot place a spacecraft on a suitable trajectory is obviously useless no matter how well-behaved its transient response.

Accuracy is generally expressed in terms of acceptable steady-state errors for specified inputs. The error E is defined as the difference between the desired output, represented by r for a command input, and the actual output c. For command inputs,

$$E(t) = r(t) - c(t) \qquad (5\text{–}2\text{–}1)$$

and

$$E(s) = r(s) - c(s) \qquad (5\text{–}2\text{–}2)$$

We shall further define the steady-state condition for each of the quantities in

---

* These coefficients are known as the stability derivatives.

Eq. (5–2–1) as occurring after all transients have died out. The steady-state error $E_{ss}(t)$ can be a function of time, as we will see shortly, and can be written as

$$E_{ss}(t) = r_{ss}(t) - c_{ss}(t) \qquad (5\text{–}2\text{–}3)$$

It should be obvious that the steady-state error has significance only for stable systems.

A word of caution is in order. The error $E(t)$ (sometimes called the error signal) is not necessarily the actuating signal $\varepsilon(t)$. The two are identical only for unity feedback systems; this case will be discussed in some detail in the next section. You will never get in trouble if you remember that *the error is the difference between the desired output and the actual output*.

Let us use the electromechanical servomechanism of Fig. 4–5–1 as a tracking radar with the transfer function

$$\frac{\theta_{\text{out}}}{\theta_{\text{cmd}}} = \frac{AK_m}{\tau_m s^2 + s + AK_m} \qquad (5\text{–}2\text{–}4)$$

Examination of the characteristic equation shows this to be a stable system with a steady-state error

$$E_{ss}(t) = \theta_{\text{cmd}_{ss}} - \theta_{\text{out}_{ss}} \qquad (5\text{–}2\text{–}5)$$

If we wish the antenna to move through a specified angle θ_0, $\theta_{\text{cmd}}(t)$ will be a step and equal to $\theta_0 u(t)$; $\theta_{\text{cmd}_{ss}}(t)$ will be θ_0. Applying the final value theorem, we see that

$$\theta_{\text{out}_{ss}} = \lim_{s \to 0} [s\theta_{\text{out}}(s)] = \theta_0 \qquad (5\text{–}2\text{–}6)$$

Substituting into Eq. (5–2–5) we find

$$E_{ss}(t) = \theta_0 - \theta_0 = 0 \qquad (5\text{–}2\text{–}7)$$

The steady-state error in the angular position of the antenna for a step input is zero.

If we wish the antenna to move at a given angular velocity, $\theta_{\text{cmd}}(t)$ will be $\omega_{\text{in}}t$, where ω_{in} is a constant and $\theta_{\text{cmd}}(s) = \omega_{\text{in}}/s^2$. From Eq. (5–2–4),

$$\theta_{\text{out}}(s) = \frac{AK_m\omega_{\text{in}}}{s^2(\tau_m s^2 + s + AK_m)} \qquad (5\text{–}2\text{–}8)$$

and

$$\theta_{\text{out}_{ss}} = \lim_{s \to 0} [s\theta_{\text{out}}(s)] = \infty \qquad (5\text{–}2\text{–}9)$$

This infinite value for $\theta_{\text{out}_{ss}}$ does not necessarily mean that E_{ss} is also infinite, since $\theta_{\text{cmd}_{ss}} = \omega_{\text{in}}t$ and also becomes infinite with time. We need to know the expression for $\theta_{\text{out}_{ss}}$ as a function of time. Expand Eq. (5–2–8) in partial fractions so that

$$\theta_{\text{out}}(s) = \frac{C_1}{s} + \frac{C_2}{s^2} + \frac{C_3 s + C_4}{\tau_m s^2 + s + AK_m} \qquad (5\text{–}2\text{–}10)$$

The last term represents the transient response; thus the first two terms are the steady-state response. The coefficients C_1 and C_2 are evaluated as

$$C_2 = \left[\frac{AK_m\omega_{in}}{\tau_m s^2 + s + AK_m} \right]_{s=0} = \omega_{in} \tag{5-2-11}$$

and

$$C_1 = \left[\frac{d}{ds}\left(\frac{AK_m\omega_{in}}{\tau_m s^2 + s + AK_m} \right) \right]_{s=0} = -\frac{\omega_{in}}{AK_m} \tag{5-2-12}$$

Therefore,

$$\theta_{out_{ss}} = \omega_{in}t - \frac{\omega_{in}}{AK_m} \tag{5-2-13}$$

When we substitute into Eq. (5-2-5) we obtain

$$E_{ss} = \omega_{in}t - \left(\omega_{in}t - \frac{\omega_{in}}{AK_m} \right) = \frac{\omega_{in}}{AK_m} \tag{5-2-14}$$

We see that the error is finite and constant: the larger the input velocity, the larger the error in the angular position of the antenna. Physically, this error means that the antenna will acquire the desired angular velocity but will never catch up with the desired angular position. The steady-state error can be reduced by increasing the amplifier gain A, the motor gain K_m, or both; such increases, however, will affect the transient response by increasing ω_n and decreasing ζ.

If we introduce a constant angular acceleration input, then

$$\theta_{cmd}(t) = \frac{a_{in}t^2}{2} \qquad \theta_{cmd}(s) = \frac{a_{in}}{s^3}$$

and

$$\theta_{out}(s) = \frac{AK_m a_{in}}{s^2(\tau_m s^2 + s + AK_m)} = \frac{C_1}{s} + \frac{C_2}{s^2} + \frac{C_3}{s^3} + \cdots \tag{5-2-15}$$

Only the terms of the steady-state response are shown in Eq. (5-2-15). Two differentiations are required to evaluate the coefficients C_1, C_2, and C_3. After the proper substitutions into Eq. (5-2-5), remembering that $\theta_{cmd_{ss}} = \theta_{cmd}$, we obtain an expression for the steady-state error:

$$E_{ss}(t) = \frac{a_{in}t}{AK_m} + a_{in}\left[\frac{\tau_m}{AK_m} - \frac{1}{(AK_m)^2} \right] \tag{5-2-16}$$

The explicit presence of t in the first term shows that the steady-state error is indeed a function of time and will eventually become infinite. Physically, this error indicates that the antenna cannot accelerate as fast as the target it is tracking; it therefore cannot keep up with the target and will fall farther and farther behind. The magnitude of the error can be reduced by increasing AK_m with due regard to the concomitant effect on the transient response.

The three steady-state input-output combinations are shown in Fig. 5-2-1. Since a step in acceleration is the integral of a ramp (a step in velocity), we could have anticipated that the constant error due to a velocity input would give rise to an infinite error for an acceleration input.

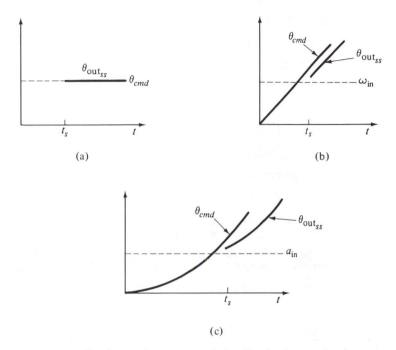

Figure 5-2-1.   Steady-state input-output relationships for the second-order system of Eq. (5-2-4): (a) step in position; (b) step in velocity; (c) step in acceleration.

Let us apply the steady-state error as an accuracy criterion to our tracking system. If the maximum allowable steady-state error in position is $\pm 0.1°$ for an expected maximum input angular velocity of $10°/\text{sec}$, Eq. (5-2-14) is relevant and can be written as the inequality

$$E_{ss_{max}} \leqslant \frac{\omega_{in_{max}}}{AK_m} \qquad (5\text{-}2\text{-}17)$$

Rearranging this inequality, we can obtain the requirement that

$$AK_m \geqslant \frac{\omega_{in_{max}}}{E_{ss_{max}}} \qquad (5\text{-}2\text{-}18)$$

In this case, $AK_m \geqslant 100$, and we have established a lower limit on the product of the motor and amplifier gains. This limit must not be violated when seeking an acceptable transient response; increasing the damping ratio will reduce AK_m. The time-dependent error due to an acceleration input should probably be considered. We shall merely state that it will generally be acceptable if the duration of the input acceleration is not unduly long.

The procedures we have followed for determining steady-state errors are straightforward and unambiguous. However, if a system has unity feedback (H equal to unity), the error constants and error coefficients of the next sections may be simpler indicators of system accuracy.

5–3 ERROR CONSTANTS

From the generalized block diagram in Fig. 5–3–1a of a disturbance-free feedback system, we see that the actuating signal $\varepsilon(s)$ is

$$\varepsilon(s) = r(s) - Hc(s) \tag{5-3-1}$$

If H is equal to unity, the system becomes the unity feedback system of Fig. 5–3–1b and the actuating signal is

$$\varepsilon(s) = r(s) - c(s) = E(s) \tag{5-3-2}$$

and becomes the error signal. We can establish a relationship between the input and the newly defined error signal with the *error transfer function*

$$W_E = \frac{E}{r}(s) = \frac{1}{1 + G} \tag{5-3-3}$$

Although the concepts and definitions we are about to develop are directly applicable only to unity feedback systems,* the unity feedback system is prevalent enough to have warranted their development and use as design criteria. Returning to Eq. (5–3–3), we can find the error for any command input from

$$E(s) = \frac{r(s)}{1 + G(s)} \tag{5-3-4}$$

If we determine that the system is stable, we can apply the final value theorem to find the steady-state error to be

$$E_{ss} = \lim_{s \to 0} \frac{sr(s)}{1 + G(s)} \tag{5-3-5}$$

Although any command input may be specified, the accuracy of a system usually may be adequately determined with one or more of the three basic step inputs—position, velocity, and acceleration. These basic inputs and their transforms are repeated for convenience in Fig. 5–3–2. Substituting each of these inputs into Eq. (5–3–5), we obtain the following expressions for the steady-state error for each type of input:

$$E_{p_{ss}} = \frac{r_0}{1 + \lim\limits_{s \to 0} G} = \frac{r_0}{1 + K_p} \tag{5-3-6}$$

$$E_{v_{ss}} = \frac{v_0}{\lim\limits_{s \to 0} sG} = \frac{v_0}{K_v} \tag{5-3-7}$$

$$E_{a_{ss}} = \frac{a_0}{\lim\limits_{s \to 0} s^2 G} = \frac{a_0}{K_a} \tag{5-3-8}$$

$E_{p_{ss}}$ is called the *steady-state position error*; K_p is called the *position error constant* and is defined in terms of the forward transfer function G as

$$K_p = \lim_{s \to 0} G \tag{5-3-9}$$

---

* A technique for non-unity feedback systems is described in Sec. 5–5.

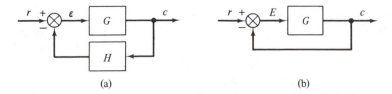

Figure 5–3–1. A disturbance-free feedback control system: (a) nonunity feedback; (b) unity feedback.

$E_{v_{ss}}$ and $E_{a_{ss}}$ are called the *steady-state velocity error* and *steady-state acceleration error*, respectively. K_v and K_a are the *velocity and acceleration error constants* and are defined as

$$K_v = \lim_{s \to 0} sG \tag{5–3–10}$$

$$K_a = \lim_{s \to 0} s^2 G \tag{5–3–11}$$

The error constants are sometimes referred to as the *static error coefficients*. Do not forget that the error is the difference between the input and the output and that the words *position*, *velocity*, and *acceleration* are used to describe the input only.

The forward transfer function G is a ratio of polynomials. If we factor out the constant terms, G can be written in either of the following two forms:

$$G = \frac{K_e(a_m s^m + a_{m-1} s^{m-1} + \cdots + a_1 s + 1)}{s^n(b_j s^j + b_{j-1} s^{j-1} + \cdots + b_1 s + 1)} \tag{5–3–12}$$

or

$$G = \frac{K_e \, \Pi \, (a_i s + 1)}{s^n \, \Pi \, (b_k s + 1)} \tag{5–3–13}$$

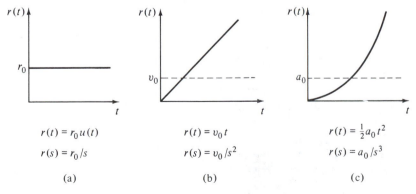

$$r(t) = r_0 u(t)$$
$$r(s) = r_0 / s$$

(a)

$$r(t) = v_0 t$$
$$r(s) = v_0 / s^2$$

(b)

$$r(t) = \tfrac{1}{2} a_0 t^2$$
$$r(s) = a_0 / s^3$$

(c)

Figure 5–3–2. Basic command inputs: (a) step in position; (b) step in velocity; (c) step in acceleration.

where K_e, the forward gain,* is independent of s and can be considered a generalized error constant. The s^n term in the denominator of the forward transfer function reflects the n number of integrations in the forward path and is used to classify unity feedback systems. If n is 0, the system is a type-0 system; if n is 1, the system is a type-1 system, etc.

Returning to Eqs. (5–3–12) and (5–3–13), we see that

$$\lim_{s \to 0} G = \lim_{s \to 0} \frac{K_e}{s^n} \tag{5–3–14}$$

If we know the system type and the value of K_e, we can quickly evaluate the error constants and determine the steady-state errors. The tracking system of Fig. 4–5–1 is a unity feedback system with

$$G = \frac{AK_m}{s(\tau_m s + 1)} \tag{5–3–15}$$

and

$$\lim_{s \to 0} G = \lim_{s \to 0} \frac{AK_m}{s} \tag{5–3–16}$$

This is a type-1 system with K_e equal to AK_m. The error constants are

$$K_p = \lim_{s \to 0} G = \lim_{s \to 0} \frac{AK_m}{s} = \infty \tag{5–3–17}$$

$$K_v = \lim_{s \to 0} sG = \lim_{s \to 0} AK_m = AK_m \tag{5–3–18}$$

$$K_a = \lim_{s \to 0} s^2 G = \lim_{s \to 0} sAK_m = 0 \tag{5–3–19}$$

Substituting these values into Eqs. (5–3–6) through (5–3–8), we find the steady-state errors in position ($\theta_{\text{cmd}_{ss}} - \theta_{\text{out}_{ss}}$) to be

$$E_{p_{ss}} = \frac{\theta_{\text{in}}}{1 + \infty} = 0 \tag{5–3–20}$$

$$E_{v_{ss}} = \frac{\omega_{\text{in}}}{AK_m} \tag{5–3–21}$$

$$E_{a_{ss}} = \frac{a_{\text{in}}}{0} = \infty \tag{5–3–22}$$

These errors agree with those of the previous section, although we do not know how rapidly $E_{a_{ss}}$ increases with time.

It is a simple matter to determine the error constants and steady-state errors for each type of system and construct Table 5–3–1. It is also possible to obtain the error constants and determine the type of a system from the Bode plots of Chap. 8. The error constants can also be used to determine the errors arising from inputs that can be represented by a polynomial in t. If $r(t) = b_0 + b_1 t +$

---

* With unity feedback, K_e is also the open-loop gain.

$b_2 t^2$, then $E_{ss} = E_{pss} + E_{vss} + E_{ass}$, so that

$$E_{ss} = \frac{b_0}{1 + K_p} + \frac{b_1}{K_v} + \frac{b_2}{K_a} \qquad (5\text{–}3\text{–}23\text{a})$$

Introducing the error constants of our type-1 second-order tracking system, Eq. (5–3–23a) becomes

$$E_{ss} = 0 + \frac{b_1}{AK_m} + \frac{b_2}{0} \qquad (5\text{–}3\text{–}23\text{b})$$

We see that the error becomes infinite with increasing time only because of the acceleration input, the error from the step input being zero and the error from the ramp error being constant and inversely proportional to AK_m.

Table 5–3–1 illustrates several points of interest. Steady-state accuracy for unity feedback systems can be expressed as a minimum acceptable error constant; the larger the value specified, the smaller the acceptable error. Since a finite error constant is the product of component gains, the requirement for an accurate system implies high gains, with upper limits established by stability and transient response considerations. Notice that an infinite steady-state error does not necessarily mean a system is unstable; *never* use the final value theorem or the magnitude of a steady-state error to detemine system stability. As a final comment, do not conclude from Table 5–3–1 that the higher the system type, the better the control system, or that a steady-state error can be eliminated simply by adding integrations. Adding an integration to raise the type may make a stable system unstable, even though a zero steady-state error is indicated. Integrations always reduce the stability of a system and are not to be added casually.

TABLE 5–3–1. ERROR CONSTANTS

Type System	Position Step		Velocity Step		Acceleration Step	
	K_p	E_{pss}	K_v	E_{vss}	K_a	E_{ass}
0	K_e	$\dfrac{r_0}{1 + K_e}$	0	∞	0	∞
1	∞	0	K_e	v_0/K_e	0	∞
2	∞	0	∞	0	K_e	a_0/K_e
3	∞	0	∞	0	∞	0

*5–4 ERROR COEFFICIENTS

The *error coefficients** provide more information than the error constants, but as might be expected, they require more effort to obtain. The convolution integral and a truncated Taylor's series can be combined to obtain a series expression

---

* The error coefficients are sometimes called the *dynamic error coefficients*.

for the steady-state actuating signal

$$\varepsilon_{ss}(s) = C_0 r(t) + C_1 r'(t) + \frac{C_2}{2!} r''(t) + \cdots + \frac{C_n}{n!} r^{(n)}(t) \qquad (5\text{-}4\text{-}1)$$

where the primes denote time derivatives. For unity feedback systems, the actuating signal is the error and the coefficients C_0, C_1 ... C_n are the error coefficients. These coefficients may be obtained from successive differentiations with respect to s of the error transfer function W_E. Thus,

$$C_0 = \lim_{s \to 0} W_E \qquad (5\text{-}4\text{-}2)$$

$$C_1 = \lim_{s \to 0} \frac{dW_E}{ds} \qquad (5\text{-}4\text{-}3)$$

.

.

.

$$C_n = \lim_{s \to 0} \frac{d^n W_E}{ds_n} \qquad (5\text{-}4\text{-}4)$$

where

$$W_E = \frac{1}{1 + G} \qquad (5\text{-}4\text{-}5)$$

Returning to our type-1 tracking servo,

$$W_E = \frac{s(\tau_m s + 1)}{\tau_m s^2 + s + AK_m} \qquad (5\text{-}4\text{-}6)$$

The first three error coefficients are

$$C_0 = \lim_{s \to 0} W_E = 0 \qquad (5\text{-}4\text{-}7)$$

$$C_1 = \lim_{s \to 0} \frac{dW_E}{ds} = \frac{1}{AK_m} \qquad (5\text{-}4\text{-}8)$$

and

$$C_2 = \lim_{s \to 0} \frac{d^2 W_E}{ds^2} = 2\left[\frac{\tau_m}{AK_m} - \frac{1}{(AK_m)^2} \right] \qquad (5\text{-}4\text{-}9)$$

Substituting into Eq. (5–4–1) yields the *steady-state error series* expression for this particular type-1 system:

$$E_{ss}(t) = 0r(t) + \frac{1}{AK_m} r'(t) + \left[\frac{\tau_m}{AK_m} - \frac{1}{(AK_m)^2} \right] r''(t) + \cdots \qquad (5\text{-}4\text{-}10)$$

If the input is a step in position, $r(t) = r_0$ and the derivatives r' and r'' are both equal to zero. Therefore, $E_{p_{ss}} = 0$. If the input is a step in velocity, $r(t) = \omega_{in}t$, $r'(t) = \omega_{in}$, and all higher derivatives are zero; thus, $E_{v_{ss}} = \omega_{in}/AK_m$. If the input is a step in acceleration, $r(t) = a_{in}t^2/2$, $r'(t) = a_{in}t$, $r''(t) = a_{in}$, and all higher derivatives are zero. The error series contains two terms and is

$$E_{a_{ss}} = \frac{a_{in}t}{AK_m} + a_{in}\left[\frac{\tau_m}{AK_m} - \frac{1}{(AK_m)^2}\right] \tag{5-4-11}$$

These steady-state errors agree with the results obtained in Sec. 5-2 [see Eq. (5-2-16)] where we used the more basic approach that is applicable to both unity and non-unity feedback systems.

If we apply a polynomial command input to our type-1 tracking servo:

$$r(t) = b_0 + b_1t + \frac{b_2}{2}t^2 \tag{5-4-12}$$

then

$$r'(t) = b_1 + b_2t \tag{5-4-13}$$

$$r''(t) = b_2 \tag{5-4-14}$$

The errors series becomes

$$E_{ss}(t) = \frac{b_1}{AK_m} + b_2\left[\frac{\tau_m}{AK_m} - \frac{1}{(AK_m)^2}\right] + \frac{b_2}{AK_m}t \tag{5-4-15}$$

Equation (5-4-15) confirms that the steady-state acceleration error is indeed infinite but shows how this instantaneous error grows with time. This knowledge can be important in the design of an actual system, because an acceleration input cannot be maintained indefinitely and the maximum acceleration error may not necessarily exceed an allowable limit. Implicit in the useful application of the error series and error constant concepts are the convergence of the series and the decay of the transient response within an acceptable period of time. With regard to the latter condition, a steady-state acceleration error of zero for a radar system tracking an incoming ballistic missile would be meaningless if the settling time were 30 min or longer.

5-5 NON-UNITY FEEDBACK ERROR CONSTANTS

When the feedback transfer function H is other than unity, as indicated in Fig. 5-5-1a, the actuating signal ε is equal to $(r - HC)$ and is *not* the error signal E. By introducing an inner feedback loop with the transfer function $(H - 1)$ as shown in Fig. 5-5-1b, the non-unity feedback system can now be represented by the equivalent unity feedback system of Fig. 5-5-1c. The new forward transfer function G^* is related to G and H by the expression

$$G^* = \frac{G}{1 + G(H - 1)} \tag{5-5-1}$$

and the error transfer function of Eq. (5-3-3) becomes

$$W_E^* = \frac{1}{1 + G^*} = \frac{1 + G(H - 1)}{1 + GH} \tag{5-5-2}$$

It is now possible to find error constants and error coefficients for systems with

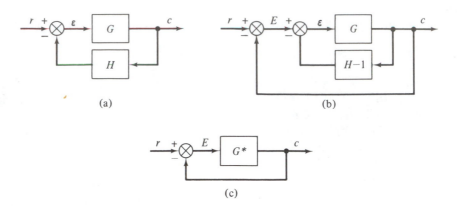

(a)   (b)

(c)

Figure 5–5–1. A non-unity feedback system: (a) with single outer loops; (b) with an added inner loop; (c) with equivalent unity feedback.

non-unity feedback by substituting G^* for G and W_E^* for W_E in the appropriate unity feedback expressions, thus determining the effects of non-unity feedback upon the steady-state accuracy of a system.

The effects of non-unity feedback upon both the steady-state accuracy and stability of a system can be considerable, and each case should be individually and carefully checked. Furthermore, as will be discussed in Sec. 5–8, the performance of a system is much more sensitive to parameter changes in the feedback path than in the forward path.

Let us look at two examples that show the effects of scalar feedback ($H = b_0$). First consider a system with the type-0 transfer function

$$G = \frac{K}{s^2 + s + 1} \tag{5-5-3}$$

With unity feedback, $K_p = K$ so that $E_{p_{ss}} = r_0/(1 + K)$, and $K_v = K_a = 0$ so that $E_{v_{ss}} = E_{a_{ss}} = \infty$. Now let $H = b_0$, resulting in

$$G^* = \frac{K}{s^2 + s + K(b_0 - 1) + 1} \tag{5-5-4}$$

If b_0 is set equal to $(1 - 1/K)$, G^* becomes

$$G^* = \frac{K}{s(s + 1)} \tag{5-5-5}$$

We see that the system type has been increased to type 1 with the result that $E_{p_{ss}}$ becomes zero and $E_{v_{ss}}$ becomes finite and equal to v_0/K. It can also be shown that the damping ratio has been increased and the natural undamped frequency decreased, while the same settling time has been maintained.

Let us now consider a system with a type-1 or higher forward transfer function. If, for example,

$$G = \frac{K(0.5s + 1)}{s^2(0.1s + 1)} \tag{5-5-6}$$

then the system is type 2 with unity feedback. Consequently, $E_{p_{ss}} = E_{v_{ss}} = 0$

and $E_{a_{ss}} = a_0/K$. Setting $H = b_0$ yields

$$G^* = \frac{K(0.5s + 1)}{0.1s^3 + s^2 + 0.5K(b_0 - 1)s + K(b_0 - 1)} \tag{5-5-7}$$

Any value of b_0 other than unity reduces the type of the system from 2 to 0 with a corresponding reduction in the steady-state accuracy.

With the addition of derivative and/or integral feedback we speak of non-scalar feedback, the effects of which can be examined in a similar manner. Non-scalar feedback is often introduced to improve the stability and transient response of a system; do not forget to consider the possible impact upon the steady-state accuracy.

Consider, for example, a unity feedback system with the forward transfer function

$$G = \frac{K}{s^2(s + 10)} \tag{5-5-8}$$

This system would be type 2 if it were stable, which it is not. Adding a zero to the forward path would make the system stable and maintain the system type at 2. If, however, rate feedback were introduced into the outer loop and combined with the unity feedback so that $H = 0.5s + 1$, the zero would be introduced into the feedback path. The system would be stable, but the feedback would be non-scalar. $(H - 1)$ becomes equal to $0.5s$, and

$$G^* = \frac{K}{s(s^2 + 10s + 0.5K)} \tag{5-5-9}$$

As a consequence, the type of the system has been reduced from 2 to 1. Furthermore, with non-scalar feedback the type of the system cannot be raised higher than 1, no matter how many integrators are added to the forward path.

5-6 DISTURBANCE ERRORS

We have been describing the accuracy of a system in terms of its response to command or desirable inputs. Systems are also subjected to undesirable inputs, such as noise in command inputs and disturbances arising from changes in the plant parameters or changes in the environment in which the plant is operating. A control system may be idealized as a perfect filter that faithfully passes all desirable inputs and suppresses all undesirable inputs. Needless to say, this goal is never completely realized.

Noise inputs that enter the system with the command inputs require filtering techniques to remove or suppress them without affecting the command input itself. We shall limit our discussion to disturbance inputs which enter the system at the plant rather than at the controller, as indicated in the generalized block diagram of Fig. 5-1-1. This diagram is drawn in more detail in Fig. 5-6-1a to show the disturbance inputs entering between the controller and the plant. In Fig. 5-6-1b the diagram is redrawn with the disturbance d as the principal input. Since the system is linear, the principle of superposition holds, and we can

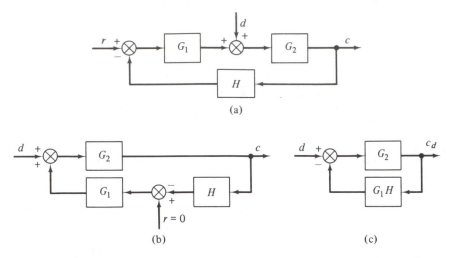

Figure 5–6–1. A feedback control system with disturbance inputs: (a) general block diagram; (b) disturbance as principal input; (c) negative feedback.

assume r equal to zero. The negative sign change associated with the command input comparator can be shifted to the disturbance input comparator. We now have the negative feedback system of Fig. 5–6–1c, where the symbol c_d is used to distinguish the output due to a disturbance input from the output c due to a command input. The disturbance input transfer function can be written as

$$\frac{c_d}{d} = \frac{G_2}{1 + G_1G_2H} \tag{5–6–1}$$

When this transfer function is compared with the usual input-output transfer function with $d = 0$

$$\frac{c}{r} = \frac{G_1G_2}{1 + G_1G_2H} \tag{5–6–2}$$

we see that characteristic equations are identical, as is to be expected, but that the numerator functions are different. A disturbance input, therefore, will not affect the stability of the system but may change the shape of the transient response and introduce steady-state errors that must be considered in determining the overall accuracy of the system.

Since any change in the output in response to a disturbance is undesirable, the disturbance error E_d is the actual output itself, represented by c_d,* and is not $(d - c)$. Consequently, Eq. (5–6–1) is the error transfer function for a disturbance input. Notice that it is not necessary for H to be equal to unity in order to establish the error transfer function, as is the case for a command input. Do *not* use error constants or error coefficients to evaluate disturbance errors. Instead, use the error transfer function for the disturbance input of interest.

---

* Technically, E_d is equal to $-c_d$.

The total steady-state error is the sum of the command error and the disturbance error. With unity feedback, the total error can be written as

$$E_{ss} = \lim_{s \to 0} \frac{sr(s) + sG_2 d(s)}{1 + G_1 G_2} \qquad (5\text{-}6\text{-}3)$$

It is often difficult to minimize both components of the error simultaneously. Obviously, it is necessary to have some knowledge as to the nature of probable disturbance inputs.

Let us return once more to our tracking radar system and consider the effect upon the accuracy of a disturbance torque T_d arising from wind loading of the antenna. To introduce the disturbance torque properly requires the detailed block diagram of Fig. 5-6-2a, which is reduced to the two block diagrams of

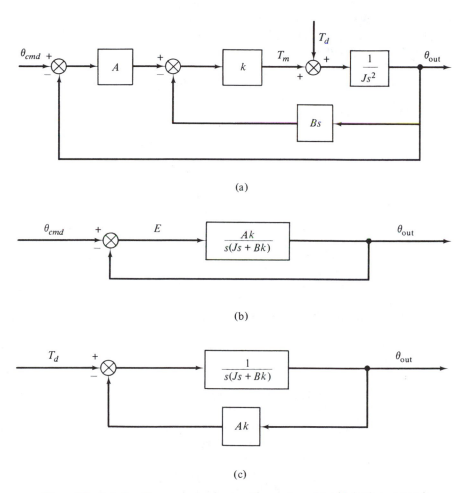

(a)

(b)

(c)

Figure Fig. 5-6-2. The second-order tracking system: (a) with both command and disturbance inputs; (b) with command input only; (c) with disturbance input only.

Fig. 5–6–2b and c, one for each type of input. In the Laplace domain the combined error can be written as

$$E(s) = \frac{s(Js + Bk)\theta_{cmd}(s) + T_d(s)}{Js^2 + Bks + kA} \qquad (5-6-4)$$

Let us assume that a constant disturbance torque, $T_d = T_{do}(u(t)$, and a ramp command input, $\theta_{cmd}(t) = \omega_{cmd}t$, are applied simultaneously. The combined steady-state error, as determined by applying the final value theorem with $\theta_{cmd}(s) = \omega_{cmd}/s^2$ and $T_d(s) = T_{do}/s$, is

$$E_{ss} = \frac{T_{do}}{kA} + \frac{B\omega_{cmd}}{A} \qquad (5-6-5)$$

Both error components can be reduced by increasing the amplifier gain A; the disturbance error alone can be reduced by increasing the motor torque constant k. In either case, the stability will be decreased and the transient response changed. In determining the component of the steady-state error due to a disturbance, do not use error constants or coefficients, since the actuating signal ε is not a measure of the disturbance error.

Both error terms of Eq. (5–6–5) can be set equal to zero by introducing an integrator into the controller block represented by the amplifier gain A. This additional integrator increases the type of the system from 1 to 2, thus eliminating the velocity error, and by being introduced *ahead* of the point of entry of the disturbance into the system, eliminates the steady-state error resulting from a step in the disturbance input. This additional integrator must be accompanied by at least one zero if the system is to remain stable. Do not forget that an unstable system does not have a steady-state response and that higher-order inputs can result in finite steady-state errors.

Very often disturbance inputs are cyclic in nature, such as gusty winds. Then the frequency response of the system becomes relevant and useful. The relationship between the frequency response and the accuracy of the system will be discussed in the next section.

5–7 ACCURACY AND THE FREQUENCY RESPONSE

If a stable linear system is subjected to a sinusoidal input of known amplitude and frequency, the output response will undergo a decaying transient response and settle down to a sinusoidal steady-state response with the same frequency as the input, but not necessarily with the same amplitude; there may also be a phase angle displacement. These relationships are illustrated in Fig. 5–7–1 for a command input. Do not confuse the forcing frequency ω_f and the phase angle Φ with the damped frequency ω and the phase angle ψ associated with an underdamped transient response.

Since sinusoids and their derivatives cannot be precisely defined as time becomes large, the concepts of error constants, coefficients, and series are not applicable. We can, however, specify the input-output relationship by specifying

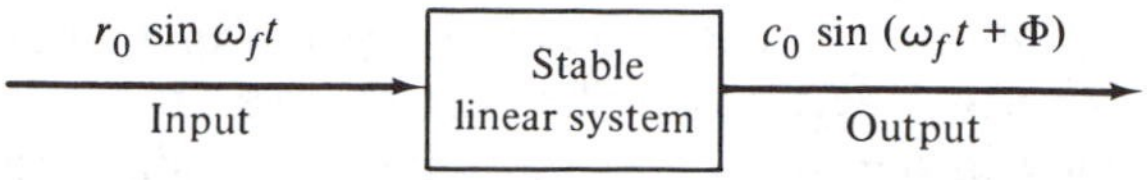

$r_0 \sin \omega_f t$

Input

| Stable
linear system |

$c_0 \sin (\omega_f t + \Phi)$

Output

Figure 5–7–1. Steady-state input-output relationship for a sinusoidal input.

the phase angle and the ratio of the output amplitude to that of the input as a function of the forcing frequency. Knowledge of the amplitude ratio M^* is sufficient for an examination of steady-state accuracy. It is customary to plot the amplitude ratio versus the forcing frequency on logarithmic scales; these are known as *Bode plots* and will be described in detail in subsequent chapters.

To illustrate the use of amplitude ratio plots, we shall return to our tracking system and consider sinusoidal inputs entering the controller. The transfer function is written as

$$\frac{\theta_{\text{out}}}{\theta_{\text{cmd}}} = \frac{kA/J}{s^2 + \dfrac{Bk}{J}s + \dfrac{kA}{J}} \tag{5–7–1}$$

where $\omega_n = (kA/J)^{1/2}$. The amplitude ratio Bode plot for this system is shown in Fig. 5–7–2. For sinusoidal inputs with forcing frequencies less than approximately $0.1\,\omega_n$ rad/sec, the amplitude ratio is unity; this would obviously be the desired range for command inputs. Inputs with frequencies greater than the undamped natural frequency ω_n are greatly reduced in amplitude (attenuated); we obviously would like to have all undesirable inputs in this frequency range. In the vicinity of the natural frequency, the amplitude ratio is greater than unity; inputs are magnified. We do not want inputs of any kind in this region.

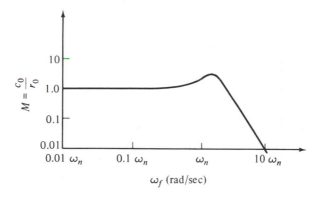

Figure 5–7–2. Bode plot of the amplitude ratio of the second-order tracking radar for a command input.

We can also draw some conclusions from Fig. 5–7–2 as to the steady-state accuracy of the system for step inputs. A step in position can be considered a zero-frequency input for which our Bode plot shows an amplitude ratio of unity

---

* The symbol M denotes the system amplitude ratio, i.e., $M = c_0/r_0$.

or a zero steady-state error, implying that the system is type 1 or higher. The errors due to steps in velocity and acceleration are less discernible. If we consider the latter to be high frequency in nature in that the input is rapidly changing, the Bode plot indicates large acceleration errors. The larger the frequency range of a unity amplitude ratio region, the greater the possibility of small velocity errors; in other words, the larger the bandwidth, the smaller the velocity error. This generalization is confirmed by Eq. (5–2–14) (which shows that increasing the amplifier gain A reduces the velocity error) and by Eq. (5–7–1) (which shows that ω_n increases as A increases). Increasing ω_n and thus increasing the bandwidth also reduces the settling and rise times but increases the possibility of accepting undesirable high-frequency inputs.

Inputs that enter at points other than the controller will have different Bode plots. In the preceding section, the wind-loading torque T_d produces an output error θ_d represented by

$$\frac{\theta_d}{T_d} = \frac{1/J}{s^2 + \dfrac{Bk}{J}s + \dfrac{kA}{J}} \tag{5–7–2}$$

The corresponding Bode plot is shown in Fig. 5–7–3 for kA equal to 10. Our objective in this case is a low amplitude ratio for all frequencies. At low frequencies, the error amplitude ratio is equal to 0.1 or $1/kA$. Increasing kA decreases steady-state errors and increases ω_n and thus the bandwidth.

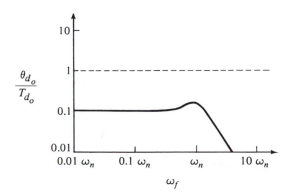

Figure 5–7–3. Bode plot of the amplitude ratio of the second-order tracking radar for a disturbance input with $kA = 10$.

The frequency response need not be limited to pure sinusoidal inputs but can be used with any rapidly varying input. Such inputs may be periodic, random, or a combination thereof. If the frequency distribution of these inputs can be calculated, measured, or even estimated, the frequency response will give us a good idea as to the steady-state accuracy of the system.

*5–8 THE SENSITIVITY FUNCTION AND ROBUSTNESS

The *sensitivity function* or simply the *sensitivity* of one function A with respect to a second function B is given the symbol S_B^A and defined as the percentage change in A resulting from a percentage change in B. It may be expressed as

$$S_B^A = \frac{\Delta A/A}{\Delta B/B} = \frac{dA/A}{dB/B} = \frac{d \ln A}{d \ln B} = \frac{B}{A}\frac{dA}{dB} \qquad (5\text{–}8\text{–}1)$$

We shall use the last of these expressions, namely,

$$S_B^A = \frac{B}{A}\frac{dA}{dB} \qquad (5\text{–}8\text{–}2)$$

When used in conjunction with the transfer function, the sensitivity will be a function of the Laplace variable s. If s is set equal to zero, the sensitivity becomes the *steady-state* or *static sensitivity*. If s is replaced by $j\omega_f$, where ω_f is the forcing frequency, then the sensitivity is called the *dynamic sensitivity*. The larger the magnitude of the sensitivity, the greater the effect on A of changes in B.

The sensitivity function can be used to determine the effects of changes in system parameters and operating conditions upon the transfer function and upon the input-output relationships. Obviously, we should like such sensitivities to be as low as possible, preferably zero. If a particular sensitivity is high, the sensitivity function will indicate what steps can be taken to reduce it. A system with low sensitivity functions is often referred to as a *robust* system.

Let us first consider the simple open-loop system of Fig. 5–8–1a and calculate the sensitivity of the system transfer function

$$W(s) = \frac{c}{r} = G_1 G_2 \qquad (5\text{–}8\text{–}3)$$

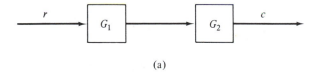

(a)

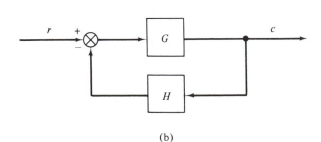

(b)

Figure 5–8–1. (a) Simple open-loop system; (b) simple closed-loop system.

to changes in the component transfer function G_1. Using Eq. (5–8–2), we find that

$$S_{G_1}^W = \frac{G_1}{W} \frac{dW}{dG_1} = \frac{G_1 G_2}{W} = 1 \qquad (5\text{–}8\text{–}4)$$

This sensitivity of unity reveals a one-to-one relationship between changes in G_1 (or G_2) and errors in the input-output relationship; a 10 percent change in G_1 produces a 10 percent error in c.

If we add a negative feedback loop, as in Fig. 5–8–1b, the system transfer function can be written as

$$W(s) = \frac{c}{r} = \frac{G}{1 + GH} \qquad (5\text{–}8\text{–}5)$$

Now

$$S_G^W = \frac{G}{W} \frac{dW}{dG} = \frac{1}{1 + GH} \cong \frac{1}{GH} \qquad (5\text{–}8\text{–}6)$$

From Eq. (5–8–6) we see that increasing $|GH|$ reduces the sensitivity. Since $|GH|$ is normally much greater than unity, the sensitivity approaches zero and the input-output relationship becomes essentially insensitive to changes in the forward transfer function G.

We need to examine the sensitivity of the transfer function with respect to the feedback transfer function H:

$$S_H^W = \frac{H}{W} \frac{dW}{dH} = \frac{-GH}{1 + GH} \cong -1 \qquad (5\text{–}8\text{–}7)$$

This high sensitivity indicates that feedback transfers the sensitivity from the forward elements to the feedback elements. The designer, however, can specify and control the feedback elements, whereas he or she generally cannot change plant specifications which are forward transfer functions. Furthermore, the feedback transfer function is often unity, implying only a transducer in the feedback loop.

Let us examine the sensitivity of our tracking radar system to changes in the parameters τ_m and AK_m. Since the system transfer function is

$$W(s) = \frac{\theta_{\text{out}}}{\theta_{\text{cmd}}} = \frac{AK_m}{\tau_m s^2 + s + AK_m} \qquad (5\text{–}8\text{–}8)$$

then

$$S_{\tau_m}^W = \frac{\tau_m}{W} \frac{dW}{d\tau_m} = \frac{-\tau_m s^2}{\tau_m s^2 + s + AK_m} \qquad (5\text{–}8\text{–}9)$$

and

$$S_{AK_m}^W = \frac{AK_m}{W} \frac{dW}{dAK_m} = \frac{s(\tau_m s + 1)}{\tau_m s^2 + s + AK_m} \qquad (5\text{–}8\text{–}10)$$

If s is set equal to zero in these two sensitivity equations, the static sensitivities are both found to be zero. What does zero static sensitivity mean? It means only that the steady-state accuracy for a position step input is unaffected by any changes in that particular parameter. We can extend the static sensitivity to include other step inputs by taking the limit of $1/s^n$ times the sensitivity function as s goes to zero. If n is zero, the static sensitivity pertains to a position step input, if $n = 1$ to a velocity step (ramp) input, if $n = 2$ to an acceleration step, etc.

From Eq. (5–8–9) we find the limit of the sensitivity function for τ_m to be zero when $n = 1$, $-\tau_m/AK_m$ when $n = 2$, and infinite when $n \geqslant 3$. The zero for $n = 1$ indicates that a change in τ_m does not affect the steady-state accuracy for a velocity input; the finite value for $n = 2$ indicates that τ_m contributes a constant error for an acceleration input and that an increase in τ_m increases that error; the infinite value for $n \geqslant 3$ indicates only that τ_m appears in an error term that increases with time.

The sensitivity function for AK_m in Eq. (5–8–10) yields the limits: zero for $n = 0$, $1/AK_m$ for $n = 1$, and infinity for $n \geqslant 2$. We interpret these values as meaning that AK_m does not affect the steady-state accuracy for a step input, that AK_m does contribute a constant error for a velocity input and the error is reduced as AK_m is increased, and that for acceleration and higher-order inputs AK_m appears in time-dependent error terms. These interpretations can be verified by examining Eqs. (5–2–14) and (5–2–16) which are expressions for the velocity and acceleration errors.

The two dynamic sensitivities are found by replacing s by $j\omega_f$ in Eqs. (5–8–9) and (5–8–10) and are

$$S^W_{\tau_m} = \frac{\tau_m \omega_f^2}{AK_m - \tau_m \omega_f^2 + jAK_m \omega_f} \tag{5–8–11}$$

$$S^W_{AK_m} = \frac{-\tau_m \omega_f^2 + j\omega_f}{AK_m - \tau_m \omega_f^2 + jAK_m \omega_f} \tag{5–8–12}$$

When plotted against the input frequency, these equations will show how the amplitude ratios and phase angles will be affected by changes in the parameters of interest for sinusoidal inputs of varying forcing frequencies. These plots will also show the effects of parameter changes upon the gain margin and phase margin, which are also measures of the robustness of a system and which will be defined and discussed in Sec. 7–5.

Without attempting to explore all the possible uses of the sensitivity function concept, let us conclude this section with the statement that sensitivity functions can be useful in indicating how well a system will perform when operated off the design point, i.e., when the parameters and operating conditions take on values other than those used in the design of the control system.

PROBLEMS

5–1. The motion of the center of mass of an aircraft in the vertical plane can be described by the two nonlinear equations:

$$T - D - W \sin \gamma = m\dot{V}$$
$$L - W \cos \gamma = mV\dot{\gamma}$$

where T is the thrust, D the drag, L the lift, and W the weight, all in pounds. V is the true airspeed in feet per second, and γ is the flight path angle in degrees.

(a) Write the equilibrium (initial steady-state) equations.

(b) If an SST weighs 400,000 lbs and has a maximum lift-to-drag ratio $(L/D)_{max}$ of 8, what is the minimum thrust required to climb at a constant flight path angle of 10°? To fly at a constant altitude, $\gamma_0 = 0$°?

(c) For an aircraft in level flight ($\gamma_0 = 0$), find the thrust-to-weight ratio (T/W) in terms of its instantaneous lift-to-drag ratio (L/D).

5-2. If a liquid is to be heated in a well-mixed insulated tank to a desired temperature, the differential equations can be written as

$$\frac{Cd(MT_{out})}{dt} = Q_{in} + W_{in}CT_{in} - W_{out}CT_{out}$$

$$\frac{dM}{dt} = W_{in} - W_{out}$$

where T_{out} is the desired temperature (°F), M is the mass of liquid in the tank (lb), C is the heat capacity (assumed constant) of the liquid [BTU(lb-°F)], W is the mass flow rate (lb/min), Q_{in} is the heat flow rate from the heater (BTU/min), and T_{in} is the temperature of the liquid entering the tank.

(a) Write the equilibrium (initial steady-state) equations and find equations for T_{out_0} and Q_{in_0}.

(b) If the liquid is water ($C = 1$), the mass of the water in the tank is 400 lb, and the temperature of the input water is 60°F, how large must the heater be to heat 25 lb/min at 145°F?

(c) If the maximum output of the heater for the nominal conditions stated in (b) is 2000 BTU/min, find the maximum steady-state temperature if
(1) The flow rate increases to 35 lb/min
(2) The input temperature decreases to 45°F.
(3) The mass of water in the tank increases to 1500 lb.

5-3. For each of the following systems, use the final value theorem to find the steady-state output and steady-state error for the unit step input $r(t) = u(t)$. System must be stable to use the FVT.

(a) $$\frac{c}{r} = \frac{2(s^2 + 2s + 3)}{s^4 + 2s^3 + 6s^2 + 10s + 6}$$

(b) $$\frac{c}{r} = \frac{10(s^2 + 3s + 3)}{s^4 + 2s^3 + 6s^2 + 10s + K}$$

(c) $$\frac{c}{r} = \frac{2(s^2 + 2s + 2)}{s^4 + 2s^3 + 6s^2 + 10s + 8}$$

(d) $$\frac{c}{r} = \frac{5(s + 2)}{s^3 + 20s^2 + 10s + 15}$$

5-4. For each of the following systems find $c_{ss}(t)$ and then $E_{ss}(t)$ for $r(t) = 3t$. Sketch the steady-state input-output relationships.

(a) $$\frac{c}{r} = \frac{5(s + 4)}{s^3 + 25s^2 + 10s + 25}$$

(b) $$\frac{c}{r} = \frac{5}{s^3 + 4s^2 + 15s + 5}$$

(c)
$$\frac{c}{r} = \frac{3(s + 6)}{s^3 + 10s^2 + 5s + 30}$$

(d)
$$\frac{c}{r} = \frac{24(s + 2)}{s^2 + 10s + 48}$$

(e)
$$\frac{c}{r} = \frac{3}{s^2 + s - 3}$$

5–5. For each of the unity feedback systems whose forward transfer function is given below, find the system type, the error constants, and the corresponding steady-state errors. Be sure to check the stability.

(a)
$$G(s) = \frac{20}{s(s + 5)(s + 4)}$$

(b)
$$G(s) = \frac{20}{s^2(s + 5)(s + 4)}$$

(c)
$$G(s) = \frac{20(s + 1)}{s^2(s + 5)(s + 4)}$$

(d)
$$G(s) = \frac{10}{s^3 + 8s^2 + 6s + 20}$$

(e)
$$G(s) = \frac{10(s^3 + 4s^2 + 5s + 2)}{s^5 + 15s^4 + 50s^3}$$

5–6. Add scalar feedback in the form of $H = b_0$ to each of the stable systems in Prob. 5–5. Find G^* and determine the system type. Then find the error constants and corresponding steady-state errors. Find the value of b_0, if any, that minimizes the errors.

5–7. The forward transfer function of a second-order control system with unity feedback is

$$G = \frac{25}{s(s + 5)}$$

(a) Find the error constants and steady-state errors.
(b) Find the steady-state error for $r(t) = 1 + 2t + 0.5t^2$.
(c) What is the error at 10 sec for the input in (b)?

5–8. For the system in Prob. 5–7,
(a) Write the first three terms of the steady-state error series.
(b) What is the steady-state error at the end of 10 sec when $r(t) = 1 + 2t + 0.5t^2$?

5–9. Same as Prob. 5–7 but with

$$G = \frac{25}{s^2 + 1.06s + 5}$$

5–10. Same as Prob. 5–8 but with

$$G = \frac{25}{s^2 + 1.06s + 5}$$

5–11. A system has the type-1 forward transfer function

$$G = \frac{K}{s(\tau s + 1)}$$

(a) For $H = 1$ find the error constants and corresponding steady-state errors.

(b) Repeat (a) for $H = 0.1$.

(c) Repeat (a) for $H = 0.3s + 0.3$.

(d) Repeat (a) for $H = 0.3s + 1$.

(e) Discuss the effects of non-unity feedback upon the accuracy of the system.

5–12. For the system in Prob. 5–11 find the first two terms of the error series for the various feedback transfer functions given and discuss the results.

5–13. Consider the system of Fig. P5–13.

(a) Let $k_2 = A = H = 1$. Find the error constants and steady-state errors.

(b) With $k_2 = A = 1$ and $H = 0.2$ repeat (a) and compare the results.

(c) With $H = 0.2$ find the values of k_2 and A that will reduce the errors of (b) to those of (a).

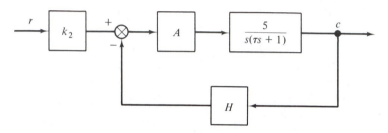

Figure P5–13

5–14. A system has a forward transfer function

$$G = \frac{10}{s(0.1s + 1)}$$

(a) If $H = 1$, find ζ and ω_n of the system, as well as the error constants and corresponding steady-state errors.

(b) If $H = 0.1s + 1$ repeat (a) and compare the results. What are the effects of non-scalar feedback?

5–15. The system of Fig. P5–15 has a type-0 plant with

$$G_p = 10/(s + 5)(s + 10)$$

(a) For $G_c = 1$ (proportional control) find the total steady-state error due to a step in position in both r and d when $A = 10$.

(b) For $G_c = 1/s$ (integral control) repeat (a).

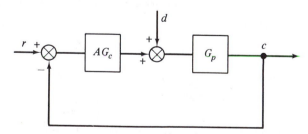

Figure P5–15

5–16. The system of Fig. P5–15 has a type-1 plant with $G_p = 2/s(s + 5)$.

    (a) Same as (a) in Prob. 5–15.

    (b) With $G_c = (s + 1)/s$ (proportional plus integral control) repeat (a).

5–17. A negative feedback system has a type-1 $G = K/s(s + a)$.

    (a) With $H = 1$, find the error constants and steady-state errors.

    (b) With $H = b_0$, but not equal to 1, find the error constants and steady-state errors.

    (c) With $H = b_0$ find the static sensitivities of the transfer function to K and to b_0. Interpret these sensitivities in terms of the results for (b).

6

The Root Locus

6–1 INTRODUCTION

The three basic performance criteria for a control system are stability, acceptable steady-state accuracy, and an acceptable transient response. With the system transfer function known, the Routh-Hurwitz criterion will tell us whether or not a system is stable. If it is stable, the steady-state accuracy can be determined for various types of inputs, as can the sensitivity functions if we so desire. We have seen that good accuracies and low sensitivities imply high gains. To determine the nature of the transient response, we need to know the location in the s plane of the roots of the characteristic equation. Unfortunately, the characteristic equation is normally unfactored and of high order.

The root locus technique* is a graphical method of determining the location of the roots of the characteristic equation as any single parameter, such as a gain or time constant, is varied from zero to infinity. The root locus, therefore, provides information not only as to the absolute stability of a system but also as to its degree of stability, which is another way of describing the nature of the transient response. If the system is unstable or has an unacceptable transient response, the root locus indicates possible ways to improve the response and is a convenient method of depicting qualitatively the effects of any such changes. The root locus technique is extremely useful when used in conjunction with the frequency response techniques of the next chapters. As a point of interest, the root locus can be used to factor polynomials (see App. B) or to plot frequency response functions (see Sec. 7–4), although such uses are neither convenient nor efficient.

* This technique was developed by Walter R. Evans.

130

The root locus can obviously be plotted by a computer, as were many in this book, and programs that can easily do so are now available for use on microcomputers. The root locus can also be plotted manually by the designer using a Spirule (if one can be found) or with a protractor and ruler. We, however, will emphasize rapid sketching techniques with the assumption that plots will be obtained by a computer. Sketching gives insight into the concept of the root locus and the effects of poles and zeros upon the stability of a system. Many times a sketch is sufficient in itself; if not, it can be helpful in verifying and interpreting the actual plot.

6–2 THE ANGLE AND MAGNITUDE CRITERIA

Without transport lag the transfer function of a system can be reduced to a ratio of polynomials such that

$$W(s) = \frac{c}{r} = \frac{N(s)}{D(s)} \tag{6–2–1}$$

The root locus technique is developed by expressing the characteristic function $D(s)$ as the sum of the integer unity and a new ratio of polynomials in s. The characteristic equation will be written as

$$D(s) = 1 + K\frac{Z(s)}{P(s)} = 0 \tag{6–2–2}$$

where K is the parameter of interest. K is independent of s and must not appear in the polynomials $Z(s)$ and $P(s)$. The form of $KZ(s)/P(s)$ is important; the ratio should be written as

$$K\frac{Z(s)}{P(s)} = \frac{K(s + z_1)(s + z_2) \cdots (s + z_i)}{s^n(s + p_1)(s + p_2) \cdots (s + p_j)} \tag{6–2–3}$$

where $-z_1, -z_2, \ldots$ are the (open-loop) zeros and $-p_1, -p_2, \ldots$ are the (open-loop) poles; these poles and zeros may be real or complex conjugates. Note in Eq. (6–2–3) that the coefficient of s, wherever it may appear, is always set equal to unity for root locus operations.

A *zero* is a value of s that makes $Z(s)$ equal to zero and is given the symbol $\bigcirc$. Do not automatically assume that this zero is also a closed-loop zero that makes $N(s)$ equal to zero in the system transfer function; it may be, but is not necessarily so. A *pole* is a value of s that makes $P(s)$ equal to zero and is given the symbol X. The s^n term represents n poles, all equal to zero and located at the origin of the s plane; there may be cases where s^n is in the numerator, indicating n zeros at the origin. A root of the characteristic equation has previously been defined as a value of s that makes $D(s)$ equal to zero and retains its symbol $\square$.

Since s is a complex variable and the poles and zeros may be complex, $KZ(s)/P(s)$ is a complex function* and may, therefore, be handled as a vector

---

* See App. A for a treatment of complex variables.

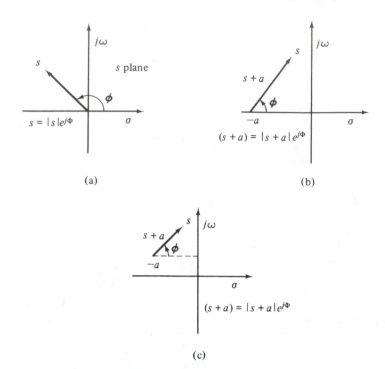

Figure 6–2–1. Vector treatment of complex numbers: (a) the Laplace variable s; (b) the factor $(s + a)$ with a real; (c) the factor $(s + a)$ with a complex.

having a magnitude and an associated angle or argument. Each of the factors on the right side of Eq. (6–2–3) can also be treated as a vector with an individual magnitude and associated angle, as shown in Fig. 6–2–1. Notice that the angle ϕ is measured from the horizontal and is positive in the counterclockwise direction.

If we express each factor in Eq. (6–2–3) in polar form, then

$$K\frac{Z(s)}{P(s)} = \frac{K|s + z_1|e^{j\phi_{z1}}|s + z_2|e^{j\phi_{z2}} \cdots}{|s^n|e^{jn\phi_n}|s + p_1|e^{j\phi_{p1}}|s + p_2|e^{j\phi_{p2}} \cdots} \qquad (6\text{–}2\text{–}4)$$

If we now collect the magnitudes together and multiply the exponentials together, we can write

$$K\frac{Z(s)}{P(s)} = \frac{K|s + z_1||s + z_2| \cdots}{|s^n||s + p_1||s + p_2| \cdots} e^{j\Sigma\phi_i} \qquad (6\text{–}2\text{–}5)$$

where

$$\sum\phi_i = \phi_{z1} + \phi_{z2} + \cdots - n\phi_n - \phi_{p1} - \phi_{p2} - \cdots \qquad (6\text{–}2\text{–}6)$$

Notice that the signs of the angles associated with the poles are changed as they are moved from the denominator to the numerator.

Returning to the characteristic equation of Eq. (6–2–2) and solving for $KZ(s)/P(s)$ yields

$$K\frac{Z(s)}{P(s)} = -1 = 1e^{j(2k+1)180°} \qquad k = 0, 1, 2 \ldots \qquad (6\text{–}2\text{–}7)$$

since -1 can be represented by a vector of unity magnitude and an angle that is an odd multiple of 180°. In order to satisfy Eq. (6–2–7) the magnitude of $KZ(s)/P(s)$ must be equal to unity, and its associated angle must be equal to an odd multiple of 180°; these two criteria are known as the *magnitude criterion* and the *angle criterion,* respectively. The root locus is plotted by finding those values of s that satisfy the angle criterion. In terms of Eq. (6–2–6) the angle criterion can be expressed as

$$\sum \phi_i = \sum \text{ angles of the zeros } - \sum \text{ angles of the poles} \qquad (6\text{–}2\text{–}8)$$
$$= (2k + 1) \, 180° \qquad k = 0, 1, 2 \, ...$$

where the angle of a zero or pole is the angle of the vector drawn to a point on the root locus from the zero or the pole. With the angle criterion satisfied, the value of K at each point on the locus is adjusted to satisfy the magnitude criterion,* which from Eq. (6–2–5) can be written as

$$\frac{K|s + z_1||s + z_2| \cdots}{|s^n||s + p_1||s + p_2| \cdots} = 1 \qquad (6\text{–}2\text{–}9)$$

so that

$$K = \frac{\text{product of the pole distances}}{\text{product of the zero distances}} \qquad (6\text{–}2\text{–}10)$$

A pole or zero distance is the distance from the pole or zero to the point on the root locus. If there are no zeros, the product of the zero distances is set equal to unity.

To plot a root locus, we need only to measure the angles from the poles and zeros to a trial point. If the sum of these angles, with due regard to the signs of the angles of the poles, is an odd multiple of 180°, the trial point is a root of the characteristic equation and lies on the root locus. If the angle criterion is not satisfied, the trial point is moved until it does satisfy the criterion. This trial point search is the basis of the Spirule operation, which simplifies the measurement and addition of the angles. Fortunately, it is possible to derive rules to make the sketching of a root locus simple and rapid. To determine graphically the value of K associated with a specific root location requires a plot of the root locus; a sketch is not sufficient. When a plot is needed, the existence of a sketch will aid and speed the plotting process.

The rules for sketching will be described in the next section.

6–3 ROOT LOCUS SKETCHING

We shall develop and demonstrate the basic rules for root locus sketching by applying them to the system shown in Fig. 6–3–1. We wish to find the locus of the roots as the gain K is varied from zero to plus infinity. The transfer

---

* The magnitude criterion requires the coefficient of the highest power of s in each factor to be unity.

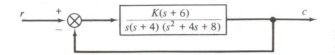

Figure 6–3–1. A unity feedback fourth-order control system.

function can be written either as

$$\frac{c}{r} = \frac{G}{1 + GH} = \frac{\dfrac{K(s + 6)}{s(s + 4)(s^2 + 4s + 8)}}{1 + \dfrac{K(s + 6)}{s(s + 4)(s^2 + 4s + 8)}} \tag{6–3–1}$$

or as

$$\frac{c}{r} = \frac{N(s)}{D(s)} = \frac{K(s + 6)}{s^4 + 8s^3 + 24s^2 + (32 + K)s + 6K} \tag{6–3–2}$$

Since K is our parameter of interest, the characteristic equation must be written in the form

$$D(s) = 1 + K\frac{Z(s)}{P(s)} = 0 \tag{6–3–3}$$

In Eq. (6–3–1) we see that the unreduced characteristic function is automatically in the correct form, so that

$$K\frac{Z(s)}{P(s)} = GH = \frac{K(s + 6)}{s(s + 4)(s^2 + 4s + 8)} \tag{6–3–4}$$

This will always be the case when we are interested in the effects of gain changes, especially of the amplifier gain, upon the stability and transient behavior of the system. Since this is one of the most frequent uses of the root locus, most authors do not use the $KZ(s)/P(s)$ function as such, but instead use the open-loop transfer function. If the parameter of interest is not a gain, GH is not the correct function to use. The $KZ(s)/P(s)$ function is a more general form and is always correct.

In Eq. (6–3–2), the characteristic equation in polynomial form can be written as

$$s^4 + 8s^3 + 24s^2 + 32s + Ks + 6K = 0 \tag{6–3–5}$$

To put this or any characteristic equation into the form of Eq. (6–3–3), divide both sides of the equation by the sum of all terms not containing K to obtain

$$1 + \frac{K(s + 6)}{s(s + 4)(s^2 + 4s + 8)} = 0 \tag{6–3–6}$$

so that

$$K\frac{Z(s)}{P(s)} = \frac{K(s + 6)}{s(s + 4)(s^2 + 4s + 8)} \tag{6–3–7}$$

There are one zero ($n_z = 1$) and four poles ($n_p = 4$) in this function. The real zero is located at $s = -6$; there is one pole at the origin, a second at $s = -4$, and a conjugate pair at $s = -2 \pm j2$. These poles and zeros are located in the s plane in Fig. 6-3-2a.

We are now ready to state the rules for sketching; proofs will be less than rigorous and sometimes omitted.

Rule 1: *There is a single-valued branch of the root locus for each root of the characteristic equation, and the total number of branches is equal to either the number of poles or the number of zeros, whichever is larger.* In our example, there are more poles than zeros and the number of branches will be four. This rule can easily be verified by rewriting Eq. (6-3-3) as

$$D(s) = P(s) + KZ(s) = 0 \qquad (6-3-8)$$

Equation (6-3-8) shows that the order of the characteristic equation and thus

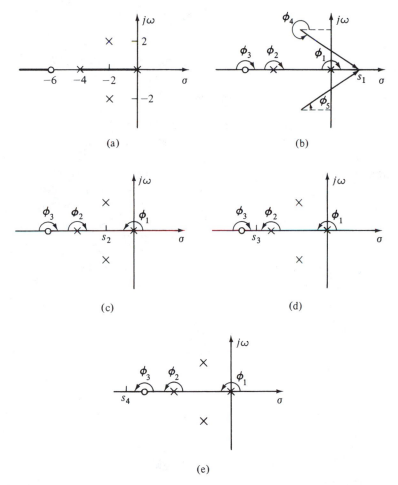

(a)

(b)

(c)

(d)

(e)

Figure 6-3-2. Application of angle criterion to locate the root locus along the real axis: (a) real axis locus; (b) trial point not on locus; (c) trial point on locus; (d) not on locus; (e) on locus.

the number of roots is established by the order of $P(s)$ or $Z(s)$, whichever is higher.

Rule 2: *Each branch of the root locus starts at a pole, where $K = 0$, and ends at a zero, where $K = +\infty$. If the number of poles exceeds the number of zeros, there will be zeros at infinity equal in number to the excess. Excess zeros similarly mean poles at infinity.* In our example, there are three more poles than zeros. There will be three zeros at infinity which means that three branches will go to infinity. Equation (6–3–8) will be used to verify this rule. First, set K equal to zero, then

$$D(s) = P(s) = 0 \qquad (6\text{–}3\text{–}9)$$

and we see that the poles are the roots. Now divide Eq. (6–3–8) by K so that

$$D(s) = \frac{P(s)}{K} + Z(s) = 0 \qquad (6\text{–}3\text{–}10)$$

and set K equal to plus infinity. Then

$$D(s) = Z(s) = 0 \qquad (6\text{–}3\text{–}11)$$

and the zeros are the roots for this particular value of K.

Rule 3: *Along the real axis the locus includes all points to the left of an odd number of real poles and zeros: no distinction is made between poles and zeros, and complex poles and zeros are neglected.* Rule 3 is applied to our example in Fig. 6–3–2a; the locus exists between the two poles on the real axis and to the left of the zero, extending to infinity. This rule is developed and proved by applying the angle criterion to a series of trial points along the real axis. Let us first take a trial point s_1 anywhere on the positive real axis and draw a vector from each pole and zero to the trial point as shown in Fig. 6–3–2b. We see first of all that the sum of the two angles associated with the complex poles is zero, and the angles associated with the real poles and zeros to the left of the trial point are also zero. Consequently, all complex poles and zeros can be neglected, as well as any real poles and zeros to the left of the trial point. For trial point s_1, there are no poles or zeros to the right of s_1 and the sum of the angles is zero; the angle criterion is not satisfied, and the positive real axis is not part of the root locus. A second trial point s_2 located anywhere between the two real poles, as shown in Fig. 6–3–2c, has a real pole to the right with an angle of $-180°$, which is equivalent to $+180°$, and the angle criterion is satisfied. Additional trial points are located in Fig. 6–3–2d and e and tested; s_3 does not meet the criterion, but s_4 does. From these trials we deduce Rule 3.

Rules 4 and 5 furnish information as to the behavior of any branches of the root locus that become infinite.

Rule 4: *If the number of poles n_p exceeds the number of zeros n_z, then as K approaches infinity, $(n_p - n_z)$ branches will become asymptotic to straight lines intersecting the real axis at angles given by*

$$\theta_k = \frac{(2k + 1)180°}{n_p - n_z} \qquad k = 0, 1, 2 \ldots \qquad (6\text{–}3\text{–}12)$$

If n_z exceeds n_p, then as K approaches zero, $(n_z - n_p)$ branches behave as above and originate at infinity along the asymptotes.

In our example,

$$\theta_k = \frac{(2k + 1)180°}{4 - 1} = (2k + 1)60° \qquad k = 0, 1, 2 \qquad (6\text{--}3\text{--}13)$$

There will be three asymptotes, one for each infinite zero, intersecting the real axis at angles of 60°, 180°, and 300°.

In developing this rule for the case of excess poles, it can be shown that as K becomes infinite, so does s. If we greatly reduce the scale of the s plane, as in Fig. 6–3–3, the cluster of poles and zeros essentially reduces to a point at the origin. Thus, all the vectors drawn from the poles and zeros to a trial point s will have equal angles designated by θ. Applying the angle criterion with due regard to signs we have

$$n_z\theta - n_p\theta = (2k + 1)180° \qquad (6\text{--}3\text{--}14)$$

or

$$\theta_k = \frac{(2k + 1)180°}{n_z - n_p} = \frac{(2k + 1)180°}{n_p - n_z} \qquad k = 0, 1, 2 \ldots \qquad (6\text{--}3\text{--}15)$$

There is no need to worry about the upper limit on k because the angles will start repeating themselves after the required number of asymptotic angles are calculated.

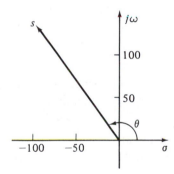

Figure 6–3–3.   Pole and zero angles for a very large value of s.

Rule 5: *The asymptotes of Rule 4 will intersect the real axis at a point* given by*

$$cg = \frac{\Sigma \text{ poles} - \Sigma \text{ zeros}}{n_p - n_z} \qquad (6\text{--}3\text{--}16)$$

For our example,

$$cg = \frac{-4 - 2 + j2 - 2 - j2 - (-6)}{3} = \frac{-2}{3} \qquad (6\text{--}3\text{--}17)$$

Notice that only the real part of the complex poles (and zeros) contributes to the cg location. The proof of Rule 5 is based on the theory of polynomials; it is tedious and therefore has been omitted. Branches of the root locus can cross

---

* This point is called the centroid or center of gravity of the roots.

the asymptotes when s is not approaching infinity. Note that if $n_z > n_p$, then $(n_p - n_z)$ in Eq. (6–3–16) becomes negative, affecting only the sign of the cg.

The result of applying Rules 1 through 5 to the example is shown in Fig. 6–3–4; arrows are drawn indicating the only possible movement of the roots along the real axis as K is increased. It is apparent that the two roots departing the real poles will come together and then, since each branch is single-valued, will leave the real axis at a *breakaway point*. Similarly, two branches must arrive together at an *arrival point* on the real axis to the left of the zero. Upon arrival one root will go to the right to the finite zero, and the other will go to the left to infinity.

When sketching we are not normally interested in the precise location of these breakaway and arrival points, which can easily be found from computer solutions and plots. Consequently, the next rule is included primarily for completeness and understanding and for the occasional case when you may wish to determine if two or more branches do indeed reach the real axis.

Rule 6: *A breakaway point on the real axis occurs when the value of K is at a maximum with respect to the values of K at points on either side; an arrival point occurs when the value of K is at a minimum with respect to the values of K of points on the real axis on either side of the arrival point.*

A closed-form method for finding any breakaway and arrival points is to solve Eq. (6–2–2) for K $[K = -P(s)/Z(s)]$, set the derivative of K with respect to s equal to zero $(dK/ds = 0)$, and solve for the breakaway and arrival points. Applying Rule 6 to our example, Eq. (6–3–6) yields

$$K = \frac{-s(s + 4)(s^2 + 4s + 8)}{s + 6} \qquad (6–3–18)$$

and taking the derivative of K with respect to s, setting it equal to zero, and simplifying results in the fourth-order equation

$$s^4 + 13.33s^3 + 56s^2 + 96s + 64 = 0 \qquad (6–3–19)$$

This equation has two real roots at -3.08 and -7.3, which can be identified by our knowledge of the acceptable and appropriate regions of the real axis as the

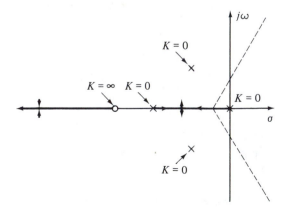

Figure 6–3–4. Application of Rules 1 through 8 for root locus of Eq. (6–3–5).

breakaway and arrival points, respectively. The corresponding values of K (K_b and K_a) can be found, either by substitution into Eq. (6–13–18) or by application of the magnitude criterion of Eq. (6–2–10), to be equal to 5 and 594.6, respectively.

Since real roots represent arrival and departure points on the real axis, the presence (or absence) of real root(s) would be a test for determining if branches originating at complex conjugate poles actually reach the real axis (and the transient modes become overdamped).

Taking the derivative of the expression for K and solving the resulting high-order polynomial is a lot of work and seems somewhat ridiculous since we are developing the root locus to avoid the necessity of factoring higher-order polynomials. It is often easier to use an iterative method of substituting trial values of s into the K expression to find the maximum and minimum values of K. For this example, we know that the breakaway point lies between 0 and -4, that the arrival point lies beyond -6, and that other values of s, therefore, need not be considered. The partial table below shows how this approach might be used.

s_b:	-2	-2.5	-2.75	$\underline{-3.0}$	-3.25
K:	4	4.55	4.82	$\underline{5.0}$	4.93
s_a:	-6.5	-6.75	-7.0	$\underline{-7.25}$	-7.5
K:	788	657	609	$\underline{594}$	599

Since K_{max} in the breakaway region is 5, -3 is the approximate location of the breakaway point. Since K_{min} in the arrival region is 594, -7.25 is the approximate location of the arrival point.

This iterative approach gives both s and K at the same time. Since monotonically increasing values of K along the real axis indicate no arrival (or departure) points on the real axis, this method can also be used as an arrival test for branches originating at complex conjugate poles.

In the next section we will look at the qualitative effects of real poles and zeros upon the breakaway point. Now let us move on to the next rule.

Rule 7: *Branches of the root locus are symmetrical with respect to the real axis since all complex roots appear in conjugate pairs.*

Rule 8: *Two branches of a root locus break away from or arrive at the real axis at angles of $\pm 90°$.* For three or more branches use Rule 9 below to calculate the departure or arrival angles.

Figure 6–3–4 shows the application of Rules 1 through 8. To complete our sketch we need to know which pair of roots approaches the pair of asymptotes to the right. Since the branches of a root locus are relatively well-behaved, the angle of departure of the branches from the complex poles should indicate whether these two branches approach the asymptotes or come together at the arrival point on the real axis.

Rule 9: *The angle criterion can be used to find the angle of departure from a complex pole and the angle of arrival at a complex zero by selecting a trial point so close to the complex pole or zero that the angles from the other poles and zeros are unaffected.* Let us redraw our poles and zeros in Fig. 6–3–5 and choose a trial point so close to the pole at $-2 + j2$ as to be indistinguishable

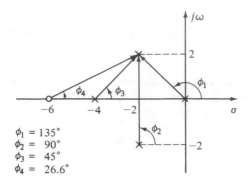

$\phi_1 = 135°$
$\phi_2 = 90°$
$\phi_3 = 45°$
$\phi_4 = 26.6°$

Figure 6–3–5. Use of the angle criterion to find the departure angle from a complex pole.

from it. Draw the vectors from the remaining poles and zeros and apply the angle criterion to obtain

$$-\phi_d - \phi_1 - \phi_2 - \phi_3 + \phi_4 = 180° \qquad (6\text{–}3\text{–}20)$$

where ϕ_d is the angle of departure we are seeking. The other angles can be calculated or measured. Substituting values into Eq. (6–3–20) and solving yields a ϕ_d equal to $-63.4°$ or $296.6°$. This departure angle and the resulting root locus are sketched in Fig. 6–3–6.

The completed root locus sketch in Fig. 6–3–6 shows how the roots of the characteristic equation vary as the gain K changes and thus gives some qualitative insight into the probable transient (dynamic) behavior of the system. For example, the two branches originating at the complex poles represent an underdamped second-order transient mode that becomes less stable as K increases and becomes unstable when the two branches enter the right half of the s plane. The other pair of roots starts out as an overdamped second-order mode (two first-order time lags) and begins to oscillate at the breakaway point, becoming underdamped. As K continues to increase, the damping ratio decreases but reaches a minimum value and then increases until the mode becomes overdamped again when reaching the arrival point.

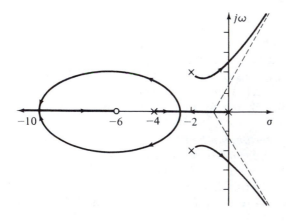

Figure 6–3–6. Root locus sketch of Eq. (6–3–5) as K is varied from 0 to $+\infty$.

Since two branches cross the imaginary axis and enter the right-half plane, this is a *conditionally stable* system. K_c, the value of K at the crossing, can be found graphically from a plot by using the magnitude criterion, or mathematically, as is done here, by using the Routh-Hurwitz criterion. With the use of the characteristic equation of Eq. (6–3–5), the Routhian array can be written as

$$
\begin{array}{llll}
s^4: & 1 & 24 & 6K \quad 0 \\
s^3: & 8 & 32 + K & 0 \quad 0 \\
s^2: & 160 - K & 48K & 0 \quad 0 \\
s^1: & -K^2 - 256K + 5120 & 0 & 0 \\
s^0: & 48K & 0 &
\end{array}
$$

K_c is the value of K that makes the s^1 row all zeros and is equal to 18.64. The corresponding roots are found from the auxiliary equation (the s^2 row set equal to zero with $K = K_c$) to be $\pm j2.52$.

There is one more rule that is limited in its usefulness but may be helpful at times, particularly with lower-order systems.

Rule 10: *If $n_p - n_z \geq 2$, then the sum of the roots of the characteristic equation is constant and is equal to the sum of the poles.* For our example, the sum of the poles is equal to -8; therefore, the sum of the four roots is always equal to -8. When $K = K_c$, the two imaginary roots cancel each other and the sum of the remaining two roots is equal to -8. Since $K_b < K_c < K_a$, these two roots are complex conjugates with the real part equal to -4. The imaginary part can be determined by applying the magnitude criterion at the point along the ($s = -4$) vertical line that either intersects a branch of a root locus plot or satisfies the angle criterion.

Rule 10 is most useful for third-order systems, such as the one represented by

$$
\frac{KZ(s)}{P(s)} = \frac{K}{s(s + 2)(s + 4)} \tag{6–3–21}
$$

The corresponding root locus is sketched in Fig. 6–3–7. When $K = K_c$, the

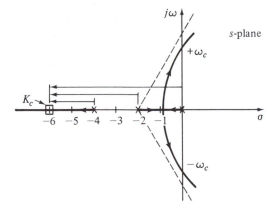

Figure 6–3–7. The use of Rule 10 to find r when $K = K_c$.

pair of imaginary roots $\pm j\omega_c$ plus the real root r must equal the sum of the poles
or

$$+j\omega_c - j\omega_c + r = -2-4 \qquad (6\text{--}3\text{--}22)$$

We find r to be equal to -6 when $K = K_c$. Now we use the magnitude criterion
to find

$$K_c = \frac{2 \cdot 4 \cdot 6}{1} = 48. \qquad (6\text{--}3\text{--}23)$$

6–4 ADDITIONAL SKETCHES AND CONSIDERATIONS

The control system designer should be adept at sketching the root locus and
should have a good feel for changes in the stability and dynamic response of a
system whenever poles and zeros are added. Let us consider a simple second-
order system with unity feedback and of type 1 as represented by the tracking
system of Fig. 6–4–1a, whose system (closed-loop) transfer function is

$$\frac{\theta_{\text{out}}}{\theta_{\text{cmd}}} = \frac{\dfrac{AK_m}{s(\tau_m s + 1)}}{1 + \dfrac{AK_m}{s(\tau_m s + 1)}} \qquad (6\text{--}4\text{--}1)$$

with a characteristic equation

$$1 + \frac{AK_m}{s(\tau_m s + 1)} = 0 \qquad (6\text{--}4\text{--}2)$$

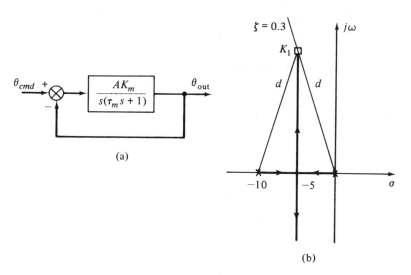

(a)

(b)

Figure 6–4–1. The second-order tracking radar: (a) block diagram; (b) root locus
for variable amplifier gain.

If the parameter of interest is the amplifier gain, then

$$K\frac{Z(s)}{P(s)} = \frac{AK_m/\tau_m}{s(s + 1/\tau_m)} \tag{6–4–3}$$

If we select a motor with $K_m = 5$ and $\tau_m = 0.1$,

$$K\frac{Z(s)}{P(s)} = \frac{K}{s(s + 10)} \tag{6–4–4}$$

where $K = AK_m/\tau_m = 50A$.

There are no zeros ($n_z = 0$) and two poles ($n_p = 2$); there will be two branches and two infinite zeros. The next step is to locate the poles in the s plane and apply Rule 3 to find that on the real axis the locus exists only between the two poles. Since there are two infinite zeros, there will be two asymptotes with angles

$$\theta_k = \frac{(2k + 1)180°}{2} = (2k + 1)90° \qquad k = 0, 1 \tag{6–4–5}$$

or 90° and 270°. The cg or intersection of the asymptotes and the real axis is

$$cg = \frac{-0 - 10}{2} = -5 \tag{6–4–6}$$

There obviously will be a breakaway point between the two poles. With

$$K = -s(s + 10) \qquad \frac{dK}{ds} = -2s - 10 = 0 \tag{6–4–7}$$

the breakaway point is located at -5, and the corresponding value of K is 25. Using iterative values of s produces the partial table

$$s: \quad -3 \quad -4 \quad \underline{-5} \quad -6$$

$$K: \quad 16 \quad 24 \quad \underline{25} \quad 24$$

which yields identical values of -5 and 25 for s_b and K_b. Notice that the breakaway point is exactly halfway between the two poles. This is always the case when there are only two poles on the real axis. Since two branches leave at the breakaway point, the breakaway angles are $\pm 90°$. The sketch is shown in Fig. 6–4–1b.

If done to scale, this sketch is a plot and is the classic root locus of a simple second-order system with a variable gain. Think of the amplifier gain as a variable spring constant in a simple spring-mass-dashpot system. When the gain is small, the roots are real and negative and the system is overdamped. As the gain is increased, the damping ratio will decrease until the system passes through the critical damping point and becomes underdamped. The system is always stable, although as the gain approaches infinity, the damping ratio approaches zero and the undamped natural frequency approaches infinity; the settling time, however, remains constant.

The shape of the transient response is a design criterion and is controlled by the damping ratio ζ. If we wish a ζ of 0.3, we can use the root locus to find the appropriate value of A. Figure 4–5–2b shows that the 0.3 damping ratio makes an angle θ with the real axis, where $\theta = \cos^{-1} \zeta$ or 72.6°. When the $\zeta = 0.3$ line is drawn in Fig. 6–4–1b to intersect the root locus, the value of K that will place the root at that intersection can be found from the magnitude criterion to be

$$K_1 = d^2 = 277.8 \qquad (6\text{–}4\text{–}8)$$

Since $K_1 = 50A$, an amplifier gain of 5.556 will produce a transient mode with $\zeta = 0.3$. Do not forget that the steady-state accuracy requirements must also be met; be sure that this value of A is equal to or larger than the minimum value of A for an acceptable steady-state error.

In order to determine the influence of poles and zeros on a root locus let us first add a pole to the $KZ(s)/P(s)$ function of Eq. (6–4–4):

$$K\frac{Z(s)}{P(s)} = \frac{K}{s(s + 10)(s + 15)} \qquad (6\text{–}4\text{–}9)$$

This additional pole could arise from time lags in the amplifier or other parts of the system. The excess of poles with respect to zeros has been increased from two to three; consequently, there will be three infinite zeros and three asymptotes making angles of 60°, 180°, and 300° with the real axis and with

$$cg = \frac{-10 - 15}{3} = -8\frac{1}{3} \qquad (6\text{–}4\text{–}10)$$

With the use of either method of Rule 6, the breakaway point is located at -3.92 with a corresponding gain of 264, a shift to the right and an increase in the gain, both resulting from the addition of the pole. The root locus is sketched in Fig. 6–4–2a and shows the system to be conditionally stable; if K is sufficiently increased, the roots will move into the right-half plane and the system becomes unstable. Notice that the added pole bends the asymptotes toward the imaginary axis and shifts the breakaway point toward the imaginary axis. This system is less stable than the preceding one. We conclude that poles added to the $KZ(s)/P(s)$

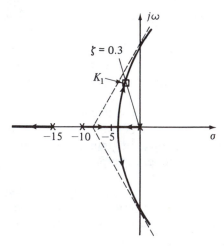

Figure 6–4–2. Root locus sketch for the conditionally stable system of Eq. (6–4–9).

function destabilize a system; the closer the pole to the imaginary axis, the more destabilizing its influence.

To find the value of K that will give the second-order transient mode a damping ratio of 0.3, draw the 0.3 line in Fig. 6–4–2 at an angle of 72.6°. Since this root locus is a sketch only, use a Spirule or protractor to search along the 0.3 line until the angle criterion is satisfied; then determine K_1 from the magnitude criterion.

As the added pole is moved to the right toward the imaginary axis, both the cg and the breakaway point move toward the imaginary axis and the stability of the system continues to decrease; it becomes unstable for all positive values of K when the pole reaches the origin. With the added pole at the origin,

$$K\frac{Z(s)}{P(s)} = \frac{K}{s^2(s + 10)} \qquad (6\text{–}4\text{–}11)$$

With unity feedback, as is the case here, the system becomes type 2 but is now unstable for all values of gain, as can be seen in the root locus sketch of Fig. 6–5–5a and as verified by the Routh-Hurwitz criterion.

We conclude that *poles tend to destabilize a system and that the closer they are to the imaginary axis, the more destabilizing they are.*

Let us now add a zero to our original second-order system of Eq. (6–4–

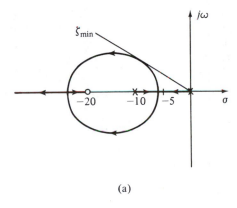

(a)

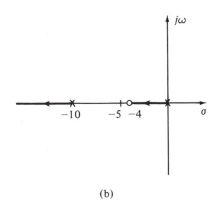

(b)

Figure 6–4–3. Effect of a zero added to the second-order system of Fig. 6–4–1: (a) to left of poles; (b) between the poles.

4) to obtain

$$K\frac{Z(s)}{P(s)} = K\frac{(s + z)}{s(s + 10)} \tag{6–4–12}$$

If the zero is located to the left of the two poles, say at -20, then the root locus as sketched in Fig. 6–4–3a is that of a stable system. This system is more stable than the original one of Fig. 6–4–1 in that the damping ratio can only be reduced to ζ_{min} as indicated in the sketch. Notice that the breakaway point has been shifted away from the imaginary axis to approximately -6. If the zero is moved closer to the imaginary axis, say to -4, so that it is between the two poles, the system becomes even more stable, as shown in Fig. 6–4–3b. This system is overdamped for all values of K, behaving more and more like a first-order system as K is increased.

We conclude that *zeros added to the* $KZ(s)/P(s)$ *function stabilize the system.* Remember that a system can be too stable to be a useful system in that it may be too slow to respond to a command.

Since most real systems are conditionally stable with respect to gain, let us add a zero to the three-pole system of Eq. (6–4–9) and sketch the root locus for four locations of the zero as shown in Fig. 6–4–4. Although the system

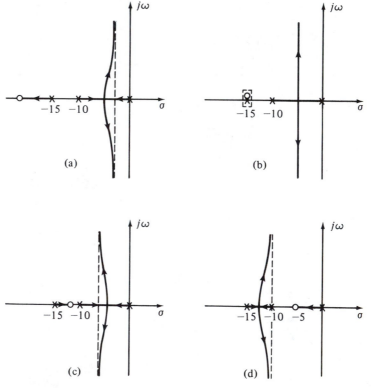

Figure 6–4–4.   Effects of adding a zero to the conditionally stable system of Fig. 6–4–2.

becomes more stable as the zero moves toward the imaginary axis, it is not obvious which system is the most acceptable; that is dependent upon the performance specifications. Incidentally, a zero in the right-half plane will make this system unstable for all positive values of K.

In Fig. 6-4-4b the zero coincides with one of the poles; resist the temptation to cancel the pole in the $KZ(s)/P(s)$ function. Although the pole-zero coincidence appears to restore the system to its original second-order configuration with the root locus of Fig. 6-4-1b, it represents a root of the characteristic equation that is independent of K. It is a *stationary* root (thus the dashed square around the pair in Fig. 6-4-4b) that will be canceled in the system transfer functions *only* if there is a closed-loop zero at that location. *Never attempt to cancel a pole in the right half of the s plane* since the pole-zero combination forms an unstable stationary root that may not be canceled at all for certain input-input combinations. Furthermore, perfect cancellation is virtually impossible inasmuch as precise locations are not known for the pole and zero and each can shift.

If a zero is deliberately added to a system by means of a passive network, it will be accompanied by a pole. If the pole is farther from the imaginary axis than the zero, the combination is the equivalent of the passive lead network of Sec. 3-3 and is stabilizing. If, on the other hand, the zero is farther away, we have a lag network that in spite of being destabilizing can allow increased gains. The use of such pole-zero combinations is discussed in the chapter on system compensation.

6-5 OTHER ROOT LOCUS USES

The root locus technique is not limited to system characteristic equations or to gain variations only. It may be used to determine the dynamic characteristics of plants and components and to examine the effects of changes in other parameters. Consider, for example, the automobile suspension system of Eq. (3-2-9) whose characteristic equation is

$$ms^2 + Bs + k = 0 \qquad (6\text{-}5\text{-}1a)$$

The effect of the shock absorber damping coefficient upon the dynamic (transient) performance can be examined by rewriting the characteristic equation as

$$ms^2 + k + Bs = 0 \qquad (6\text{-}5\text{-}1b)$$

so that

$$1 + K\frac{Z(s)}{P(s)} = 1 + \frac{(B/m)s}{s^2 + k/m} \qquad (6\text{-}5\text{-}1c)$$

Leaving B unspecified but using the values of Sec. 5-1 for m and k, namely, 31.1 slugs (453.7 kg) and 31.1 lb/ft (453.7 N/m), respectively,

$$K\frac{Z(s)}{P(s)} = \frac{Ks}{(s^2 + 1)} \qquad (6\text{-}5\text{-}1d)$$

where $K = B/31.1$ ($B/0.0022$).

The root locus is shown in Fig. 6-5-1 with the arrival point at $s = -1$,

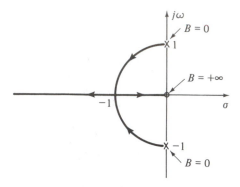

Figure 6–5–1. Root locus for suspension system of Eq. (6–5–1a) for variable B.

where $K = 2$ and the damping coefficient of the shock absorber is 62.2 1b-sec/ft (907.4 N-sec/m). At that point, the suspension system is critically damped. For lower values, it is underdamped, becoming marginally stable when B goes to zero, and is overdamped for larger values, becoming limitedly stable as B approaches infinity.

Let us return to the tracking system of Eq. (6–4–1) and see how its performance is affected by the motor time constant τ_m. After writing the characteristic equation in polynomial form,

$$\tau_m s^2 + s + AK_m = 0 \qquad (6\text{–}5\text{–}2a)$$

we isolate the terms containing τ_m:

$$s + AK_m + \tau_m s^2 = 0 \qquad (6\text{–}5\text{–}2b)$$

and divide both sides of the equation by the terms that do not contain τ_m to obtain

$$1 + \frac{\tau_m s^2}{s + AK_m} = 0 \qquad (6\text{–}5\text{–}2c)$$

which is in the form of $1 + KZ(s)/P(s)$. The root locus is sketched with

$$K\frac{Z(s)}{P(s)} = \tau_m \frac{s^2}{s + AK_m} \qquad (6\text{–}5\text{–}2d)$$

where the values of A and K_m (or of their product) must be fixed. This particular $KZ(s)/P(s)$ is not a physical transfer function, as has been the case when a gain is varied, but rather is a purely mathematical function, as indicated by the fact that the numerator is of higher order than the denominator. Consequently, for the first time there will be excess zeros, which means that there will be poles at infinity. In this case there is one excess zero which locates the infinite pole on the single asymptote represented by the negative real axis.

The root locus is sketched in Fig. 6–5–2 and shows that the system is essentially first-order when $\tau_m = 0$, and approaches instability when τ_m becomes infinite. The Routh-Hurwitz criterion is probably the simplest method for determining that the root locus does not cross the imaginary axis.

A more realistic case would be the tracking system of Fig. 6–5–3 with a system time lag represented by an amplifier time lag. To determine the effects

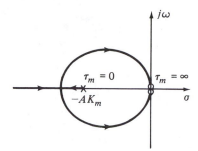

Figure 6–5–2. Root locus sketch of second-order system of Fig. 6–4–1 for variable τ_m and fixed AK_m.

of τ_a, the system transfer function is written as

$$\frac{\theta_{\text{out}}}{\theta_{\text{cmd}}} = \frac{AK_m}{\tau_a\tau_m s^3 + (\tau_a + \tau_m)s^2 + s + AK_m} \qquad (6\text{–}5\text{–}3)$$

The characteristic equation is rearranged to identify and isolate the terms containing τ_a:

$$\tau_m s^2 + s + AK_m + \tau_a s^2(\tau_m s + 1) = 0 \qquad (6\text{–}5\text{–}4)$$

Equation (6–5–4) is put into the form

$$1 + \frac{\tau_a s^2(\tau_m s + 1)}{\tau_m s^2 + s + AK_m} = 0 \qquad (6\text{–}5\text{–}5)$$

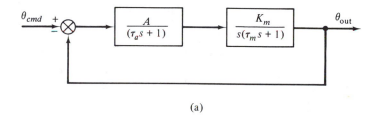

(a)

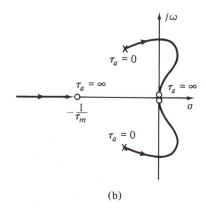

(b)

Figure 6–5–3. Root locus sketch of a third-order system for variable τ_a and fixed AK_m and τ_m.

For root locus purposes,

$$K\frac{Z(s)}{P(s)} = \frac{\tau_a s^2 \left(s + \dfrac{1}{\tau_m}\right)}{s^2 + \dfrac{1}{\tau_m}s + \dfrac{AK_m}{\tau_m}} \qquad (6\text{-}5\text{-}6)$$

where τ_m and AK_m must be specified. With $\tau_m = 0.1$ sec and $AK_m = 25.0$, there will be a pair of complex poles (the roots of the underdamped system without a time lag), three finite zeros, and a pole at minus infinity. The root locus is sketched in Fig. 6–5–3b and shows the system to be conditionally stable with respect to τ_a for this combination of τ_m and AK_m. The crossing of the imaginary axis was confirmed by applying the Routh-Hurwitz criterion (see Sec. 4–3 where the stability boundary for this system was examined).

When there are excess zeros, $(n_p - n_z)$ becomes negative. This has no affect on the angles of the asymptotes, since $-180°$ is equal to $+180°$, but does affect the sign of the cg and can cause trouble if one is not careful. Consider a characteristic equation with

$$K\frac{Z(s)}{P(s)} = K\frac{s^2(s + 1)}{s + 10} \qquad (6\text{-}5\text{-}7)$$

There are two excess zeros, two asymptotes making angles of $\pm 90°$ with the real axis, $n_p - n_z = -2$, and the

$$cg = \frac{-10 - (-1)}{-2} = +4.5$$

Consequently, the asymptotes and the two poles at infinity are in the right-half plane and the system is unstable for all positive values of K, as can be seen from the root locus sketch of Fig. 6–5–4 and verified by application of the Routh-Hurwitz criterion.

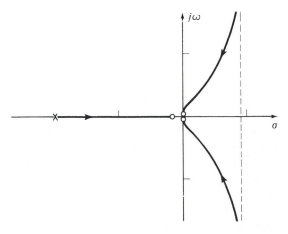

Figure 6–5–4. Root locus of Eq. (6–5–7) with excess zeros.

Some examples of root locus sketches for various pole-zero combinations are shown in Fig. 6–5–5. The reader is advised to reproduce these sketches as an exercise in applying the sketching rules.

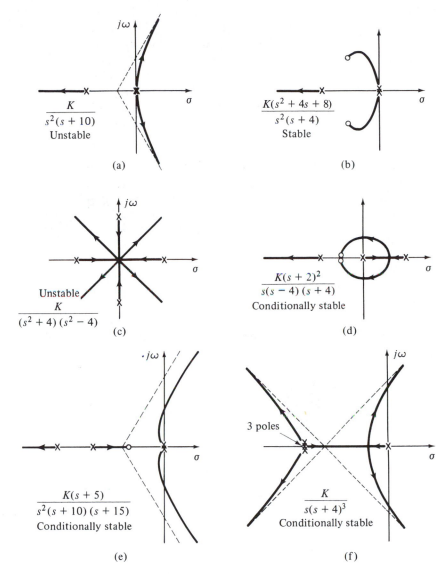

Figure 6–5–5. Some root locus sketches.

*6–6 ROOT CONTOURS

A disadvantage of the root locus method is that a single root locus can only show the effects of varying one parameter at a time. Consequently, the analysis and design of multiparameter and multiple-loop systems may require several individual root locus plots or the superimposition of more than one root locus in a single plot. The latter technique results in a series of root loci for varying parameters that are referred to as *root contours*.

The stability and performance of the third-order system of Fig. 6–5–3a are determined by the values of the two time constants and the product of the amplifier and motor gains, three parameters in all. Figure 6–5–3b is a root locus plot for a variable τ_a, with AK_m and τ_m fixed at 25 and 0.1 sec, respectively. To construct root contours for variable values of AK_m as well as for a variable τ_a (with τ_m still held fixed), the locus of the roots for a variable AK_m with τ_a set equal to zero is embedded in the root locus of Fig. 6–5–3b. With $\tau_a = 0$ and $\tau_m = 0.1$ sec, and the characteristic equation written as

$$s^2 + 10s + 10AK_m = 1 + \frac{10AK_m}{s(s + 10)} = 0 \qquad (6\text{–}6\text{–}1)$$

the root locus for positive values of AK_m is as shown in Fig. 6–6–1a. These roots will be the poles for the root contours that will be constructed using the complete fourth-order characteristic equation written (with $\tau_m = 0.1$ sec) as

$$1 + \frac{\tau_a s^2(s + 10)}{s^2 + 10s + 10AK_m} = 0 \qquad (6\text{–}6\text{–}2)$$

Note that the $P(s)$ expression in Eq. (6–6–2) is identical to the system characteristic equation of Eq. 6–6–1, with $\tau_a = 0$.

Root contours for varying AK_m and varying τ_m are shown in Fig. 6–6–1b. Notice that each contour starts on the root locus of Fig. 6–6–1a, which indeed is embedded in this figure. Also note that the system is stable for all positive values of τ_a when AK_m is equal to or less than 10 but conditionally stable for

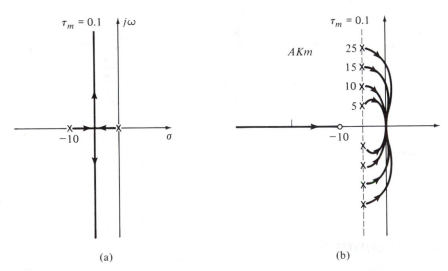

Figure 6–6–1. Root contours for Eq. (6–6–1) with $\tau_m = 0.1$: (a) variable AK_m with $\tau_a = 0$; (b) variable τ_a for various values of AK_m.

greater values of AK_m, as can be determined or verified by use of the Routh-Hurwitz criterion.

Examination of Fig. 6–6–1a and b shows that a reduction in τ_m is accompanied by a shift in the locus of the poles in Fig. 6–6–1b as the pole represented by $-1/\tau_m$ moves. A decrease in τ_m (a motor with a faster response time) shifts the poles to the left, away from the imaginary axis, and makes the system more stable, whereas an increase in τ_m moves the poles toward the imaginary axis and the system becomes less stable. The effects of a varying τ_m can be incorporated into the root contour representation by embedding the root locus for a variable τ_m for various values of AK_m, with $\tau_a = 0$, using the reduced-order characteristic equation written in the form

$$1 + \frac{\tau_m s^2}{s + AK_m} = 0 \qquad (6\text{–}6\text{–}3)$$

Design criteria can also be superimposed upon the root contour plots and used to establish regions of acceptable component parameters. For example, roots with a minimum damping ratio of 0.5 would lie in the region to the left of a radial line making an angle of 60° with the negative real axis; roots with a minimum undamped natural frequency of 5 rad/sec would lie above a circular arc with its center at the origin and a radius of 5; and roots with AK_m greater than 10 would lie above the corresponding AK_m parameter line. Any combination of τ_m and τ_a within the region established by these three lines would result in a system that meets these minimum design criteria.

*6–7 THE ZERO-ANGLE (NEGATIVE) ROOT LOCUS

The root locus rules and examples of the preceding sections were based on the two assumptions that feedback was negative and that the variable parameter was positive. These two assumptions lead to a characteristic equation in the form of

$$1 + K\frac{Z(s)}{P(s)} = 0 \qquad (6\text{–}7\text{–}1)$$

and

$$K\frac{Z(s)}{P(s)} = -1 \qquad (6\text{–}7\text{–}2)$$

From Eq. (6–7–2) we developed the criteria that the magnitude of $KZ(s)/P(s)$ be unity and that its associated angle be an odd multiple of 180°.

If either of these assumptions is not valid, then the characteristic equation takes the form

$$1 - K\frac{Z(s)}{P(s)} = 0 \qquad (6\text{–}7\text{–}3)$$

and

$$K\frac{Z(s)}{P(s)} = +1 \qquad (6\text{--}7\text{--}4)$$

The two sets of conditions for Eqs. (6–7–3) and (6–7–4) are negative feedback with a negative parameter and positive feedback with a positive parameter. Since these conditions usually result in an unstable system, they are not deliberately introduced into a control system but may be encountered in a plant or in the inner loop.

A third and possibly more significant condition is the presence of an odd number (usually one) of zeros in the right half of the s plane. This condition, which may be encountered in plant transfer functions, will be examined in more detail at the end of this section.

To satisfy Eq. (6–7–4), the magnitude of $KZ(s)/P(s)$ must be equal to unity as before, but the associated angle must be zero. This is a different angle criterion that can be written as

$$\sum \phi_i = \sum \text{ angles of the zeros}$$
$$- \sum \text{ angles of the poles} = k360° \qquad k = 0, 1 \ldots \qquad (6\text{--}7\text{--}5)$$

This new angle criterion modifies some, but not all, of the rules used previously. For obvious reasons, the resulting root locus is usually referred to as a *zero-angle root locus* and sometimes as a *negative root locus* or a *complementary root locus*.

Let us use the example of Eq. (6–3–4) to illustrate the differences in the rules when K takes on negative values. The $KZ(s)/P(s)$ function remains unchanged as

$$K\frac{Z(s)}{P(s)} = \frac{K(s + 6)}{s(s + 4)(s^2 + 4s + 8)} \qquad (6\text{--}7\text{--}6)$$

but now K will vary from 0 to $-\infty$. Rules 1 and 2 remain unchanged except that $K = -\infty$ at the zeros. Rule 3 is changed to state that the locus on the real axis includes all points to the left of an *even* number of poles and zeros (zero is considered an even number). These and succeeding rules are illustrated in Fig. 6–7–1.

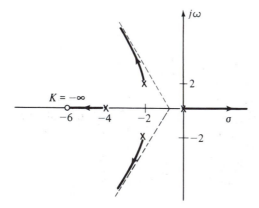

Figure 6–7–1. Root locus sketch of Eq. (6–3–4) for negative values of K.

The angles of the asymptotes are found from the expression

$$\theta_k = \frac{k360°}{n_p - n_z} \qquad k = 0, 1, 2, \ldots \qquad (6\text{-}7\text{-}7)$$

The three asymptotes for our example will have angles of 0°, 120°, and 240°. The rule for the cg is unchanged, and the cg remains at $-2/3$.

The departure angle from the complex pole is found by summing all the angles and setting the sum equal to zero. For this example,

$$-\phi_d - 135° - 90° - 45° + 26.6° = 0 \qquad (6\text{-}7\text{-}8)$$

or $\phi_d = -243.4° = +116.6°$. The root locus sketch is now complete and shows the system to be unstable for all negative values of K since there is always a positive real root.

When the root locus sketch for positive values of K is superimposed on the sketch for negative values, as in Fig. 6-7-2, it becomes apparent that the branches are continuous contours that close at the zeros. In our example, one of the branches closes at the finite zero and the other three close at infinity. Without knowledge of the zero-angle root locus rules, this continuity of the branches can be used to sketch the root locus for positive feedback with a positive K, or the root locus for negative feedback with positive K.

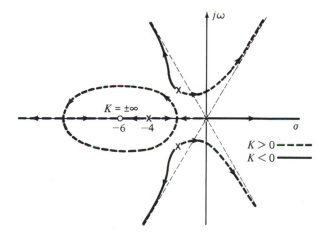

Figure 6-7-2. Root locus sketch of Eq. (6-3-4) for all values of K, both positive and negative.

With zeros in the right-half plane the $KZ(s)/P(s)$ function is a nonminimum phase function, and it is not necessarily obvious what type of root locus is appropriate or what the sign of the feedback should be. In fact, any confusion that might arise from the presence of right-half plane zeros results from the interaction of the sign of the open-loop steady-state value and the feedback sign.

Let us look at the unity-feedback second-order system of Fig. 6-7-3a with negative feedback and with a single zero in the right-half plane. The characteristic equation is

$$1 + \frac{K(s - 3)}{s(s + 10)} = 0 \qquad (6\text{-}7\text{-}9)$$

The regular root locus of Fig. 6–7–3b with positive K shows that the system is unstable for all positive K, whereas the zero-angle root locus of Fig. 6–7–3c with negative K shows the system to be conditionally stable. To decide which is correct, use the Routh-Hurwitz criterion. The rewritten characteristic equation is

$$s^2 + (10 + K)s - 3K = 0 \qquad (6\text{–}7\text{–}10)$$

The quick check shows that, if all the coefficients are to be of the same sign, K must be negative. The Routhian array is

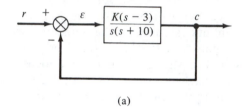

$$
\begin{array}{llll}
s^2: & 1 & -3K & 0 \\
s^1: & 10 + K & 0 & 0 \\
s^0: & -3K & 0 &
\end{array}
$$

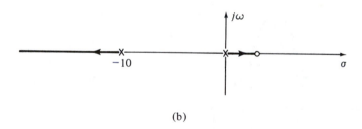

(a)

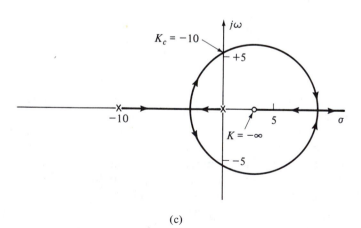

(b)

(c)

Figure 6–7–3. A nonminimum phase transfer function: (a) system block diagram; (b) regular root locus for positive K; (c) zero-angle root locus for negative K.

The s^0 row shows that K must be negative, and the s^1 row shows that K must be greater than -10. Consequently, the system is conditionally stable and the conditions for stability are that $-10 < K < 0$.

From a physical point of view, the steady-state value of the open-loop transfer function is negative, so that with negative feedback the actuating signal ε is the sum of the command and feedback variables rather than their difference, as is the usual case. Therefore, changing the feedback sign of Fig. 6–7–3a from negative to positive, as shown in Fig. 6–7–4a, establishes an equivalent system but one with positive values of K. Since this is a positive feedback system with a positive K, the root locus must be the zero-angle root locus of Fig. 6–7–3c.

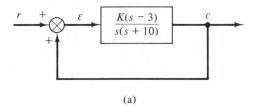

(a)

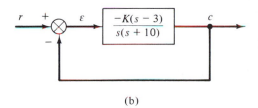

(b)

Figure 6–7–4. Equivalent representations of Fig. 6–7–3a: (a) positive feedback with positive K; (b) negative feedback with forward path sign change and positive K.

Negative feedback and positive values of K can be retained by introducing a negative sign into the open-loop transfer function (by reversing the polarity of the amplifier, for example) as shown in Fig. 6–7–4b. For both representations, the correct root locus is the zero-angle root locus shown in Fig. 6–7–3c for positive values of K. Sometimes the negative sign is inherent in the plant transfer function.

Whenever there is any doubt as to the type of root locus that is appropriate, first examine the characteristic equation in the form $1 + KZ(s)/P(s) = 0$ to determine the appropriate relationship between the angle criterion and the sign of K. Then apply the Routh-Hurwitz criterion to the characteristic equation in polynomial form to confirm the sign of K. For example, for each of the system representations of Fig. 6–7–4, the characteristic equation can be written as

$$1 - \frac{K(s - 3)}{s(s + 10)} = 0 \qquad (6\text{–}7\text{–}11\text{a})$$

or

$$\frac{K(s - 3)}{s(s + 10)} = +1 \qquad (6\text{–}7\text{–}11\text{b})$$

and in polynomial form as

$$s^2 + (10 - K)s + 3K = 0 \qquad (6\text{--}7\text{--}11c)$$

Equation (6–7–11b) shows that the zero-angle criterion is appropriate for positive values of K. The quick Routh-Hurwitz check, applied to Eq. (6–7–11c), confirms that K must be positive (and less than 10) as a minimum requirement for stability.

An odd number of zeros in the right-half plane has the same effect as a single zero, whereas an even number of zeros in the right-half plane has no effect since the corresponding steady-state value is positive.

*6–8 THE INVERSE ROOT LOCUS

The root locus method of the preceding sections is sometimes called the *direct root locus* method. Essentially, it is a technique for analysis, in that it locates the roots of the characteristic equation when the poles and zeros of the individual elements of the system are known. There is an *inverse root locus* method that is a synthesis technique, because it indicates the poles and zeros required to produce specified roots; in other words, it shows what elements are required to create a desired system. This method is interesting and under certain conditions may be useful in the design of a controller. This design technique is sometimes referred to as *pole placement*.

To develop the inverse root locus, the system transfer function

$$W(s) = \frac{c}{r}(s) = \frac{G}{1 + GH} \qquad (6\text{--}8\text{--}1)$$

is solved for G in terms of the transfer function W, so that

$$G = \frac{W}{1 - HW} \qquad (6\text{--}8\text{--}2)$$

If the feedback transfer function H is not unity, it must be known or specified; this is a definite disadvantage. If, however, the feedback is unity, Eq. (6–8–2) can be written as

$$G = \frac{W}{1 - W} = K\frac{Z(s)}{P(s)} \qquad (6\text{--}8\text{--}3)$$

The zeros of G are the values of s that make $W(s) = 0$, and the poles of G are the roots of the equation

$$1 - W = 0 \qquad (6\text{--}8\text{--}4)$$

If W is written as the ratio of two polynomials

$$W = K_w \frac{N(s)}{D(s)} \qquad (6\text{--}8\text{--}5)$$

where $N(s)$ and $D(s)$ are in the proper form for root locus sketching, then Eq. (6–8–4) becomes

$$1 - K_w \frac{N(s)}{D(s)} = 0 \qquad (6\text{--}8\text{--}6)$$

Solving for

$$K_w \frac{N(s)}{D(s)} = +1 \qquad (6\text{–}8\text{–}7)$$

we find that the magnitude and angle criteria are those for the zero-angle root locus of the preceding section. Applying the angle criterion will plot the locus of the poles of G as K_w takes on positive values. The zeros of G will be the closed-loop zeros of W, and the K of Eq. (6–8–3) will be K_w. These last two relationships can be verified by substituting Eq. (6–8–5) into Eq. (6–8–3) to obtain

$$G = K \frac{Z(s)}{P(s)} = \frac{K_w \dfrac{N(s)}{D(s)}}{1 - K_w \dfrac{N(s)}{D(s)}} = \frac{K_w N(s)}{D(s) - K_w N(s)} \qquad (6\text{–}8\text{–}8)$$

We shall illustrate the use of the inverse root locus method with two examples. The first example is quite simple. We wish to design a unity feedback control system that behaves as a simple second-order system. The steady-state error for a position step input is to be zero, and the transient response is to have a percent overshoot less than 40 percent and a settling time of approximately 0.6 sec. The transient response requirements lead to a desired damping ratio ζ of approximately 0.3 and an undamped natural frequency ω_n of 16.67 rad/sec. The roots of the characteristic equation will, therefore, be located at $s = -5 \pm j15.9$. The desired system transfer function can now be written as

$$W(s) = K \frac{N(s)}{D(s)} = \frac{K}{s^2 + 10s + 277.8} \qquad (6\text{–}8\text{–}9)$$

The two roots are located in the s plane in Fig. 6–8–1; there are no zeros. Since there are no roots or zeros on the real axis, the entire real axis satisfies the zero-angle criterion and is part of the locus. The absence of finite zeros means that there are two at infinity; as might be expected, their angles are $0°$ and $180°$. The departure angle from the upper complex root is $-90°$; it can easily be seen that every point on the line joining the complex roots is on the locus. The final sketch in Fig. 6–8–1 shows all possible values of the poles of G for the given roots as K is increased.

In reference to the performance criteria, the zero steady-state error for a position step input implies a type-1 or higher system; i.e., one or more poles of G must be located at the origin of the s plane. The pole locus sketch shows a single pole at the origin when $K = K_1$. The value of K_1 can be found, either numerically or graphically, from the magnitude criterion, or from Eq. (6–8–6).

$$1 - K \frac{N(s)}{D(s)} = s^2 + 10s + (277.8 - K) = 0 \qquad (6\text{–}8\text{–}10)$$

For s to be equal to zero in Eq. (6–8–10), $277.8 - K_1$ must also be equal to zero, or $K_1 = 277.8$. K_1 could also have been evaluated in this particular case by applying the final value theorem to Eq. (6–8–9).

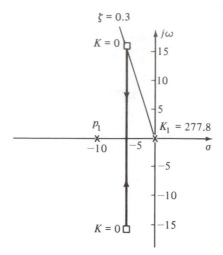

Figure 6–8–1. The inverse root locus of the transfer function of Eq. (6–8–9).

With one pole located at the origin, the other pole can be determined graphically* or by use of Rule 10 of Sec. 6–3 which says that

$$0 + p_1 = -5 + j15.95 - 5 - j15.95 \qquad (6\text{–}8\text{–}11)$$

or $p_1 = -10$. We can now find G from

$$G = K_1\frac{N(s)}{P(s)} = \frac{277.8}{s(s + 10)} \qquad (6\text{–}8\text{–}12)$$

where G is the product of all the forward element transfer functions. As such, G must include the plant transfer function as well as any controller transfer functions. If we compare Eq. (6–8–12) with Eq. (6–4–4) and Fig. 6–8–1 with Fig. 6–4–1a, we see that the direct and inverse root locus sketches are of the same system with the same transient response and accuracy.

A more complicated and realistic example might entail a transfer function with three roots and a single closed-loop zero; H is taken to be unity. Such a configuration gives the system designer greater freedom in incorporating the plant dynamics, reducing the velocity error, controlling the bandwidth, etc. Without going into the details of how to determine the desired transfer function—not a simple matter by any means—let us consider the tracking system of Fig. 6–8–2a with a desired transfer function

$$W(s) = K\frac{N(s)}{D(s)} = \frac{K(s + 20)}{(s + 50)(s^2 + 40s + 600)} \qquad (6\text{–}8\text{–}13)$$

There are one zero and three roots; therefore, the number of asymptotes is two, as before. The inverse root locus sketch is shown in Fig. 6–8–2b. If we assume a type-1 or higher system, the value of K is found to be equal to 2000 for a pole

---

* The other pole can also be determined directly from Eq. (6–8–10).

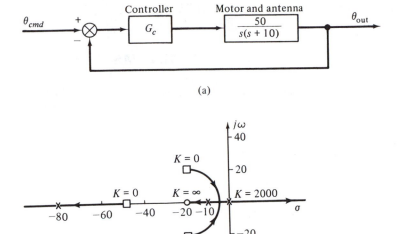

(a)

(b)

Figure 6–8–2. (a) Block diagram of a tracking radar; (b) inverse root locus of Eq. (6–8–13).

at the origin. The remaining two poles are found from the equation $1 - W = s(s^2 + 90s + 800) = 0$ to be located at -10 and -80.

The total forward transfer function G is thus

$$G = \frac{2000(s + 20)}{s(s + 10)(s + 80)} \qquad (6\text{–}8\text{–}14)$$

but

$$G = \frac{50G_c}{s(s + 10)} \qquad (6\text{–}8\text{–}15)$$

Solving Eq. (6–8–15) for the required controller function yields

$$G_c = \frac{40(s + 20)}{(s + 80)} \qquad (6\text{–}8\text{–}16)$$

This controller transfer function could be obtained by combining an amplifier with a gain of 40 and a passive lead network of Sec. 3–3 with $\alpha = 4$ and $\tau = 1/80$ sec.

The principle drawbacks to this synthesis technique are twofold. The first is the difficulty of uniquely specifying a desired transfer function. The second is that the controller required is often not physically realizable or is expensive in terms of sensors and signal processing. This frequently is the case with other techniques that directly synthesize a controller for satisfying a predetermined performance index or set of criteria.

PROBLEMS

6–1. For each of the negative and unity feedback systems whose forward transfer function G is given below, sketch the root locus for positive values of K and describe the system stability.

(a) $\dfrac{K}{(s + 10)(s + 15)}$

(b) $\dfrac{K(s + 20)}{(s + 10)(s + 25)}$

(c) $\dfrac{K(s + 20)}{s(s + 10)(s + 25)}$

(d) $\dfrac{K}{s(s - 3)(s + 25)}$

(e) $\dfrac{K(s + 15)}{s(s - 3)(s + 15)}$

(f) $\dfrac{K(s + 2)^2}{s^3(s + 10)}$

6–2. Sketch the root locus for positive values of K when $H = 1$ and G is as given. Find the approximate values of any points of departure from and arrival at the real axis. Describe the system stability.

(a) $G = \dfrac{K(s + 3)}{s^2}$

(b) $G = \dfrac{K(s + 7)(s + 10)}{s(s + 5)}$

(c) $G = \dfrac{Ks(s + 5)}{(s + 15)(s + 10)}$

(d) $G = \dfrac{Ks(s + 8)}{s(s + 5)(s + 10)}$

6–3. Sketch the root locus for positive K. Describe the system stability and find departure angles from complex poles and arrival angles at complex zeros when $H = 1$ and

(a) $G = \dfrac{K(s^2 + 2s + 4)}{s^2(s + 5)(s + 10)}$

(b) $G = \dfrac{K}{s(s + 1)(s^2 + 4s + 8)}$

(c) $G = \dfrac{K(s^2 + 2s + 4)}{(s^2 + 8s + 25)(s^2 + 2s + 2)}$

(d) $G = \dfrac{K(s + 5)(s + 8)}{s(s + 1)(s + 10)(s^2 + 6s + 30)}$

6–4. Sketch the root locus for positive K and describe the stability of the systems with the following characteristic equations:

(a) $s^3 + 10s^2 + (30 + K)s + 3K = 0$

(b) $Ks^3 + (1 + 3K)s^2 + 6s + 9 = 0$

(c) $Ks^3 + 5Ks^2 + (1 + 12K)s + 2 = 0$

(d) $s^4 + 4s^3 + (13 + K)s^2 + 9Ks + 20K = 0$

6–5. Sketch the root locus for positive K for the systems with the $KZ(s)/P(s)$ functions listed below. Use the Routh-Hurwitz criterion to locate the point(s) where any branches cross the imaginary axis and to find the corresponding values of K_c.

(a) $\dfrac{K}{s(s + 3)(s + 5)^2}$

(b) $\dfrac{K(s + 1)}{s^2(s + 5)(s + 12)}$

(c) $\dfrac{K(s - 5)(s - 10)}{(s + 5)(s + 10)}$

(d) $\dfrac{Ks^2(s + 5)}{(s^2 + 6s + 9)}$

6–6. Sketch the root locus with $KZ(s)/P(s)$ as shown below and find the K required to yield a damping ratio of 0.5 for the dominant pair of roots.

(a) $\dfrac{K}{s(s + 4)}$

(b) $\dfrac{K}{s(s + 4)(s + 8)}$

(c) $\dfrac{K(s + 1)}{s(s + 4)(s + 8)(s + 16)}$

(d) $\dfrac{K(s + 10)}{s(s + 5)(s + 10)(s + 15)}$

6–7. (a) With $H = G_c = 1$ plot the root locus of the system of Fig. P6–7. Find the amplifier gain A for $\zeta = 1.0$, as well as the settling time t_s and natural frequency ω_n of the system.

(b) To reduce the settling time we "cancel" the plant pole at -3 and replace it with a pole at -20 by letting $G_c = (s + 3)/(s + 20)$. Plot the root locus and find A, approximate t_s, and ω_n for $\zeta = 1.0$.

(c) Write the transfer functions for c/r and c/d. Has the transient mode arising from the plant pole at -3 been canceled?

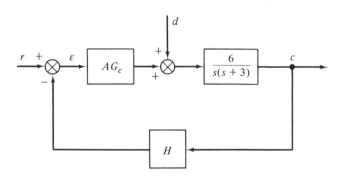

Figure P6–7

6–8. (a) For the system of Fig. P6–7 with $G_c = 1$ and $H = (s + 3)/(s + 30)$, plot the root locus for positive values of A. Find the value of A that yields $\zeta = 1.0$.

(b) With $d = 0$ find $c(t)$ for a unit step in $r(t)$. Has the transient mode due to the plant pole at -3 been canceled?

6–9. (a) With $B = 0$, i.e., no rate feedback in the inner loop, write the system TF c/r

for the system of Fig. P6–9. Sketch the root locus for positive values of A and describe the system stability.

(b) With $B > 0$, write the inner loop TF c/e_1. Sketch the root locus for variable B. Locate the inner-loop roots for $B = 0.5$.

(c) With $B = 0.5$, write the system TF c/r. Sketch the root locus for positive A and describe the system stability.

(d) What is the relationship between the inner-loop roots and the system poles?

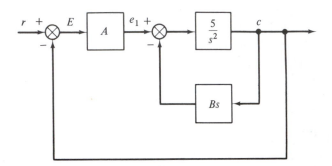

Figure P6–9

6–10. (a) With $H = 0$ sketch the root locus of the system of Fig. P6–10 and determine if the specification $\zeta = 0.7$ can be met by the proper choice of A and K.

(b) Let $H = 0.5s + 1$. Sketch the root locus of the inner loop and locate the roots for $K = 15$.

(c) With the inner loop of (b) sketch the system root locus and determine if $\zeta = 0.7$ can now be achieved.

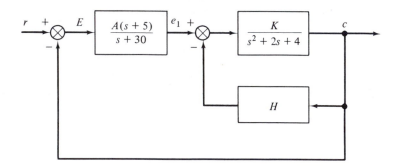

Figure P6–10

6–11. A unity feedback system has a forward transfer function

$$G = \frac{A(s + 1)}{s^2(\tau s + 1)}$$

(a) Sketch on one diagram the root locus for positive A for $\tau = 0.1$, 0.5, and 2. Describe the effects of increasing τ upon the stability of the system.

(b) For $A = 6$ sketch the root locus for positive τ and describe the system stability.

6–12. A unity feedback system has a forward transfer function

$$G = \frac{K}{s^2 + a_1 s + a_0}$$

The design values of the system parameters are $K = 20$, $a_1 = 6$, $a_0 = 6$.
(a) Locate the roots for the design values and describe the system response.
(b) By means of root locus sketches, show and describe the effect upon the system response of changes in each of the design values, one at a time.

6–13. Do Prob. 6–1 for negative values of K.

6–14. Do Prob. 6–3 for positive values of K but with positive rather than negative feedback.

6–15. For each of the negative feedback systems with $H = 1$ and G as shown, sketch on one figure the root locus for both positive and negative values of K.

(a) $G = \dfrac{K(s^2 + 2s + 8)}{s^2(s^2 + 6s + 26)}$

(b) $G = \dfrac{K(s + 3)(s^2 + 3s + 13)}{s^2(s + 10)(s + 15)}$

6–16. In the unity-feedback system of Fig. P6–16,

$$G_p = \frac{10}{(s + 1)(s + 2)}$$

By means of the inverse root locus find the controller transfer function G_c required to obtain the desired system transfer function

$$W = \frac{c}{r} = \frac{K}{(s + 50)(s + 100)}$$

when the steady-state error for a unit step input is to be
(a) Zero.
(b) 0.1.
(c) Sketch the direct root locus for $G_c G_p$ for positive values of K for each case above. Is pole cancellation required?

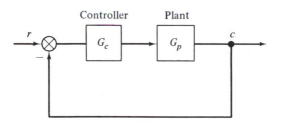

Figure P6–16

6–17. In the system of Fig. P6–16 the plant is unstable with a transfer function

$$G_p = \frac{10}{s(s - 3)}$$

(a) By means of the inverse root locus find the controller transfer function G_c that will produce a system transfer function

$$W = \frac{c}{r} = \frac{K(s + 5)}{(s + 20)(s^2 + 2.4s + 9)}$$

(b) Sketch the direct root locus for positive K of G_cG_p. Is pole cancellation required? Would you recommend this controller?

6–18. Sketch the root locus for the following $KZ(s)/P(s)$ functions and determine if any branches originating at the complex poles touch the real axis.

(a) $\dfrac{K(s + 0.4)}{(s + 10)(s^2 + 1.3s + 2.5)}$

(b) $\dfrac{K(s + 2)}{s(s + 0.5)(s + 10)(s + 20)}$

6–19. In the negative feedback system of Fig. P6–16, the proportional controller TF is $G_c = A$. For each of the plant transfer functions listed below (i) name and sketch the appropriate root locus for positive values of A, (ii) determine the stability, and (iii) verify with the Routh-Hurwitz criterion.

(a) $\dfrac{10(s - 1)}{(s + 5)(s + 15)}$

(b) $\dfrac{10(s - 1)(s - 3)}{s(s + 5)(s + 15)}$

(c) $\dfrac{10(s - 1)}{(s - 3)(s + 10)}$

(d) $\dfrac{10(s - 3)}{(s - 1)(s + 10)}$

(e) $\dfrac{10(s + 1)}{s(s - 3)(s + 10)}$

7

The Frequency Response:
Nyquist Diagrams

7–1 INTRODUCTION

There are times when it is necessary or advantageous to work in the frequency domain rather than in the Laplace domain of the root locus.

First of all, the root locus method requires a transfer function, which may be difficult or even impossible to obtain for certain components, subsystems, or even systems. In many of these cases, the frequency response can be determined experimentally for sinusoidal test inputs of known frequency and amplitude. These experimental data may then be used to obtain an approximate transfer function or may be used directly in the frequency response techniques to be described in this and subsequent chapters.

The nature of the input also influences the choice of techniques to be used for system analysis and design. Many command inputs merely instruct a system to move from one steady-state condition to a second steady-state condition. This type of input can be described adequately by suitable steps in position, velocity, and acceleration, and the Laplace domain is appropriate for this purpose. If, however, the interval between such step inputs is decreased so that the system never has time to reach the corresponding steady state (consider the evading bomber or maneuvering fighter), the step representation and Laplace domain are no longer adequate. Such rapidly varying command inputs may be periodic, random, or a combination thereof. Disturbance inputs are even more likely to be varying and random in nature. The wind loading of a tracking radar antenna, for example, results from a mean velocity component that varies with time plus superimposed random gusts. Atmospheric turbulence, particularly clear air turbulence, is even more random in nature. If the frequency distribution of these inputs can be calculated, measured, or even estimated, the frequency response can be used to determine their effects upon the system output, as was done in

Sec. 5–7. Purely random disturbance inputs that lie in the frequency and amplitude ranges of command inputs require very sophisticated techniques of analysis and suppression. It should be apparent that the control problem becomes increasingly more difficult as the frequencies and amplitudes of command and disturbance inputs approach each other.

Two situations have been touched upon that require the use of the frequency response. In Chap. 9, we shall see that the selection of controller parameters can often be accomplished more easily with frequency response rather than root locus techniques. These two basic techniques and their associated domains have inherent advantages and limitations which should be well understood when attacking a particular problem.

Remember that the frequency response is a steady-state response. Although some information can be obtained about the transient response, it is only approximate and is subject to misinterpretation. Explicit transient behavior can best be found by other means. Furthermore, the Nyquist and Bode diagrams used to determine stability are based on functions other than the system transfer function. Separate diagrams are usually required to determine the frequency response of the system.

7–2 THE FREQUENCY TRANSFER FUNCTION

It is necessary to develop an input-output relationship that can be used in the frequency domain, i.e., a frequency transfer function. Consider a linear system with a known transfer function $W(s)$ and apply the sinusoidal input

$$r(t) = r_0 \sin \omega_f t \qquad (7\text{–}2\text{–}1)$$

where r_0 is the amplitude and ω_f the input or forcing frequency.* The Laplace transform of the input is

$$r(s) = \frac{\omega_f r_0}{s^2 + \omega_f^2} \qquad (7\text{–}2\text{–}2)$$

and the transformed output is

$$c(s) = \frac{W(s)\omega_f r_0}{s^2 + \omega_f^2} \qquad (7\text{–}2\text{–}3)$$

The partial fraction expansion of $c(s)$ yields

$$c(s) = \frac{C_1}{s - j\omega_f} + \frac{C_2}{s + j\omega_f} + \frac{C_3}{s + r_2} + \frac{C_4}{s + r_2} + \cdots \qquad (7\text{–}2\text{–}4)$$

where $-r_1, -r_2, \ldots$ are the roots of the characteristic equation of the transfer function. The inverse transform of Eq. (7–2–4) is

$$c(t) = C_1 e^{j\omega_f t} + C_2 e^{-j\omega_f t} + C_3 e^{-r_1 t} + C_4 e^{-r_2 t} + \cdots \qquad (7\text{–}2\text{–}5)$$

The first two terms in Eq. (7–2–5) represent an undamped oscillation resulting from the sinusoidal input, and the remaining terms are the transient response.

* The dimensions of ω_f are radians per unit time.

If the system is unstable, at least one of the terms of the transient response will increase exponentially with time. Consequently, *there is no steady-state response and no frequency response as such for an unstable system.* If the system is stable, the transient response will disappear with time, leaving as the steady-state response

$$c_{ss} = C_1 e^{j\omega_f t} + C_2 e^{-j\omega_f t} \tag{7-2-6}$$

The coefficients C_1 and C_2 are evaluated by the Heaviside expansion theorem as

$$C_1 = \left[\frac{(s - j\omega_f) W(s) \omega_f r_0}{s^2 + \omega_f^2} \right]_{s = +j\omega_f} = \frac{W(j\omega_f) r_0}{2j} \tag{7-2-7}$$

$$C_2 = \left[\frac{(s + j\omega_f) W(s) \omega_f r_0}{s^2 + \omega_f^2} \right]_{s = -j\omega_f} = \frac{-W(-j\omega_f) r_0}{2j} \tag{7-2-8}$$

With these values for C_1 and C_2, Eq. (7-2-6) becomes

$$c_{ss} = \frac{r_0}{2j} W(j\omega_f) e^{j\omega_f t} - \frac{r_0}{2j} W(-j\omega_f) e^{-j\omega_f t} \tag{7-2-9}$$

Since they are complex functions,

$$W(j\omega_f) = Re + jIm = |W(j\omega_f)| e^{j\Phi}$$

$$W(-j\omega_f) = Re - jIm = |W(j\omega_f)| e^{-j\Phi} \tag{7-2-10}$$

where the angle Φ is the argument of $W(j\omega_f)$ and is equal to $\tan^{-1} Im/Re$. Equation (7-2-9) can now be written as

$$c_{ss} = r_0 |W(j\omega_f)| \left(\frac{e^{j(\omega_f t + \Phi)} - e^{-j(\omega_f t + \Phi)}}{2j} \right) \tag{7-2-11}$$

Since the bracketed terms are equal to $\sin(\omega_f t + \Phi)$, the steady-state response can finally be written as

$$c_{ss}(j\omega_f) = c_0 \sin(\omega_f t + \Phi) \tag{7-2-12}$$

where

$$c_0 = |W(j\omega_f)| r_0 \tag{7-2-13}$$

From these equations we see that a sinusoidal input to a linear stable system produces a steady-state response that is also sinusoidal, having the same frequency as the input but displaced through a phase angle Φ and having an amplitude that may be different. This steady-state sinusoidal response is called the *frequency response* of the system. Since the *phase angle* is the angle associated with the complex function $W(j\omega_f)$ and the *amplitude ratio* (c_0/r_0) is the magnitude of $W(j\omega_f)$, knowledge of $W(j\omega_f)$ specifies the steady-state input-output relationship in the frequency domain. $W(j\omega_f)$ is called the *frequency transfer function* and can be obtained from the transfer function $W(s)$ by replacing the Laplace variable s by $j\omega_f$. Consequently, if $W(j\omega_f)$ can be determined from experimental data, the transfer function $W(s)$ is found by replacing $j\omega_f$ by s.

In the stability analyses to follow, complex functions other than the system frequency transfer function will be used, and the terms *phase angle* and *amplitude ratio* can be confusing. When a gain is the parameter of interest, the function used will be the open-loop transfer function, and the phase angle and amplitude ratio will describe the open-loop response but not the system or closed-loop response. If the variable parameter is other than gain, the phase angle and amplitude ratio will have no physical significance whatsoever.

For a given system, the frequency response is completely specified if the amplitude ratio and phase angle are known for the range of input frequencies from 0 to $+\infty$ radians per unit time. Consider the stable first-order system of Fig. 7–2–1a with a transfer function

$$W(s) = \frac{1}{\tau s + 1} \tag{7-2-14}$$

The frequency transfer function is

$$W(j\omega_f) = \frac{1}{j\omega_f \tau + 1} \tag{7-2-15}$$

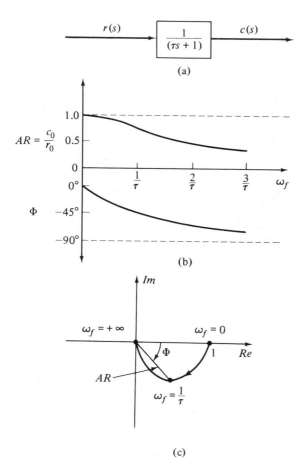

(a)

(b)

(c)

Figure 7–2–1. A simple first-order system: (a) transfer function; (b) *AR* and Φ as a function of ω_f; (c) polar plot.

The amplitude ratio* is

$$AR(j\omega_f) = M(j\omega_f) = \frac{c_0}{r_0} = |W(j\omega_f)| = \frac{1}{\sqrt{(\omega_f\tau)^2 + 1}} \qquad (7\text{-}2\text{-}16)$$

and the phase angle is

$$\Phi(\omega_f) = \underline{/W(j\omega_f)} = \underline{/1} - \underline{/(j\omega_f\tau + 1)} = -\tan^{-1}\omega_f\tau \qquad (7\text{-}2\text{-}17)$$

When the amplitude ratio and phase angle are plotted against the frequency in Fig. 7-2-1b, two plots are required. Notice that at very low input frequencies the amplitude ratio is almost unity and the phase angle is almost zero, indicating that the steady-state response is essentially following the input. As the input frequency increases, the amplitude ratio decreases and the phase angle becomes increasingly negative; the output is lagging the input with a smaller amplitude. As the input frequency becomes very large, the amplitude ratio approaches zero and the phase angle approaches $-90°$; the system in essence ceases to respond to the input.

The same information can be presented on one plot that traces the tip of the vector representing the frequency transfer function as ω_f is varied from 0 to $+\infty$ rad/sec. Such a presentation is called a *polar plot* and is illustrated in Fig. 7-2-1c.

Polar plots and AR and Φ versus ω_f plots are used to represent different types of complex functions in the frequency domain; this can be confusing. We must always keep in mind what a specific function represents and the physical significance, if any, of the amplitude ratio and phase angle.

7-3 THE NYQUIST STABILITY CRITERION

In the frequency domain, the theory of residues can be used to detect any roots in the right half of a plane. As with the root locus method, the characteristic equation in the form $1 + KZ(s)/P(s)$ is used, where again the function $KZ(s)/P(s)$ may or may not be the open-loop transfer function. To develop the Nyquist criterion, the characteristic function itself is written as a ratio of polynomials so that

$$1 + K\frac{Z(s)}{P(s)} = \frac{P(s) + KZ(s)}{P(s)} = K'\frac{(s + r_1)(s + r_2)\cdots}{(s + p_1)(s + p_2)\cdots} = 0 \qquad (7\text{-}3\text{-}1)$$

Comparing the identities of Eq. (7-3-1), we see that $-r_1$, $-r_2$, ... are the roots of the characteristic equation and that $-p_1$, $-p_2$, ... are the poles of both the characteristic function and $KZ(s)/P(s)$. Poles and roots at the origin have been omitted in the interests of simplicity.

Some roots and poles are arbitrarily located in the s plane, as indicated in Fig. 7-3-1a. In addition, s is constrained to take on the values along the arbitrary closed contour that encloses only the root r_1 and is made to travel in a clockwise

---

* M is used to denote the system amplitude ratio; AR is used to denote the magnitude of a complex function in general.

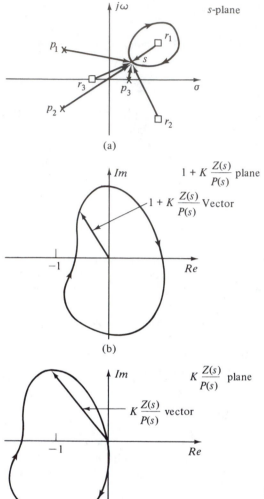

Figure 7–3–1. Encirclement of one root: (a) contour in the s plane; (b) in the $1 + KZ(s)/P(s)$ plane; (c) in the $KZ(s)/P(s)$ plane.

direction. For each value of s on the contour there will be a corresponding value of the characteristic function $1 + KZ(s)/P(s)$. The closed contour in the s plane will map into a closed contour in the $1 + KZ(s)/P(s)$ plane that is the locus of the tip of the vector representing the complex function $1 + KZ(s)/P(s)$. The magnitude of this vector will be the product of the magnitudes of the individual vectors representing the factors on the right side of Eq. (7–3–1), and the angle of this vector is the sum of the angles of the individual vectors.

In Fig. 7–3–1a, draw the vectors from the poles and roots to a point s on the contour. When s has made one complete clockwise revolution about r_1, the vector $(s + r_1)$ will have also made one clockwise revolution for a change in angle of 360°. Since the other roots and poles are outside the contour, their vectors will not make complete revolutions but will merely nod up and down as s travels around the contour. The clockwise revolution of the $(s + r_1)$ vector will produce a clockwise revolution about the origin in the $1 + KZ(s)/P(s)$ plane, as shown in Fig. 7-3-1b, since the root r_1 is in the numerator of the function.

The characteristic function, unfortunately, is rarely in the factored form of Eq. (7-3-1); if it were, the location of the roots would be immediately known and further investigation would be unnecessary. Mapping of the contour in the $KZ(s)/P(s)$ plane is more appropriate and convenient and is shown in Fig. 7-3-1c. The shape of the contour is unchanged, but now the clockwise encirclement of the origin in the $1 + KZ(s)/P(s)$ plane corresponds to a clockwise encirclement of the -1 point in the $KZ(s)/P(s)$ plane.

If the contour in the s plane is redrawn so as to encircle both a root and a pole, as shown in Fig. 7-3-2, the vectors from both r_1 and p_3 will each make

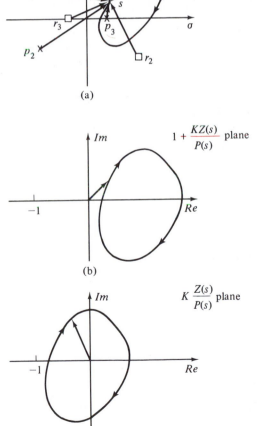

(a)

(b)

(c)

Figure 7-3-2. Encirclement of one root and one pole: (a) in the s plane; (b) in the $1 + KZ(s)/P(s)$ plane; (c) in the $KZ(s)/P(s)$ plane.

one complete revolution as s travels the contour. Since the $(s + r_1)$ vector is in the numerator, it will contribute $+360°$ to the change in angle of the $1 + KZ(s)/P(s)$ vector, whereas the $(s + p_3)$ vector, being in the denominator, will contribute $-360°$ to the change. Consequently, there will be no net change in the angle of the $1 + KZ(s)/P(s)$ vector; it will not encircle the origin in the $1 + KZ(s)/P(s)$ plane or the -1 point in the $KZ(s)/P(s)$ plane as shown in Fig. 7–3–2b and c.

The closed contour in the s plane may be drawn to include any region desired. For the sake of consistency, this contour will always be traveled in a clockwise direction. Each root within the contour will produce a clockwise encirclement of the -1 point in the $KZ(s)/P(s)$ plane; each pole within the contour will produce a counterclockwise encirclement of the -1 point. The results of the development to this point can be summarized by the relationship

$$N_n = N_r - N_p \qquad (7\text{–}3\text{–}2)$$

where N_n is the number of encirclements of the -1 point, N_r is the number of roots within the contour in the s plane, and N_p is the number of poles within the s contour. If N_n is positive, the -1 point is encircled in a clockwise manner and $N_r > N_p$. If N_n is negative, the -1 point is encircled in a counterclockwise manner and $N_r < N_p$.

Equation (7–3–2) can be used to detect the presence of roots in the right half of the s plane by selecting a contour that completely encloses the right-half plane. Such a contour is shown in Fig. 7–3–3a and is divided into three regions. In Region I, $s = j\omega$ and ω takes on values from $\omega = 0+$ to $\omega = +\infty$ along the positive imaginary axis. In Region II, $s = Re^{j\theta}$, where R is infinite and θ changes from $+90°$ to $-90°$. Finally, in Region III, $s = j\omega$ again and ω goes from $-\infty$ to $0-$ along the negative imaginary axis. Since complex factors always occur as conjugate pairs, the map of Region III will be the mirror image about

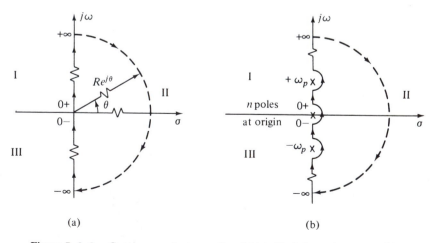

(a)   (b)

Figure 7–3–3. Contours enclosing entire right half of the s plane: (a) without singularities; (b) with singularities.

the real axis of Region I. The map of Region II is obtained by setting s equal to $Re^{j\theta}$ in $KZ(s)/P(s)$ and taking the limit as R goes to infinity. If $P(s)$ is of higher order than $Z(s)$ (the usual case), the map of Region II reduces to a point at the origin of the $KZ(s)/P(s)$ plane. If $Z(s)$ is of higher order, the map is one clockwise semicircle of infinite radius for each difference of order. For example, if $Z(s)$ is of fifth order and $P(s)$ is of third order, the map of Region II would be two clockwise semicircles or one complete clockwise revolution about the origin of the $KZ(s)/P(s)$ plane. Discussion of the mapping of Region I will be deferred to the next section.

Poles at the origin and on the imaginary axis of the s plane produce singularities in $KZ(s)/P(s)$ and must be bypassed by the small semicircles of Fig. 7–3–3b. The map of the semicircle about the pole(s) at the origin as ω goes from $0-$ to $0+$ is found by setting s equal to $\epsilon e^{j\theta}$ in $KZ(s)/P(s)$ and letting θ go from $-90°$ to $90°$ as ϵ goes to zero; each pole at the origin of the s plane maps a clockwise semicircle of infinite radius in the $KZ(s)/P(s)$ plane. The map around the complex poles at $\pm j\omega_p$ is obtained in a similar manner by setting s equal to $j\omega_p + \epsilon e^{j\theta}$ and taking the limit as ϵ goes to zero and θ goes from $-90°$ to $+90°$. Each pole on the imaginary axis maps into an infinite semicircle in passing from $\omega = \omega_p-$ to ω_p+, resulting in a deformed contour in the $KZ(s)/P(s)$ plane.

The map of the s-plane contour of Fig. 7–3–3 into the $KZ(s)/P(s)$ plane is known as a *Nyquist diagram*. The application of Eq. (7–3–2) to the Nyquist diagram is called the *Nyquist stability criterion* and is written as

$$N_r = N_n + N_p \qquad (7\text{–}3\text{–}3)$$

If we know $KZ(s)/P(s)$, we know the number of poles in the right-half plane. If we count the number of times the Nyquist diagram encircles the -1 point, we know the number of roots in the right-half plane. If the encirclements are clockwise, N_n is positive; if counterclockwise, N_n is negative. Any root in the right-half plane denotes an unstable system.

One of the several expressions for the Nyquist stability criterion states that the *Nyquist diagram of a stable system must encircle the -1 point in the counterclockwise direction as many times as there are poles in the right half of the s plane.*

7–4 THE NYQUIST DIAGRAM

A Nyquist diagram is a map in the $KZ(s)/P(s)$ plane of a closed contour that encloses the entire right half of the s plane. It may also be considered to be the polar plot of the complex function $KZ(j\omega_f)/P(j\omega_f)$ as the input frequency ω_f is varied throughout the entire frequency range from 0 to $\pm\infty$. This polar plot has a physical significance only when ω_f is positive and when $KZ(s)/P(s)$ is the open-loop transfer function.

The Nyquist diagram is constructed by sketching or plotting the map of Region I, where ω_f varies from $0+$ to $+\infty$, and then using our knowledge of Regions II and III to complete the diagram. The Region I map can be plotted point by point, calculating the amplitude ratio AR and phase angle Φ for each frequency, or can be obtained more easily from the Bode plots of the next chapter. It is even easier to sketch the Nyquist diagram with a knowledge of the behavior of the pole and zero factors at very low and very high input frequencies. A sketch is often adequate, particularly when supplemented by other techniques or simple calculations. In fact, a not-to-scale sketch is the only way to show the overall shape of the Nyquist diagram in Region I; computer plots must be limited in their frequency ranges because of scaling problems. The sketching and use of the Nyquist diagram will be illustrated by example. It is useful to remember that the phase angle of a real pole in the left-half plane monotonically decreases from $0°$ when ω_f is $0+$ to $-90°$ when ω_f is $+\infty$, and the phase angle of a complex pole (a second-order term) monotonically decreases from $0°$ to $-180°$. The phase angle of real zeros and complex zeros increases in the left-half plane from $0°$ to $+90°$ and to $+180°$, respectively.

After sketching a Nyquist diagram, the reader is encouraged to sketch the corresponding root locus and note the relationship between these two graphical techniques used to examine the stability of a plant or system.

Let us consider first an open-loop transfer function with poles in the right half of the s plane. Poles in the right-half plane indicate either an unstable plant or an unstable inner loop. Functions with poles or zeros in the right-half plane are called *nonminimum phase functions*. If a system has the characteristic equation

$$6s^2 - 5s + 1 + K = 0 \tag{7–4–1}$$

then

$$K\frac{Z(s)}{P(s)} = \frac{K}{(2s - 1)(3s - 1)} \tag{7–4–2}$$

where, in this case, $KZ(s)/P(s)$ is the open-loop transfer function. It is apparent that there are two poles in the right-half plane ($N_p = 2$) at $s = +\frac{1}{2}$ and $s = +\frac{1}{3}$.

The frequency function is obtained by replacing s in Eq. (7–4–2) by $j\omega$* so that

$$K\frac{Z(j\omega)}{P(j\omega)} = \frac{K}{(j2\omega - 1)(j3\omega - 1)} \tag{7–4–3}$$

Notice that *the constant term in each factor is set equal to unity when working in the frequency domain,* whereas in the Laplace domain the coefficient of the highest power of s is set equal to unity. The expressions for the amplitude ratio AR and the phase angle Φ of the complex function of Eq. (7–4–3) are

$$AR(\omega) = \frac{K}{\sqrt{4\omega^2 + 1}\sqrt{9\omega^2 + 1}} \tag{7–4–4}$$

$$\Phi(\omega) = -\tan^{-1}\frac{2\omega}{-1} - \tan^{-1}\frac{3\omega}{-1} \tag{7–4–5}$$

In Eq. (7–4–5) the angle associated with K is zero and the two angles shown are negative since they are the angles of the poles and appear in the denominator.

---

* The subscript f will now be dropped and used only when, to avoid confusion, it is necessary.

Since these angles are in the second quadrant by virtue of the negative real parts, Eq. (7–4–5) can be written as

$$\Phi(\omega) = \tan^{-1} 2\omega + \tan^{-1} 3\omega \tag{7–4–6}$$

When ω is equal to $0+$, Eq. (7–4–4) shows that the amplitude ratio is equal to K and Eq. (7–4–6) that the phase angle is equal to $0°$. When ω approaches $+\infty$, AR approaches zero and Φ approaches $+180°$. Between these two frequencies, AR decreases and Φ increases, both monotonically. It is helpful in sketching to prepare a brief table as follows:

ω_f (rad/sec)	AR	Φ (deg)
$0+$	K	0
$+\infty$	0	$+180$
1	$0.14K$	$+135$

where the intermediate point at 1 rad/sec is used as a checkpoint even though it is not particularly useful for this simple sketch. The Nyquist diagram is sketched in Fig. 7–4–1a using the symmetry about the real axis to close the map from $-\infty$ to $0-$. We see that the -1 point will never be encircled no matter what the value of K. Since N_n will always be zero, the number of roots in the right-half plane can be found from Eq. (7–3–3) to be

$$N_r = N_n + N_p = 0 + 2 = 2 \tag{7–4–7}$$

The two roots of the characteristic equation in the right-half plane mean that the system is unstable for all values of K; this statement can be easily verified by the Routh-Hurwitz criterion or by a root locus sketch (see Fig. 7–4–1b). In order for any system with two poles in the right-half plane to be stable, there must be two counterclockwise encirclements of the -1 point.

Adding a single real zero in the left-half plane to Eq. (7–4–3) yields the function

$$K\frac{Z(j\omega)}{P(j\omega)} = \frac{K(j\omega + 1)}{(j2\omega - 1)(j3\omega - 1)} \tag{7–4–8}$$

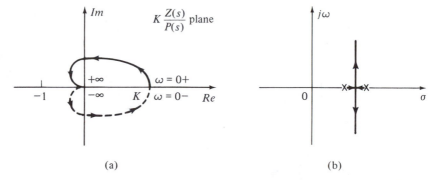

(a)   (b)

Figure 7–4–1. (a) Nyquist diagram of nonminimum phase function of Eq. (7–4–2); (b) root locus for positive K.

The AR and Φ become

$$AR = \frac{K\sqrt{\omega^2 + 1}}{\sqrt{4\omega^2 + 1}\sqrt{9\omega^2 + 1}} \qquad (7\text{–}4\text{–}9)$$

$$\Phi = + \tan^{-1}\omega + \tan^{-1}2\omega + \tan^{-1}3\omega \qquad (7\text{–}4\text{–}10)$$

The table for this sketch is

ω_f (rad/sec)	AR	Φ (deg)
$0+$	K	0
$+\infty$	0	$+270$
1	$0.2K$	$+180$

The intermediate checkpoint shows that the curve for Region I crosses the negative real axis and thus has the potential for encircling the -1 point and is, therefore, conditionally stable. Crossing the negative real axis is analogous to a root locus branch crossing the imaginary axis and leads to the statement that *crossing the negative real axis denotes a conditionally stable system.*

The Nyquist diagram is sketched in Fig. 7–4–2 for two values of K. For the smaller value of K, the curve of Fig. 7–4–2a does not encircle the -1 point, and the system is unstable. If K is increased sufficiently, the curve of Fig. 7–4–2b will encircle the -1 point twice in a counterclockwise direction and $N_n = -2$. Substituting into Eq. (7–3–3), we find

$$N_r = N_n + N_p = -2 + 2 = 0 \qquad (7\text{–}4\text{–}11)$$

Since the number of roots in the right-half plane is now zero, the system is stable for this and larger values of K. This is a conditionally stable system. We conclude that adding a zero in the left-hand plane makes a system more stable. Incidentally, the number of encirclements of the -1 point can easily be determined by drawing a straight line from -1 in any direction, as is done in Fig. 7–4–2b. Then N_n is the number of times the curve crosses this line in a clockwise direction

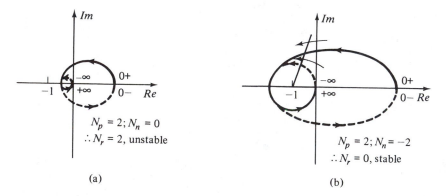

(a)   (b)

Figure 7–4–2. The Nyquist diagram of the nonminimum phase function $K(s + 1)/(2s - 1)(3s - 1)$ for a conditionally stable system: (a) unstable for this value of K; (b) stable for this larger value of K.

minus the number of times it crosses in a counterclockwise direction. In this example, $N_n = 0 - 2 = -2$.

Let us now add a single pole at the origin to our conditionally stable system of Eq. (7-4-8) to obtain

$$K\frac{Z(j\omega)}{P(j\omega)} = \frac{K(j\omega + 1)}{j\omega(j2\omega - 1)(j3\omega - 1)} \tag{7-4-12}$$

The expressions for the amplitude ratio and phase angle are

$$AR = \frac{K\sqrt{\omega^2 + 1}}{\omega\sqrt{4\omega^2 + 1}\sqrt{9\omega^2 + 1}} \tag{7-4-13}$$

$$\Phi = +\tan^{-1}\omega - 90° + \tan^{-1}2\omega + \tan^{-1}3\omega \tag{7-4-14}$$

The table for this sketch is

ω_f (rad/sec)	AR	Φ (deg)
$0+$	∞	-90
$+\infty$	0	$+180$
1	$0.2K$	$+45$

The Nyquist diagram is sketched in Fig. 7-4-3 and shows no possible encirclement of the -1 point. Notice that crossing the positive real axis does not denote conditional stability. Thus the system is unstable for all values of K, and we see that a pole destabilizes a system. Notice the effects upon the Nyquist diagram of adding a single pole at the origin. The starting point at $\omega = 0+$ has been shifted a quadrant in a clockwise direction, and the initial magnitude has been increased from a finite to an infinite value. Furthermore, the curve is no longer closed at $\omega = 0+$ and $0-$ and requires an infinite semicircle to close the contour.

More often than not, a system whose stability is being examined is composed of stable elements and the $KZ(s)/P(s)$ function has no poles or zeros in the right-

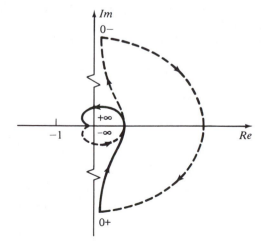

Figure 7-4-3. Nyquist diagram showing effects of adding a single pole at the origin to the function of Fig. 7-4-2; system is unstable for all values of K.

half plane. Such a function is said to be a *minimum phase function*.* When the number of poles in the right-half plane is zero, the Nyquist criterion for minimum phase functions reduces to

$$N_r = N_n \qquad (7\text{--}4\text{--}15)$$

and any encirclement of the -1 point denotes an unstable system. There is no longer any need to count encirclements or determine their direction, nor is the proper closure of the curve between $\omega = 0-$ and $0+$ necessary. Consequently, the Nyquist diagram for minimum phase functions can be simplified to a polar plot of $KZ(j\omega)/P(j\omega)$ as ω is varied from $0+$ to $+\infty$.

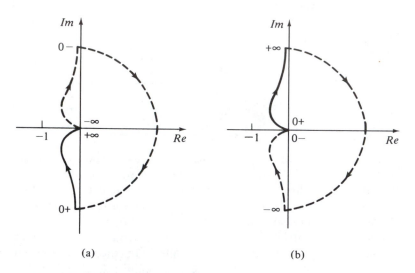

(a)   (b)

Figure 7-4-4. Two Nyquist diagrams of the minimum phase functions:

(a) $\dfrac{AK_m}{s(\tau_m s + 1)};$ (b) $\dfrac{\tau_m s^2}{s + AK_m};$ both systems are stable.

Let us consider our second-order tracking system of Fig. 6–4–1a, which has the characteristic equation

$$\tau_m s^2 + s + AK_m = 0 \qquad (7\text{--}4\text{--}16)$$

For the effect of varying the amplifier gain A with all other parameters fixed,

$$K\frac{Z(j\omega)}{P(j\omega)} = \frac{AK_m}{j\omega(j\omega\tau_m + 1)} \qquad (7\text{--}4\text{--}17)$$

The complete Nyquist diagram is sketched in Fig. 7–4–4a with the unnecessary portions dashed. The curve can never encircle the -1 point, and the system is stable for all values of A. *A simple test for the encirclement of the -1 point is to look to the right as the solid portion of the curve is traveled in the direction of increasing ω. If the -1 point cannot be directly seen, there is no encirclement and the system is stable.* This rule does *not* apply to nonminimum phase functions.

---

* K is restricted to positive values; otherwise, the function is a nonminimum phase function.

For this same system, if the motor time constant is the parameter of interest,

$$K\frac{Z(j\omega)}{P(j\omega)} = \frac{\tau_m}{AK_m}\frac{(j\omega)^2}{(j\omega/AK_m + 1)} \qquad (7\text{–}4\text{–}18)$$

where A and K_m are fixed. The Nyquist diagram is sketched in Fig. 7–4–4b. Applying the "looking to the right rule" while traveling from $0+$ to $+\infty$ shows no possible encirclement of the -1 point; the system is stable for all values of τ_m.

Adding a first-order time lag (a pole), we have the system of Fig. 6–5–3a with a characteristic equation

$$\tau_m\tau_a s^3 + (\tau_a + \tau_m)s^2 + s + AK_m = 0 \qquad (7\text{–}4\text{–}19)$$

For a varying amplifier gain with the other parameters fixed,

$$K\frac{Z(j\omega)}{P(j\omega)} = \frac{AK_m}{j\omega(j\omega\tau_m + 1)(j\omega\tau_a + 1)} \qquad (7\text{–}4\text{–}20)$$

The Nyquist diagram is sketched in Fig. 7–4–5a. The system is conditionally stable with respect to A since the curve crosses the negative real axis, where $\Phi = -180°$, and can be made to encircle -1 by increasing A. The critical value of A that causes the Nyquist diagram to pass through the -1 point can be found by expanding the right-hand side of Eq. (7–4–20) and substituting the numerical values $K_m = 5$, $\tau_m = 0.2$, and $\tau_a = 0.1$:

$$K\frac{Z(j\omega)}{P(j\omega)} = \frac{5A}{-0.3\omega^2 + j\omega(1 - 0.02\omega^2)} \qquad (7\text{–}4\text{–}21)$$

At the point where the Nyquist diagram crosses the real axis, the imaginary part of $KZ(j\omega)/P(j\omega)$ is zero; thus,

$$1 - 0.02\omega_{cr}^2 = 0 \qquad (7\text{–}4\text{–}22)$$

where ω_{cr} denotes the *crossing frequency,* which is defined as the frequency at

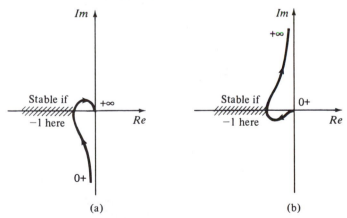

(a)   (b)

Figure 7–4–5. Conditionally stable Nyquist diagrams of the minimum phase functions:

(a) $\dfrac{AK_m}{s(\tau_m s + 1)(\tau_a s + 1)}$;   (b) $\dfrac{\tau_a s^2(\tau_m s + 1)}{\tau_m s^2 + s + AK_m}$.

which $\Phi = -180°$. Solving Eq. (7–4–22) for $\omega_{cr} = +\sqrt{50}$ rad/sec and substituting into Eq. (7–4–21), we find that

$$K\frac{Z(j\omega_{cr})}{P(j\omega_{cr})} = -\frac{A}{3} \qquad (7\text{–}4\text{–}23)$$

where $A/3$ is the AR when $\Phi = -180°$. When $A = A_c = 3$, the AR is equal to 1 and the curve will pass through the -1 point; the system will be marginally stable. When $A > A_c$, AR will be greater than unity; the curve will pass to the left of the -1 point, and the system will be unstable. This system, therefore, is conditionally stable. Obviously, the critical value of A could have been obtained by other means, such as the Routh-Hurwitz criterion.

If τ_a is the parameter of interest,

$$K\frac{Z(j\omega)}{P(j\omega)} = \frac{\tau_a}{AK_m}\frac{(j\omega)^2(j\omega\tau_m + 1)}{\left(1 - \frac{\tau_m\omega^2}{AK_m} + j\frac{\omega}{AK_m}\right)} \qquad (7\text{–}4\text{–}24)$$

The Nyquist diagram is sketched in Fig. 7–4–5b and shows the system to be conditionally stable with respect to τ_a.

Nyquist diagrams can also be plotted or sketched by a graphical technique using a pole-zero plot in the s plane. To illustrate, put the $KZ(s)/P(s)$ function of Eq. (7–4–2) in the proper form for root locus sketching:

$$K\frac{Z(s)}{P(s)} = \frac{K}{6(s - \frac{1}{2})(s - \frac{1}{3})} \qquad (7\text{–}4\text{–}25)$$

and locate the poles and zeros (none here) in the s plane as in Fig. 7–4–6a. Now, let s move from the origin up the positive imaginary axis. At any point $s = j\omega$, and AR and Φ for that particular frequency can be found from the relationships

$$AR = \left|K\frac{Z(s)}{P(s)}\right| = \frac{K}{6|s - \frac{1}{2}||s - \frac{1}{3}|} \qquad (7\text{–}4\text{–}26)$$

$$\Phi = \underline{/K\frac{Z(s)}{P(s)}} = -\underline{/\left(s - \frac{1}{2}\right)} - \underline{/\left(s - \frac{1}{3}\right)} \qquad (7\text{–}4\text{–}27)$$

For example, when $\omega = 0+$,

$$AR = \frac{K}{6(\frac{1}{3})(\frac{1}{2})} = K \qquad (7\text{–}4\text{–}28)$$

and

$$\Phi = -180° - 180° = 0° \qquad (7\text{–}4\text{–}29)$$

This point is plotted in the $KZ(s)/P(s)$ plane in Fig. 7–4–6b. When $\omega = 0.5$ rad/sec, we draw the vectors from the poles to the point where $s = j0.5$, measure (or calculate) their lengths and angles, and find the AR to be approximately $0.67K$ and Φ to be approximately $-259°$ or $+101°$. As ω increases, the vector lengths become very large and the AR goes to zero; the individual angles each approach $90°$, so that Φ goes to $-180°$ or its equivalent $+180°$. An advantage of this graphical technique is that it works for both nonminimum and minimum phase functions without distinction.

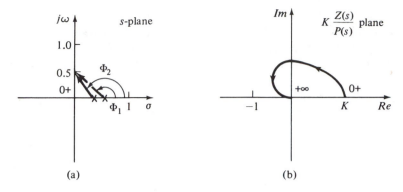

(a)   (b)

Figure 7-4-6. The Nyquist diagram of Fig. 7-4-1 by a graphical technique: (a) pole-zero plot in the s plane; (b) map in the $KZ(s)/P(s)$ plane of the imaginary axis of the s plane.

Let us now sketch the Nyquist diagram of the minimum phase function of Eq. (6-3-7) that we used to illustrate the rules for root locus sketching. In the proper form for frequency domain techniques, the KZ/P function is

$$\frac{KZ(j\omega)}{P(j\omega)} = \frac{0.1875K(j\omega/6 + 1)}{j\omega(j\omega/4 + 1)(1 - \omega^2/8 + j\omega/2)} \qquad (7\text{-}4\text{-}30)$$

The appropriate table is

ω_f (rad/sec)	AR	Φ (deg)
0+	∞	-90
$+\infty$	0	-180
1	0.18K	-225

The Nyquist diagram sketched in Fig. 7-4-7 shows that the system is conditionally stable with respect to the value of K. Notice that the curve crosses the negative real axis once from below and that the root locus of Fig. 6-3-6 crosses the

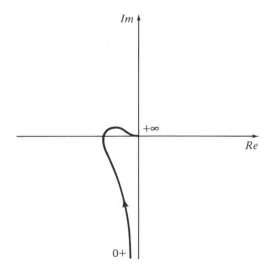

Figure 7-4-7. Nyquist diagram of the minimum phase function of Eq. (6-3-7).

imaginary axis once. The value of K that causes the Nyquist curve to pass through the -1 point is the value of K at the crossing of the imaginary axis.

Some additional examples of minimum phase Nyquist diagrams are sketched in Fig. 7–4–8. Remember that stability is determined by looking to the right for

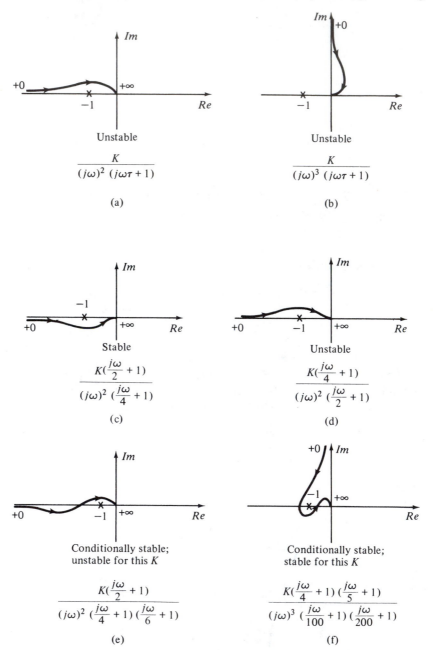

Figure 7–4–8. Some minimum phase Nyquist diagrams.

a glimpse of the -1 point; if it can be seen, the system is unstable for that particular value of K. The reader is encouraged to reproduce these sketches as well as the corresponding root locus sketches and relate the two, remembering that the Nyquist diagrams are sketched for a specific value of K.

7-5 THE GAIN MARGIN AND THE PHASE MARGIN

In addition to determining the absolute stability of a system, the Nyquist diagram provides qualitative information as to the degree of stability. The -1 point plays the same role in the Nyquist diagram as the imaginary axis does in the root locus diagram. Consider the root locus diagram of Fig. 7-5-1a for a minimum phase function; the crossing of the imaginary axis tells us that the system is conditionally stable with respect to K. When $K = K_1$, the system is stable and the corresponding Nyquist curve in Fig. 7-5-1b does not encircle the -1 point.

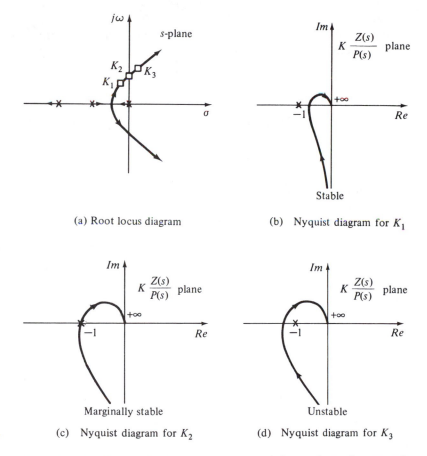

(a) Root locus diagram   (b) Nyquist diagram for K_1

(c) Nyquist diagram for K_2 (d) Nyquist diagram for K_3

Figure 7-5-1. Relationship between imaginary axis in root locus diagram and -1 point in Nyquist diagram.

It does cross the negative real axis, however, which indicates that the system is conditionally stable. When $K = K_2$, the system is marginally stable and the Nyquist curve in Fig. 7–5–1c passes through the -1 point. Finally, when $K = K_3$, the system is unstable and the Nyquist curve in Fig. 7–5–1d encircles the -1 point, passing to its left.

For a stable system, the closer the Nyquist curve approaches the -1 point, the less stable the system. The gain margin GM and the phase margin Φ_m are two quantities that give some measure of the closeness of approach to the -1 point and of the degree of stability of a system. For a conditionally stable system whose Nyquist curve crosses the negative real axis from below, the *gain margin* is the factor by which the variable parameter (usually the gain) can be increased before marginal stability occurs. Consider the Nyquist diagrams of Fig. 7–5–2 for two values of K. When $K = K_1$, the system is stable; when K is increased to K_c, the system becomes marginally stable. Increasing K further results in an unstable system. By definition,

$$GM = \frac{K_c}{K_1} \tag{7-5-1}$$

The gain margin can be evaluated by comparing the amplitude ratios for the two values of K when $\Phi = -180°$, i.e., $\omega = \omega_{cr}$, the crossing frequency. When $K = K_1$,

$$AR(\omega_{cr}) = K_1 \left| \frac{Z(j\omega_{cr})}{P(j\omega_{cr})} \right| = d \tag{7-5-2}$$

where d is the distance from the origin to the point of crossing. When $K = K_c$,

$$AR(\omega_{cr}) = K_c \left| \frac{Z(j\omega_{cr})}{P(j\omega_{cr})} \right| = 1 \tag{7-5-3}$$

From Eqs. (7–5–2) and (7–5–3), we find

$$GM = \frac{K_c}{K_1} = \frac{1}{d} = \frac{1}{\left| K_1 \dfrac{Z(j\omega_{cr})}{P(j\omega_{cr})} \right|} \tag{7-5-4}$$

For the system of Fig. 7–5–2, a GM greater than unity denotes stability, a GM equal to unity indicates marginal stability, and a GM less than unity signifies instability.

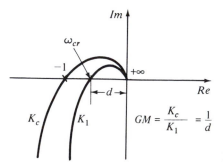

Figure 7–5–2. The gain margin GM.

Using the gain margin as a stability criterion or as a measure of the degree of stability is generally not sufficient and can easily be misleading. The Nyquist curve of the second-order system in Fig. 7–5–3a never crosses the negative real axis; d is zero and the *GM* is infinite. However, as K is increased, the curve does move toward the -1 point and the stability decreases. In Fig. 7–5–3b we have two systems with identical gain margins but different degrees of stability; B is less stable than A. These two examples illustrate the inadequacy of the gain margin in describing the degree of stability. In Fig. 7–5–3c, the Nyquist curve crosses the negative real axis from above and is stable only when the *GM* is less than unity—a reversal of the criterion just established. When the Nyquist curve crosses the negative real axis more than once, as in Fig. 7–5–3d, the gain margin becomes multivalued and loses its significance. Although the gain margin is limited in its application and usefulness, it does appear at times in design specifications.

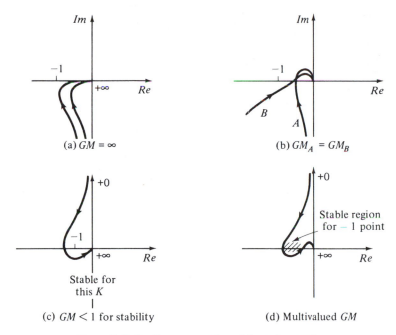

Figure 7–5–3.   Some examples of the gain margin.

The phase margin is superior to the gain margin in that it provides a consistent stability criterion for all minimum phase functions and a better measure of the degree of stability. The *phase margin* is defined as 180° plus the phase angle at the gain crossover frequency ω_{gc}, where *the gain crossover frequency is the frequency at which the AR is unity*. The expression for the phase margin can be written as

$$\Phi_m = 180° + \Phi(\omega_{gc}) \tag{7-5-5}$$

The phase margin and its application are illustrated in Fig. 7–5–4; Φ_m is measured from the negative real axis to the point where the AR is unity and is positive in the counterclockwise direction. We see that a positive Φ_m ensures stability. This fact can be used to express the Nyquist stability criterion in terms of the

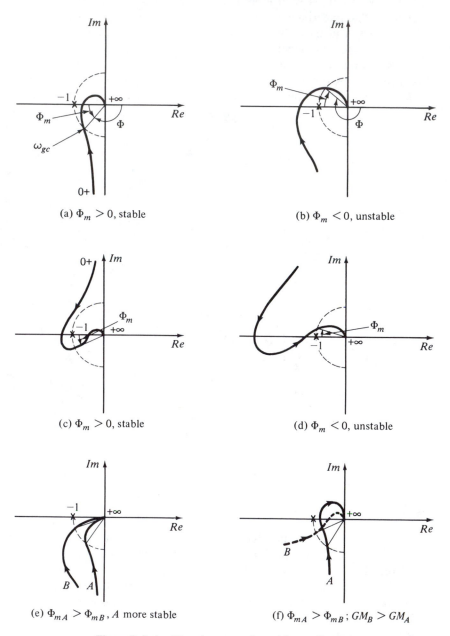

Figure 7–5–4. The phase margin and its application.

phase margin: *A system with a minimum phase $KZ(s)/P(s)$ function* is stable if $\Phi_m > 0$, marginally stable if $\Phi_m = 0$, and unstable if $\Phi_m < 0$.* Furthermore, the degree of stability increases as Φ_m increases.

This stability criterion can also be applied to systems with nonminimum phase $KZ(s)/P(s)$ functions if the systems are stable or conditionally stable. Although this criterion holds for many unstable systems as well, the necessity to identify the unstable systems for which it is not applicable negates its usefulness for nonminimum phases functions. The gain margin may also be applied to nonminimum phase functions; great care, however, is required in its interpretation and, as with the phase margin, the gain margin is not recommended for nonminimum phase functions. Count encirclements to determine the stability of nonminimum phase functions, or sketch the root locus.

In reference to *minimum phase functions only*, the phase margin (and the gain margin as well) can be obtained graphically from a precisely plotted Nyquist diagram. It can also be calculated in three steps: (1) Calculate the gain crossover frequency using the condition that the AR is unity at ω_{gc}; (2) use ω_{gc} to find the corresponding phase angle; and (3) find the phase margin from its definition. The easiest way, however, to find both the phase margin and the gain margin is to use the Bode plots of the next chapter. In fact, the easiest way to plot or sketch a Nyquist diagram itself is to construct the Bode plots first. We have treated the Nyquist diagrams first in order to establish the Nyquist stability criterion and the conditions for stability associated with the -1 point, which represents a phase angle of $-180°$ and an amplitude ratio of unity, and to define the gain and phase margins.

Since the phase margin and the (appropriate) gain margin are measures of the relative stability of a system, they can be used to indicate the degree of robustness of a system. The larger they are, the more robust the system.

PROBLEMS

7–1. Write the frequency transfer function for each of the systems represented by the differential equations of Prob. 2–6 with zero initial conditions. Also, write the expressions for the system amplitude ratio and phase angle as functions of the forcing (input) frequency ω_f.

7–2. Do Prob. 7–1 for the systems whose system (closed-loop) transfer functions are given in Prob. 4–8.

7–3. Sketch a polar plot for each of the systems of Prob. 7–2.

7–4. (a) Write expressions for the amplitude ratio and phase angle and sketch the polar plot of

$$W(s) = \frac{-10}{(s - 4)(s + 5)}$$

(b) Do the same as in (a) for

$$W(s) = \frac{10(s + 1)}{(s + 10)}$$

---

* It is also necessary that the order of $P(s)$ be greater than that of $Z(s)$.

(c) Determine the stability of each system. Did you use the polar plot? Can stability be determined from a system polar plot? Do polar plots for unstable systems have any physical significance?

(d) Discuss any similarities and differences in the polar plots of (a) and (b).

7–5. Do Prob. 7–4 for

$$W(s) = \frac{5}{(s + 3)(s + 5)}$$

and

$$W(s) = \frac{-500}{(s + 3)(s + 5)(s + 10)(s - 10)}$$

7–6. A unity feedback system has a

$$G = \frac{K}{s(0.1s + 1)(0.2s + 1)}$$

(a) Construct a brief table, sketch the Nyquist diagram, and determine the stability of the system.

(b) With $K = 5$, calculate the phase margin and the gain margin (in decibels and scalar) and locate the -1 point. Is the system stable for this value of K?

(c) Does the Nyquist diagram represent the frequency response of the system?

(d) Find the roots of the characteristic equation for $K = 5$ and determine the characteristics of the transient modes of the system.

(e) Sketch the root locus and show the approximate location of the roots for $K = 5$.

(f) Sketch the polar plot of the system transfer function for $K = 5$. Does it represent the frequency response of the system?

7–7. Do Prob. 7–6 for the forward transfer functions of Prob. 6–2 with $K = 10$.

7–8. For each of the Nyquist diagrams sketched in Fig. P7–8 determine the number of encirclements of the -1 point. With the number of poles in the right half of the s plane as shown in the figure, describe the stability of each system.

7–9. For each of the characteristic equations of Prob. 6–4 sketch the Nyquist diagram for positive K and describe the stability of the system. Verify with a root locus sketch.

7–10. Sketch the Nyquist diagram for each of the characteristic equations shown below. Locate the -1 point arbitrarily and then indicate on the sketch the phase margin Φ_m, the gain crossover frequency ω_{gc}, and the crossing frequency ω_{cr}. Describe the system stability and discuss the gain margin GM.

(a) $1 + \dfrac{K(s + 1)}{s(s + 5)(s + 10)} = 0$

(b) $1 + \dfrac{K}{s(s + 5)(s + 10)} = 0$

(c) $1 + \dfrac{K}{s^2(s + 10)} = 0$

(d) $1 + \dfrac{K(s + 1)^2}{s^2(s + 10)} = 0$

(e) $1 + \dfrac{K(s + 1)^2}{s^3(s + 10)^2} = 0$

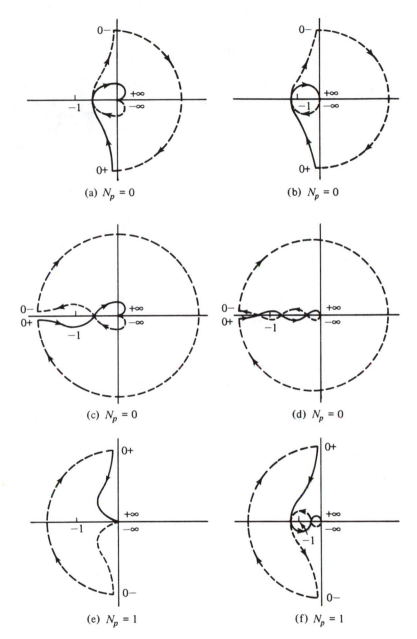

Figure P7–8

7–11. Sketch the Nyquist diagram for each of the following characteristic equations and determine the stability of the system. Verify with a root locus sketch and with the Routh-Hurwitz criterion.

(a) $1 + \dfrac{5K}{(s + 3)(s + 5)} = 0$

(b) $1 + \dfrac{500K}{(s + 3)(s + 5)(s + 10)(s - 10)} = 0$

7-12. (a) Construct a brief table and sketch the Nyquist diagram of Eq. (6-3-5); then determine the stability of the system.

(b) Does the Nyquist diagram corroborate the root locus of Fig. 6-3-6? How?

(c) Validate the conclusions of (a) and (b) using the Routh-Hurwitz criterion.

8

The Frequency Response:
Bode Plots

8–1 INTRODUCTION

The frequency transfer function of a system or of its $KZ(j\omega)/P(j\omega)$ function can be represented either by the single Nyquist diagram (a polar plot) or by plots of the amplitude ratio and the phase angle against the input (forcing) frequency. It is customary to plot the amplitude ratio in decibels and the phase angle in degrees against the common logarithm of the input frequency. In this form, the two plots are known as *Bode plots* (after H. W. Bode). There are exact Bode plots, which are best prepared with a computer, and straight-line asymptotic plots, which can be quickly and easily sketched or plotted by hand using the techniques to be developed and discussed in this chapter.

Bode plots of the $KZ(j\omega)/P(j\omega)$ function are used, in conjunction with the Nyquist stability criterion, to determine and examine the stability of a system. As with the Nyquist diagram, the value of K must be specified for Bode plots. However, only the AR plot is affected by a change in K, and it does not have to be completely redrawn; in fact, the effects of a change in K can often be visualized and determined without modifying the plot. The phase angle plot is independent of K.

Bode plots of the system transfer function are used to determine the effects of various inputs (including a step) upon the steady-state response of the system. Since the frequency response is a steady-state response, the system *must* be stable and its stability must be determined before the system Bode plots can be used (as in the application of the final value theorem).

The amplitude ratio in decibels* is defined as

$$AR_{db} = 20 \log AR \qquad (8-1-1)$$

* The decibel was originally defined as 10 times the common logarithm of a power ratio.

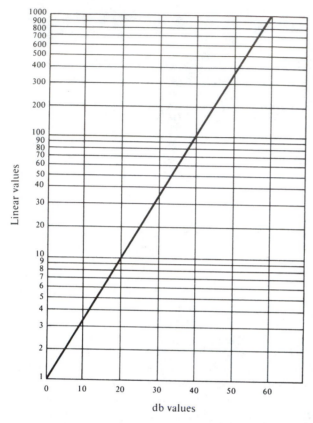

Linear values

db values

Figure 8–1–1. Linear values versus decibel values.

Figure 8–1–1 presents a chart to aid in the conversion between decibels and the linear values of the amplitude ratio. Remember that the reciprocals of numbers differ only in sign; e.g., 2 is $+6$ db, and $\frac{1}{2}$ is -6 db.

Let us start with the Bode plot of a simple first-order system having the transfer function

$$W(s) = \frac{1}{\tau s + 1} \tag{8–1–2}$$

and a frequency transfer function

$$W(j\omega) = \frac{1}{j\omega\tau + 1} \tag{8–1–3}$$

The amplitude ratio in decibels and the phase angle in degrees are plotted against the logarithm of the input frequency in Fig. 8–1–2. These Bode plots of a simple first-order system show that the AR varies from unity (0 db) to zero ($-\infty$ db) as the frequency goes from 0 to $+\infty$ and that Φ starts at $0°$ and approaches $-90°$ as the frequency increases.

A simple second-order system has the transfer function

$$W(s) = \frac{\omega_n^2}{s^2 + 2\zeta\omega_n s + \omega_n^2} \tag{8–1–4}$$

and the frequency transfer function

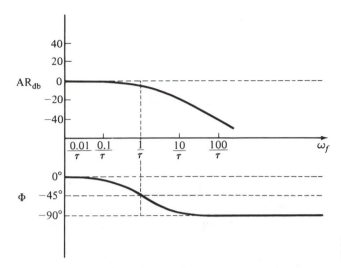

Figure 8–1–2. Bode plots of a simple first-order system.

$$W(j\omega) = \frac{\omega_n^2}{\omega_n^2 - \omega^2 + j2\zeta\omega_n\omega} \tag{8–1–5}$$

The Bode plots in Fig. 8–1–3 show that the shape of the AR and Φ curves is a strong function of the damping ratio ζ. For damping ratios less than 0.707, the AR peaks in the vicinity of the undamped natural frequency ω_n, indicating that the system output oscillates with a larger amplitude than that of the input. When the damping ratio is zero, the amplitude ratio is infinite; fortunately, a little damping does a lot to reduce this peaking effect. The phase angle starts at $0°$ for low frequencies and decreases to $-180°$ as a limit at high frequencies.

Higher-order systems may be treated as combinations of simple first- and second-order systems plus numerator factors. The frequency response of such systems can be obtained by plotting each term and then combining the individual plots. Let us assume that we are fortunate enough to have the transfer function of a higher-order system in the factored form

$$W(s) = \frac{K(\tau_1 s + 1)(\tau_2 s + 1) \cdots}{s^n(\tau_3 s + 1)(\tau_4 s + 1) \cdots} \tag{8–1–6}$$

so that

$$W(j\omega) = \frac{K(j\omega\tau_1 + 1)(j\omega\tau_2 + 1) \cdots}{(j\omega)^n(j\omega\tau_3 + 1)(j\omega\tau_4 + 1) \cdots} \tag{8–1–7}$$

The amplitude ratio can be written as

$$AR(\omega) = |W(j\omega)| = \frac{K|j\omega\tau_1 + 1||j\omega\tau_2 + 1| \cdots}{|(j\omega)^n||j\omega\tau_3 + 1||j\omega\tau_4 + 1| \cdots} \tag{8–1–8}$$

Plotting the linear magnitude of AR against the frequency requires multiplying together the individual magnitudes. If, however, the AR is expressed in decibels, Eq. (8–1–8) becomes

$$AR_{db} = 20 \log K + 20 \log|j\omega\tau_1 + | \cdots \tag{8–1–9}$$
$$- 20\,n \log \omega - 20 \log|j\omega\tau_3 + 1| - \cdots$$

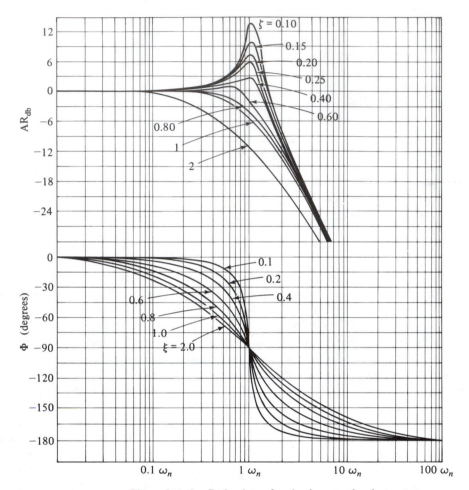

Figure 8–1–3.   Bode plots of a simple second-order system.

The value of AR_{db} is merely the sum of the individual magnitudes, or amplitude ratios, expressed in decibels. The phase angle of the frequency transfer function is the sum of the angles associated with the individual factors. The mechanism of plotting has been simplified by using logarithmic scales, as will be the process of system compensation to be discussed in Chap. 9.

8–2 BODE PLOT SKETCHING

Bode plots are most commonly used with the $KZ(j\omega)/P(j\omega)$ function to examine the stability of a system. When the function is *minimum phase*, the Bode plots can be sketched rather rapidly with a knowledge of the four basic terms that appear in the function. These terms are frequency-invariant terms K, zeros and poles at the origin $(j\omega)^{\pm n}$, first-order terms or real poles and zeros $(j\omega\tau + 1)^{\pm n}$, and second-order poles and zeros $[1 - (\omega/\omega_n)^2 + j2\zeta(\omega/\omega_n)]^{\pm n}$. The Bode plots of the first two terms are straight lines and can be plotted exactly and rapidly.

The Bode plots of the last two terms are curves; however, they can be approximated by straight-line asymptotes that are usually adequate. Semilog paper is recommended for Bode plots.

For a frequency-invariant term K, AR_{db} is simply

$$AR_{db} = K_{db} = 20 \log K \qquad (8\text{–}2\text{–}1)$$

Since K is frequency-invariant, the AR_{db} plot is a straight horizontal line as shown in Fig. 8–2–1 for K equal to 10. The phase angle associated with K is always zero and requires no plot per se.

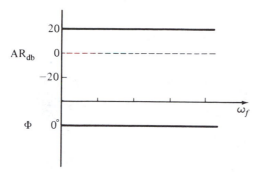

Figure 8–2–1.   Bode plots of the frequency-invariant term $K = 10$.

For poles and zeros at the origin, the frequency function is $(j\omega)^{\pm n}$. The amplitude ratio is the magnitude of $(j\omega)^n$ and is

$$AR(\omega) = \sqrt{\omega^{\pm 2n}} = \omega^{\pm n} \qquad (8\text{–}2\text{–}2)$$

The AR_{db} can be written as

$$AR_{db} = \pm 20n \log \omega \qquad (8\text{–}2\text{–}3)$$

and

$$\Phi = \pm n 90° \qquad (8\text{–}2\text{–}4)$$

where $+n$ corresponds to n zeros and $-n$ to n poles. When ω is equal to unity, the AR_{db} is zero regardless of the magnitude and sign of n, and therefore all AR_{db} plots will cross the 0-db line at $\omega = 1$. The slope of the AR_{db} curve can be found by differentiating Eq. (8–2–3) to obtain

$$\frac{d\, AR_{db}}{d \log \omega} = \pm 20n \text{ db/decade} \cong \pm 6n \text{ db/octave} \qquad (8\text{–}2\text{–}5)$$

A *decade* (dec) is a unit change in the log ω and represents an order of magnitude change in the input frequency. An *octave* represents a doubling or halving of the frequency and is approximately equal to 0.3 decade. The Bode plots for one and two poles at the origin and for one and two zeros at the origin are shown in Fig. 8–2–2; these are exact plots. The AR_{db} and Φ plots for the poles are mirror images about the 0-db and 0° lines of the plots for the zeros. This symmetry and the linearity of the plots are further advantages of using logarithmic scales.

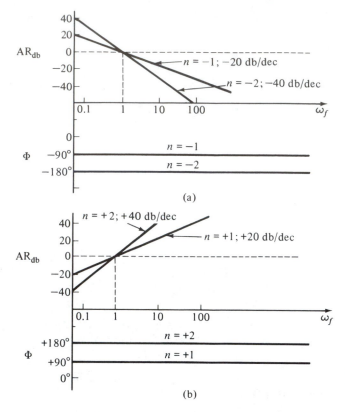

Figure 8–2–2. Bode plots of poles and zeros at the origin: (a) one and two poles; (b) one and two zeros.

Since a single real pole has a frequency function $(j\omega\tau + 1)^{-1}$ that is identical to the frequency transfer function of a simple first-order system, the Bode plots are identical and are as shown in Fig. 8–1–2. The plots are actually curves but may be approximated by straight-line asymptotes. The AR_{db} is

$$AR_{db} = 20 \log \frac{1}{\sqrt{(\omega\tau)^2 + 1}} = -20 \log \sqrt{(\omega\tau)^2 + 1} \qquad (8-2-6)$$

When $\omega\tau \ll 1$ (ω approaches zero) then

$$AR_{db} \cong -20 \log 1 = 0 \qquad (8-2-7)$$

From Eq. (8–2–7) we see that at low frequencies the AR_{db} plot coincides with the 0-db line. When $\omega\tau \gg 1$,

$$AR_{db} \cong -20 \log \omega\tau = -20 \log \omega - 20 \log \tau \qquad (8-2-8)$$

and when $\omega \gg \tau$,

$$AR_{db} \cong -20 \log \omega \qquad (8-2-9)$$

Equation (8–2–9) is the equation of a straight line with a slope of -20 db/decade or -6 db/octave. The intersection of the straight-line asymptotes can be found, by equating Eqs. (8–2–7) and (8–2–8), to occur when $-20 \log \omega\tau = 0$ or when $\omega\tau = 1$. The frequency of this intersection is called the *corner frequency*,* given

* Sometimes called the *break frequency* or *breakpoint*.

the symbol ω_c, and is equal to $1/\tau$. The AR_{db} straight-line plot for a real pole is shown in Fig. 8–2–3a. The corrections applied to this asymptotic plot to make it fit the actual curve are: at $\omega = \omega_c$, subtract 3 db; at one octave higher, $\omega = 2\omega_c$, subtract 1 db; and at one octave lower, $\omega = 0.5\omega_c$, subtract 1 db. The phase curve is an arctangent curve where

$$\Phi = -\tan^{-1} \omega\tau \tag{8–2–10}$$

when $\omega = \omega_c$, $\omega\tau = 1$, and Φ is exactly equal to $-45°$. The straight-line approximation is drawn from the $0°$ line at a point one decade below the corner frequency, through the $-45°$ point at ω_c, to the $-90°$ line at a point one decade above the corner frequency as shown in Fig. 8–2–3a. The actual phase curve is $-6°$ at $\omega = 0.1\omega_c$ and $-84°$ at $\omega = 10\omega_c$; these are the maximum deviations between the two curves. A comparison of Fig. 8–2–3a with Figs. 8–2–1 and 8–2–2 reveals that a real pole approximates an ideal amplifier with unity gain at low frequencies and a pole at the origin or an ideal integrator at high frequencies.

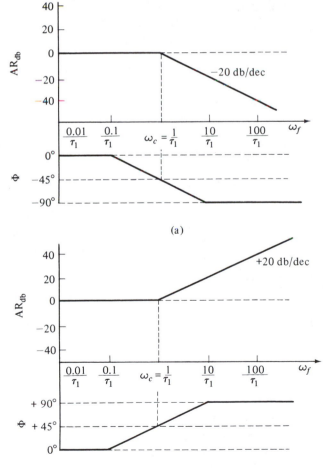

(a)

(b)

Figure 8–2–3. Asymptotic Bode plots of simple poles and zeros: (a) a simple pole, $1/(j\omega\tau + 1)$; (b) a simple zero, $j\omega\tau + 1$.

The asymptotic plots for a single real zero are shown in Fig. 8–2–3b and are simply the mirror images of those for the real pole. Naturally the corrections to the asymptotic plots to match the actual curve will change signs. A real zero approximates an ideal unity gain amplifier at low frequencies and an ideal differentiator at high frequencies.

The asymptotic plots of multiple real poles and zeros will resemble those for single poles and zeros. The corner frequency will remain unchanged, as will the plots for frequencies below ω_c; however, slopes, corrections, and high-frequency phase angles will all be changed by a factor equal to the number of multiple poles and zeros. For example, a pair of real poles will change the slope to -40 db/decade or -12 db/octave, double the corrections, and change the upper limit of the phase angle to $-180°$.

A single complex pole may also be represented by straight-line asymptotic plots, but these plots will be less accurate than those for a real pole. The frequency function is $[1 - (\omega/\omega_n)^2 + j2\zeta(\omega/\omega_n)]^{-1}$, so that

$$AR_{db} = -20 \log \sqrt{\left[1 - \left(\frac{\omega}{\omega_n}\right)^2\right]^2 + 4\zeta^2\left(\frac{\omega}{\omega_n}\right)^2} \qquad (8\text{--}2\text{--}11)$$

and

$$\Phi = -\tan^{-1} \frac{2\zeta(\omega/\omega_n)}{1 - (\omega/\omega_n)^2} \qquad (8\text{--}2\text{--}12)$$

When $\omega/\omega_n \ll 1$,

$$AR_{db} \cong -20 \log 1 = 0 \qquad (8\text{--}2\text{--}13)$$

and

$$\Phi \cong -\tan^{-1} 0 = 0° \qquad (8\text{--}2\text{--}14)$$

When $\omega/\omega_n \gg 1$,

$$AR_{db} \cong -20 \log \left(\frac{\omega}{\omega_n}\right)^2 = -40 \log \left(\frac{\omega}{\omega_n}\right) \qquad (8\text{--}2\text{--}15)$$

and

$$\Phi \cong -\tan^{-1} \frac{2\zeta}{-\omega/\omega_n} \cong -180° \qquad (8\text{--}2\text{--}16)$$

Equation (8–2–15) is the equation of a straight line with a slope of -40 db/decade or -12 db/octave. When $\omega = \omega_n = \omega_c$, the two straight lines representing the AR_{db} will intersect and Φ will be exactly equal to $-90°$. The asymptotic plots are shown in Fig. 8–2–4 and are identical to those for two real and equal poles with $\tau = 1/\omega_n$; we have in essence assumed the damping ratio of the complex pole to be unity. The actual curves are those of the simple second-order system of Fig. 8–1–3 and show the influence of the damping ratio upon the validity of the straight-line approximations. It so happens, fortunately, that it is seldom necessary to consider in our analyses the corrections due to the damping ratio; when it is necessary, they can be calculated.

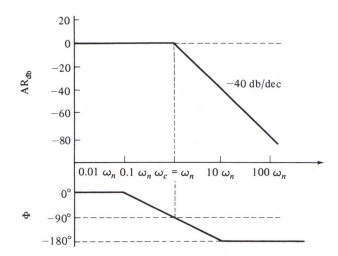

Figure 8–2–4. Asymptotic Bode plots of a complex pole.

Complex zeros, when they occur, have plots that are mirror images of those for the complex poles. Multiple complex poles and zeros are treated in the same manner as multiple real poles and zeros.

The use of these building-block Bode plots will be illustrated by example. Let us first consider a system with

$$K\frac{Z(s)}{P(s)} = \frac{K_1}{s(s+2)(s+10)} \tag{8–2–17}$$

which in the frequency domain is written as

$$K\frac{Z(j\omega)}{P(j\omega)} = \frac{K_1}{20}\frac{1}{(j\omega)(j\omega/2+1)(j\omega/10+1)} \tag{8–2–18}$$

where $K_1/20 = K$. Notice that the form of the factors in Eq. (8–2–17) has been changed in Eq. (8–2–18); it is necessary that the constant term in each factor be reduced to unity. The frequency-invariant term is $K_1/20$; before it can be plotted, K_1 must be specified. A Bode plot, like a Nyquist diagram, represents only one value of K, in contrast to the root locus diagram where K is allowed to vary. In this example, set $K_1 = 40$, so that $K = 2$ and $K_{db} = +6$ db. This horizontal line is the first thing plotted in Fig. 8–2–5. Next, the AR_{db} of the pole at the origin is drawn through the 0-db line at $\omega = 1$ with a slope of -20 db/decade. The corner frequencies associated with the two real poles are 2 and 10; these are located on the 0-db line, and a straight line with a slope of -20 db/decade is drawn from each. The total AR_{db} of the $KZ(j\omega)/P(j\omega)$ function is obtained by adding the individual amplitude ratios together; this is the solid line.

The phase shift associated with each pole is drawn as shown. The pole at the origin has a constant phase shift of $-90°$. The phase shift of the real pole with $\omega_c = 2$ is represented by a line along $0°$ for $\omega < 0.2$, by a straight line from $0°$ at $\omega = 0.2$ to $-90°$ at $\omega = 20$ (this line passes through $-45°$ at $\omega_c = 2$), and finally by a line along $-90°$ for $\omega > 20$. A similar series of lines represents the phase angle associated with the other real pole. The addition of these individual

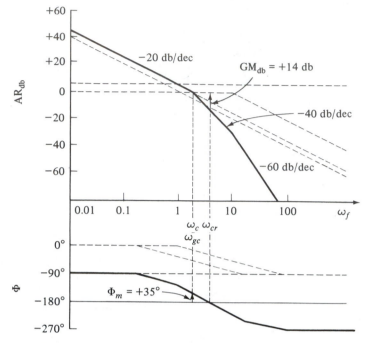

Figure 8–2–5. Bode plots of

$$\frac{K_1}{20}\frac{1}{j\omega(j\omega/2 + 1)(j\omega/10 + 1)} \quad \text{for } K_1 = 40.$$

phase plots produces the total phase angle associated with the function. Notice that increasing or decreasing the value of K moves the overall AR_{db} curve up or down with respect to the 0-db line but does not affect the phase angle plot.

For the next example, let

$$K\frac{Z(s)}{P(s)} = \frac{K_1(s + 2)}{s^2(s + 10)} \tag{8–2–19}$$

which in the frequency domain is

$$K\frac{Z(j\omega)}{P(j\omega)} = \frac{K_1}{5}\frac{j\omega/2 + 1}{(j\omega)^2(j\omega/10 + 1)} \tag{8–2–20}$$

If K_1 is set equal to 10, K is 2 and K_{db} is $+6$ db. The corner frequency of the real zero is 2, and that of the real pole is 10. Each AR_{db} curve is drawn in Fig. 8–2–6 and added to produce the overall AR_{db} plot. The phase angle of the zero passes through $+45°$ at $\omega_c = 2$, reaching $+90°$ at $\omega = 20$. This and the other individual phase angles are drawn and then added to obtain the overall phase angle plot. The overall phase angle starts at $-180°$ at low frequencies because of the double pole at the origin. The phase lead of the zero is felt first, since it has a lower corner frequency than the pole. The phase lag of the pole eventually counteracts the effect of the zero, and the phase angle returns to $-180°$ at higher frequencies.

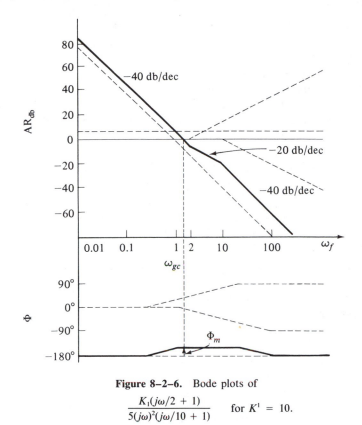

Figure 8–2–6. Bode plots of

$$\frac{K_1(j\omega/2 + 1)}{5(j\omega)^2(j\omega/10 + 1)} \quad \text{for } K^1 = 10.$$

The asymptotic Bode plots are a reasonable approximation of the exact plots except in the vicinity of the corner frequencies of lightly damped second-order terms. Fortunately, exact values in these regions usually are not needed for the analysis of well-behaved systems. However, when these regions cannot be avoided and must be considered, computer plots should be used or corrections can be made to the asymptotic plots based on a knowledge of the appropriate damping ratios.

8–3 BODE PLOTS AND STABILITY

The Nyquist stability criterion for a minimum phase system as developed in Sec. 7–5 states that a system is stable if the phase margin Φ_m is positive; furthermore, the larger the phase margin, the more stable the system. The approximate phase margin can be easily obtained from the Bode plots of the $KZ(j\omega)/P(j\omega)$ function.* Bode plots, therefore, can be used to obtain information about both the absolute and relative stability of a system.

* The open-loop frequency transfer function $GH(j\omega)$ is customarily used.

Consider the system of Eq. (8–2–17) with the Bode plots of Fig. 8–2–5. The gain crossover frequency ω_{gc} is 2 rad/sec, and the corresponding phase angle is $-145°$. Based on the definition of Eq. (7–5–5), the $\Phi_m = 180° - 145° = +35°$; therefore, this system is stable for the value of K_1 (40) for which these Bode plots were drawn.

Since the phase angle curve crosses the $-180°$ line only once from above, this system is conditionally stable in the simple sense and therefore has a gain margin with significance; examine the Nyquist diagram of Fig. 7–4–5a. If K_1 were increased to move the AR_{db} curve up so that the gain crossover occurred at the frequency where $\Phi = -180°$, the system would become marginally stable with a Φ_m equal to zero. Any further increase in K_1 would cause Φ_m to become negative and drive the system to instability. The value of GM_{db} can be found by determining the AR_{db} at the crossing frequency ω_{cr}, where the phase angle curve crosses the $-180°$ line. In this example, the AR_{db} at ω_{cr} is -14 db, which means that the overall AR_{db} curve could be shifted upward by that amount before the system would become marginally stable. In other words, the GM_{db} is equal to $+14$ db, which is equal to a linear GM of 5. Since K_1 is equal to 40, $K_c = GM \times K_1 = 5 \times 40 = 200$; this is the value of K at which the root locus would cross the imaginary axis. The gain margin and phase margin are indicated in Fig. 8–2–5. Remember that these are approximate values only[†]; more exact values require either corrections to the asymptotic plots or computer plots.

We shall now consider the system of Eq. (8–2–19) with the Bode plots of Fig. 8–2–6. At the gain crossover frequency of approximately 1.4, Φ is $-152°$ and Φ_m is $+28°$; therefore, this system is stable for $K_1 = 10$. Since there is no crossing of the $-180°$ line (see the Nyquist diagram of Fig. 7–4–8c), there is no finite GM. As K_1 is increased or decreased, the phase margin remains positive but does approach zero, or marginal stability, as a limit. The system is most stable when Φ_m is at a maximum.

If we remove the real zero from this example, we have

$$K\frac{Z(j\omega)}{P(j\omega)} = \frac{K_1}{10}\frac{1}{(j\omega)^2(j\omega/10 + 1)} \tag{8–3–1}$$

The Bode plots of this system for K_1 equal to 10 are given in Fig. 8–3–1. Since the phase margin can never be positive, the system is unstable for all values of K_1. The Nyquist diagram for this system is sketched in Fig. 7–4–8a.

As a final example, consider the system with

$$K\frac{Z(j\omega)}{P(j\omega)} = \frac{K_1}{1000}\frac{(j\omega/4 + 1)(j\omega/5 + 1)}{(j\omega)^3(j\omega/100 + 1)(j\omega/200 + 1)} \tag{8–3–2}$$

The Bode plots for $K_1 = 1000$ are shown in Fig. 8–3–2, and the Nyquist diagram for an unspecified K_1 in Fig. 7–4–8f. This system is conditionally stable but not in the simple sense, because the phase angle curve crosses the $-180°$ line twice. The Bode plots reveal a negative phase margin for this particular value of K_1 and the system is unstable. The system will be stable only when the gain crossover frequency ω_{gc} falls in the region where the phase angle curve is above

[†] The Routh-Hurwitz criterion yields a value of 240 for K_c.

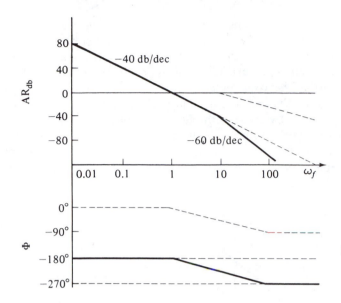

Figure 8-3-1. Bode plots of

$$\frac{K_1}{10} \frac{1}{(j\omega)^2(j\omega/10 + 1)} \quad \text{for } K_1 = 10.$$

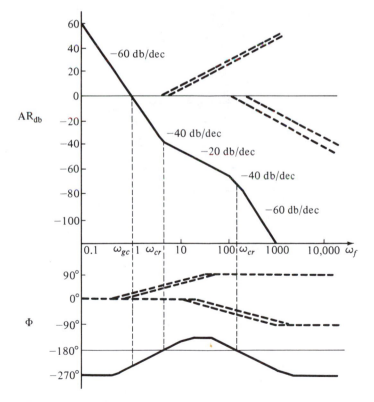

Figure 8-3-2. Bode plots of

$$\frac{K_1}{1000} \frac{(j\omega/4 + 1)(j\omega/5 + 1)}{(j\omega)^3(j\omega/100 + 1)(j\omega/200 + 1)} \quad \text{for } K_1 = 1000.$$

the $-180°$ line, i.e., between $\omega_{cr} \cong 4.47$ and $\omega_{cr} \simeq 141$. At these two frequencies, the AR_{db} is -40 and -72 db, respectively. This means that K_{db} must be increased by at least 40 db but not more than 72 db for stable operation; consequently, the system is stable only when K_1 is greater than approximately 10^5 but less than approximately 4×10^6. The root locus of this system would show two crossings of the imaginary axis; these crossings and the corresponding values of K_1 can be found, but not easily, by the Routh-Hurwitz criterion.

Bode plots are sometimes used to provide the data for AR_{db}-versus-Φ plots. Such plots do not provide any new information; they are merely another form of presentation. Two typical *gain-phase plots* are illustrated in Fig. 8–3–3,

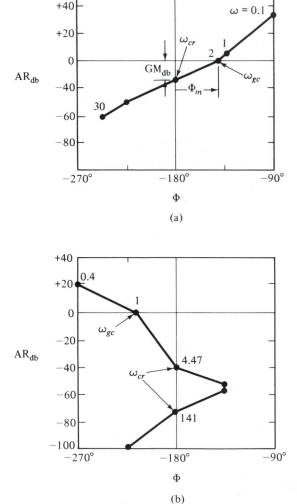

(a)

(b)

Figure 8–3–3. Gain-phase plots for two conditionally stable systems: (a) from Fig. 8–2–5 for a stable value of K; (b) from Fig. 8–3–2 for an unstable value of K.

showing that stable operation requires the gain crossover point to be to the right of the $-180°$ line.

Minimum phase functions have a unique relationship between the AR and Φ that can be utilized to indicate stability by examining the slope of the AR_{db} curve at the gain crossover frequency ω_{gc}. If the slope is -60 db/decade or more negative, the system is definitely unstable. If the slope is -40 db/decade, the system may or may not be stable; if it is stable, the phase margin will be small. If the slope is -20 db/decade or less negative, the system is stable; the smaller the slope in a negative sense, the more stable the system. Apply these relationships to the preceding examples to get a feel for the acceptable ranges of values of K_1 for stable operation.

Nonminimum phase $KZ(j\omega)/P(\omega)$ functions occur when the plant is unstable and are also possible in multiloop systems. The asymptotic plots for the real and complex poles and zeros may still be used if appropriate changes are made in the phase angle plots. The relationship between the AR_{db} and Φ plots is no longer unique. It is unwise to use the Bode plots to determine system stability; instead, use them to plot the Nyquist diagram, which then becomes the instrument for examining stability, or else sketch the root locus.

If the system (closed-loop) transfer function is in factored form, it is a simple matter to construct the Bode plots. Do *not* attempt to determine stability from Bode plots of the system frequency transfer function; an unstable system can have Bode plots identical in shape to those of a stable system. Furthermore, it is not always easy to factor the numerator and denominator of the transfer function. Therefore, the common practice is to find the phase margin Φ_m and gain crossover frequency ω_{gc} from the Bode plots of the open-loop frequency transfer function and use these two parameters with the techniques of Sec. 8–5 to obtain qualitative information on the overall system behavior.

8–4 THE EXPERIMENTAL TRANSFER FUNCTION

Experimentally obtained frequency response data can be used to obtain an approximate transfer function. The value of AR_{db} is plotted against the logarithm of the input frequency, as in Fig. 8–4–1a, and a series of straight-line asymptotes are fitted to these data. The zero slope at low frequencies tells us that there are no poles or zeros at the origin and that K_{db} is $+20$ db and K is 10. Changes in slope occur at the corner frequencies, and the sign and magnitude indicate the type of term. The change of -20 db/decade at $\omega_c = 1$ reveals a first-order pole with $\tau = 1$. The $+20$ db/decade change at $\omega_c = 10$ indicates a first-order zero with $\tau = 0.1$. Finally, the change of -40 db/decade at $\omega_c = 50$ denotes a second-order pole with $\omega_n = 50$; the slight peaking in the vicinity of 50 rad/sec is typical of a relatively high damping ratio which can be approximated by examination of Fig. 8–1–3 or can be assumed equal to unity. If the latter course is chosen, the complex pole is approximated by a pair of real and equal poles with $\tau = 0.02$. Putting this information together produces the approximate

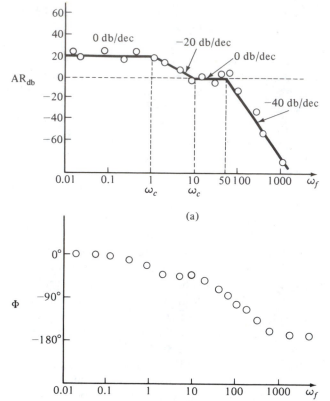

(a)

(b)

Figure 8–4–1. An approximate transfer function

$$\frac{2500(s + 10)}{(s + 1)(s + 50)^2}$$

from experimental data: (a) amplitude ratio data; (b) phase angle data.

frequency transfer function

$$TF(j\omega) = \frac{10(j\omega/10 + 1)}{(j\omega + 1)(j\omega/50 + 1)^2} \tag{8–4–1}$$

Replacing $j\omega$ by s yields a transfer function

$$TF(s) = \frac{2500(s + 10)}{(s + 1)(s + 50)^2} \tag{8–4–2}$$

The experimental phase angle data are plotted in Fig. 8–4–1b and are examined in order to verify that Eq. (8–4–1) is indeed a minimum phase function as shown. As ω becomes large, the phase angle of a minimum phase function approaches $-(n - m)90°$, where n and m are the orders of the denominator and numerator of the frequency transfer function, respectively. For nonminimum phase functions this relationship does not exist. Figure 8–4–1b is the phase plot of a minimum phase function because Φ approaches $-180°$ or $-(3 - 1)90°$ as ω becomes large.

Equation (8–4–2) can represent a component, a plant, or a system, depending upon what was tested. The fact that the output settled down to a steady-state sinusoidal output indicates that what was tested is not unstable. An unstable element or system does not have a frequency response. When experimental data are not amenable to asymptotic fitting, they may be used directly in Bode plots and Nyquist diagrams for stability determination and system compensation.

Inputs other than steady-state sinusoids can be used to obtain approximate transfer functions but not as easily nor as accurately. The response to a known pulse input can be reduced and converted to a frequency transfer function of reasonable accuracy by utilizing the theory of Fourier transformations. A step response is less accurate and is useful primarily when there is a dominant root or a dominant pair of roots; it is difficult to produce a true physical step, and there is always the possibility of driving components to saturation.

8-5 BODE PLOTS AND THE SYSTEM RESPONSE

Since the frequency domain deals solely with the steady-state response, it provides no explicit information about the transient response of the system. However, for the special case of a simple second-order system with unity feedback, exact relationships can be established between the phase margin Φ_m and the damping ratio ζ, and between the gain crossover frequency ω_{gc} and the undamped natural frequency ω_n. These relationships can then be extended with caution to higher-order systems.

The transfer function for a simple second-order system in parametric form is

$$W(s) = \frac{\omega_n^2}{s^2 + 2\zeta\omega_n s + \omega_n^2} \tag{8-5-1}$$

With unity feedback, the open-loop transfer function can be written as

$$G(s) = K\frac{Z(s)}{P(s)} = \frac{\omega_n^2}{s(s + 2\zeta\omega_n)} \tag{8-5-2}$$

The open-loop frequency transfer function becomes

$$G(j\omega) = K\frac{Z(j\omega)}{P(j\omega)} = \frac{\omega_n^2}{-\omega^2 + j2\zeta\omega_n\omega} \tag{8-5-3}$$

The ratio of the gain crossover frequency ω_{gc} to the undamped natural frequency can be found by setting the magnitude of $G(j\omega)$ equal to unity. This results in a quadratic equation in $(\omega_{gc}/\omega_n)^2$:

$$\left(\frac{\omega_{gc}}{\omega_n}\right)^4 + 4\zeta^2\left(\frac{\omega_{gc}}{\omega_n}\right)^2 - 1 = 0 \tag{8-5-4}$$

This equation can be solved for

$$\frac{\omega_{gc}}{\omega_n} = \left(\sqrt{4\zeta^4 + 1} - 2\zeta^2\right)^{1/2} \tag{8-5-5}$$

where the negative solutions have been rejected. This relationship has been plotted in Fig. 8-5-1.

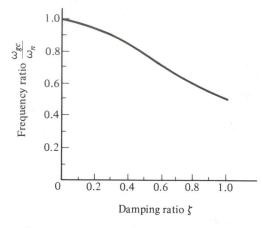

Figure 8–5–1. Ratio of the gain crossover frequency to the natural frequency as a function of the damping ratio for a simple second-order system.

From the definition of the phase margin,

$$\Phi_m = 180° - \tan^{-1} 2\zeta \frac{\omega_n}{-\omega_{gc}} = \tan^{-1} 2\zeta \frac{\omega_n}{\omega_{gc}} \qquad (8\text{--}5\text{--}6)$$

When Eq. (8–5–5) is substituted into Eq. (8–5–6), we obtain a relationship for Φ_m in terms of ζ alone:

$$\Phi_m = \tan^{-1} 2\zeta \left(\frac{1}{\sqrt{4\zeta^4 + 1} - 2\zeta^2} \right)^{1/2} \qquad (8\text{--}5\text{--}7)$$

This relationship is plotted in Fig. 8–5–2; when $\Phi_m < 40°$, the damping ratio can be approximated by

$$\zeta \cong \frac{\pi \Phi_m}{360} \qquad (8\text{--}5\text{--}8a)$$

where Φ_m is in degrees.

Once ζ is calculated, ω_{gc}/ω_n can be found from Eq. (8–5–5) or, more easily, from Eq. (8–5–6) rewritten as

$$\frac{\omega_{gc}}{\omega_n} = \frac{2\zeta}{\tan \Phi_m} \qquad (8\text{--}5\text{--}8b)$$

Figures 8–5–1 and 8–5–2, in conjunction with the Bode plots of the open-

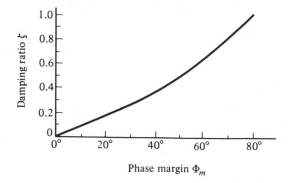

Figure 8–5–2. The damping ratio as a function of the phase margin for a simple second-order system.

loop frequency transfer function, can be extended to higher-order systems to obtain approximate values of ζ and ω_n.* The values of Φ_m and ω_{gc} are taken directly from the Bode plots. Φ_m is used first to find ζ, which in turn enables us to obtain ω_{gc}/ω_n and thus ω_n.

With ζ and ω_n known, other characteristics of the system (closed-loop) frequency response can be determined; the information is exact for simple second-order systems and approximate for other systems. These system characteristics are the amplitude ratio M for a given input frequency, the peak amplitude ratio M_p, and the peak frequency ω_p, which is the input frequency at which M_p occurs. From Eq. (8–5–1) written in the frequency domain,

$$M(j\omega) = |W(j\omega)| = \frac{\omega_n^2}{[(\omega_n^2 - \omega^2)^2 + 4\zeta^2\omega^2\omega_n^2]^{1/2}} \qquad (8\text{–}5\text{–}9)$$

The peak frequency ω_p is found by differentiating Eq. (8–5–9) with respect to ω, setting the derivative equal to zero, and solving for ω_p:

$$\omega_p = \omega_n \sqrt{1 - 2\zeta^2} \qquad (8\text{–}5\text{–}10)$$

The ratio ω_p/ω_n is plotted against ζ in Fig. 8–5–3. Note that ζ must be less than 0.707 for ω_p to be real; as we have seen previously, there is no M_p for $\zeta > 0.707$. Substituting Eq. (8–5–10) into Eq. (8–5–9) yields the relationship

$$M_p = \frac{1}{2\zeta\sqrt{1 - \zeta^2}} \qquad (8\text{–}5\text{–}11)$$

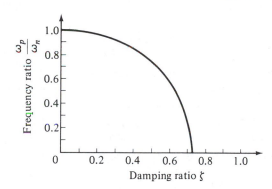

Figure 8–5–3. Ratio of the peak frequency to the natural frequency as a function of the damping ratio for a simple second-order system.

This equation is plotted in Fig. 8–5–4.

Let us now apply these procedures to the unity feedback system represented by the open-loop transfer function of Eq. (8–2–17) and the asymptotic Bode plots of Fig. 8–2–5. From these Bode plots we obtain approximate values for Φ_m and ω_{gc}, namely, $+35°$ and 2 rad/sec. Using Eqs. (8–5–8a) and (8–5–8b),

$$\zeta = \frac{35\pi}{360} = 0.305$$

$$\frac{\omega_{gc}}{\omega_n} = \frac{2 \times 0.305}{\tan 35°} = 0.87$$

* We are in effect approximating higher-order systems by a second-order system.

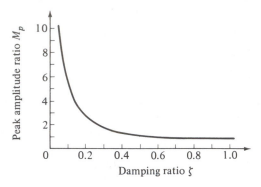

Figure 8–5–4. The peak amplitude ratio as a function of the damping ratio for a simple second-order system.

and $\omega_n = 2.3$ rad/sec. Equations (8–5–11) and (8–5–10) yield values of 1.72 for M_p and 2.07 rad/sec for ω_p. The system transfer function with $K_1 = 40$ and unity feedback is

$$\frac{c(s)}{r(s)} = \frac{40}{(s + 10.45)(s^2 + 1.55s + 3.83)}$$

This is a simple third-order system whose second-order transient mode has $\zeta = 0.396$ and $\omega_n = 1.96$ rad/sec. A Bode plot of this transfer function shows $M_p = 1.36$ at $\omega_p = 1.6$ rad/sec. If exact Bode plots are used, the exact value of Φ_m is $+43.2°$ with $\omega_{gc} = 1.56$ rad/sec. These values yield $\zeta = 0.377$ and $\omega_{gc}/\omega_n = 0.803$, resulting in $\omega_n = 1.94$ and $M_p = 1.43$ at 1.64 rad/sec, a closer match than the results using the asymptotic Bode plots. With a higher-order system with numerator dynamics do not expect as good a match.

In summary, relationships have been established among the phase margin Φ_m and the gain crossover frequency ω_{gc} (obtained from the Bode plots of the open-loop frequency transfer function) and the damping ratio ζ, the undamped natural frequency ω_n, the peak amplitude ratio M_p, and the peak frequency ω_p of the closed-loop system. These relationships are exact only for a simple second-order system with unity feedback; they are approximate for all other systems. If these other systems have a dominant pair of roots, the approximations are usually good. *If there is no dominant pair, the approximations may be poor and very misleading.*

The AR_{db} plot of the open-loop transfer frequency function of a unity feedback system also provides information on the steady-state accuracy of a system. In Fig. 8–5–5 are shown the AR_{db} plots of three minimum phase open-loop frequency transfer functions. In the first plot the zero slope at low frequencies means that this system is type 0; the position error constant in decibels K_{pdb} is the AR_{db} value at these frequencies. In the second plot the -20 db/decade slope at low frequencies reveals that there is a single pole at the origin and that the system is type 1. If the low-frequency line is extended until it intercepts the 0-db line at ω_1, the velocity error constant K_v is numerically equal to ω_1. The third system is a type-2 system since the low-frequency slope of -40 db/decade indicates two integrations or two poles at the origin. The acceleration error constant K_a is the square of the frequency ω_1 at which the extrapolated low-frequency line intercepts the 0-db line.

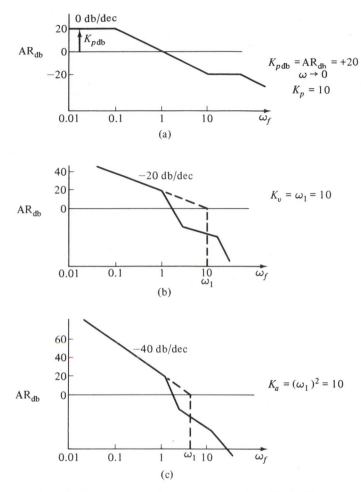

Figure 8-5-5. System type and error constants for unity feedback systems: (a) type 0; (b) type 1; (c) type 2.

*8-6 OTHER FREQUENCY RESPONSE TECHNIQUES

There are various techniques for plotting Nyquist diagrams in different ways so as to yield information on system stability and performance and to aid in the design of an acceptable system. There are such techniques as inverse polar plots [the polar plot of $P(j\omega)/KZ(j\omega)$], M and N contours, and Nichols charts. The M and N contour method is based upon contours of constant amplitude ratio M and constant phase angle N superimposed upon the Nyquist diagram, either normal or inverse. Nichols charts are similar in approach but have a rectangular coordinate system using the AR and Φ values of the open-loop frequency transfer function as the ordinates. Such specialized techniques are all based on the assumption of unity feedback and are not often used these days. In fact, the Nyquist diagram itself is seldom used; Bode plots and root locus diagrams are simpler and more effective for preliminary analysis and design. For final and precise design, numerical methods in conjunction with the computer are preferred.

There is a graphical technique for finding the system frequency response from the plotted Nyquist diagram of the open-loop frequency transfer function. If the system has unity feedback, then the closed-loop frequency transfer function can be written as

$$W(j\omega) = \frac{G(j\omega)}{1 + G(j\omega)} = Me^{j\Phi_{CL}} \tag{8-6-1}$$

where

$$M = AR_{CL} = \frac{|G(j\omega)|}{|1 + G(j\omega)|} \tag{8-6-2}$$

and

$$\Phi_{CL} = \Phi_n - \Phi_d \tag{8-6-3}$$

In Eq. (8–6–3) Φ_n is the argument of $G(j\omega)$ and Φ_d is the argument of $1 + G(j\omega)$.

Use the Bode plots of $G(j\omega)$ to plot the Nyquist diagram of $G(j\omega)$ as indicated in Fig. 8–6–1. At any particular input frequency, such as ω_1, the values necessary to calculate M and Φ_{CL} for that particular frequency can be measured directly from the Nyquist plot. $|G(j\omega)|$ is the length of the vector drawn from the origin to ω_1, and $|1 + G(j\omega)|$ is the length of the vector drawn from the -1 point to ω_1. The angles Φ_n and Φ_d are as shown, taken to be positive in the clockwise direction. By taking a sufficient number of frequency points, the system frequency response can be specified.

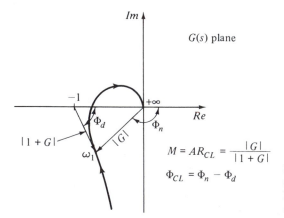

$$M = AR_{CL} = \frac{|G|}{|1 + G|}$$

$$\Phi_{CL} = \Phi_n - \Phi_d$$

Figure 8–6–1. The closed-loop frequency response from the open-loop Nyquist diagram of a unity feedback system.

This technique can be used for systems with non-unity feedback, but more effort is involved. The system is treated as a unity feedback system in series with an open-loop transfer function, as shown in Fig. 8–6–2. The system transfer function can now be written as

$$\frac{c}{r}(s) = \left(\frac{GH}{1 + GH}\right)\left(\frac{1}{H}\right) = Me^{j\Phi_{CL}} \tag{8-6-4}$$

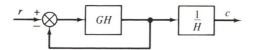

Figure 8-6-2. A non-unity feedback system represented by an equivalent unity feedback system.

The expressions for M and Φ_{CL} are

$$M = \left| \frac{GH}{1 + GH} \right| \cdot \left| \frac{1}{H} \right| \tag{8-6-5}$$

and

$$\Phi_{CL} = \Big/\frac{GH}{1 + GH} - \Big/\frac{1}{H} \tag{8-6-6}$$

The Nyquist diagram of GH, in conjunction with the graphical technique of the preceding paragraphs, can be used to find the magnitude and angle associated with the $GH/(1 + GH)$ function, which can be properly combined with the magnitude and angle of H to obtain the system frequency response.

The system frequency response of both unity and non-unity feedback systems can also be obtained graphically from a root locus diagram with the closed-loop zeros superimposed. This is an extension of the graphical method of Sec. 7-4 used to construct Nyquist diagrams. Consider the non-unity feedback system of Fig. 8-6-3a and its root locus for variable K in Fig. 8-6-3b. K must be specified in order to determine the frequency response; the value K_1 might be established by setting a minimum error constant or by specifying the damping

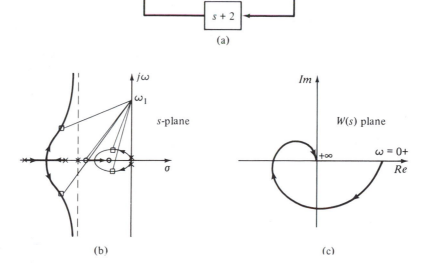

Figure 8-6-3. System (closed-loop) frequency response by graphical means; (a) block diagram; (b) root locus; (c) polar plot.

ratio of the dominant pair of roots. With K_1 specified, the four roots are located along with the single closed-loop zero at $s = -4$. As s takes on values along the positive imaginary axis, there will be five vectors to be considered. The general shape of the polar plot of the frequency response is shown in Fig. 8–6–3c. With K specified and the roots located, it would have been much easier to construct the Bode plots directly rather than use this graphical technique.

Bode plots of the frequency response of a system will frequently indicate which roots, and thus which transient modes, are dominant. Consider a fourth-order system with two underdamped transient modes, one mode with $\omega_n \cong 0.1$ and the other with $\omega_n \cong 1.0$. The M_{db} plots for two different input-output relationships are sketched in Fig. 8–6–4. The first sketch indicates that the lower-frequency mode ($\omega_n = 0.1$) is dominant for this transfer function, whereas the higher-frequency mode dominates in the second sketch.

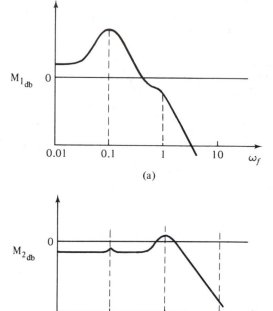

(a)

(b)

Figure 8–6–4. Two frequency responses of a fourth-order system: (a) lower-frequency mode dominant; (b) higher-frequency mode dominant.

When working in the frequency domain, it is essential to remember that the Nyquist diagrams and Bode plots of the $KZ(j\omega)/P(j\omega)$ function do *not* represent the frequency response of the system itself.

PROBLEMS

8–1. Construct asymptotic Bode plots of the $KZ(s)/P(s)$ functions listed below. Find the gain margin GM and the phase margin Φ_m and describe the stability of each system. For conditionally stable systems find the critical values of K and compare them with those obtained by the Routh-Hurwitz criterion.

(a) $\dfrac{100}{s}$

(b) $\dfrac{200}{s(s + 20)}$

(c) $\dfrac{20}{s(s + 1)(s + 10}$

(d) $\dfrac{5000(s + 5)}{s^2(s + 1)(s + 10)(s + 50)}$

(e) $\dfrac{1000(s + 1)}{s^2(s^2 + 14s + 100)}$

(f) $\dfrac{1.11(s^2 + 3s + 9)}{s^3(s^2 + 8s + 100)}$

8-2. From the asymptotic Bode plots of Prob. 8–1 construct a gain-phase plot for each systems and use it to determine the stability.

8-3. For each of the systems whose open-loop transfer function appears below, construct the asymptotic Bode plots for an arbitrary K, and sketch the Nyquist diagram and root locus. Describe the system stability in terms of each sketch. Find any critical values of K from the Bode plots.

(a) $GH = \dfrac{Ks}{(s^2 + 0.8s + 1)(s + 10)^2}$

(b) $GH = \dfrac{K(s + 2)(s + 10)}{s^2(s + 4)(s + 20)^3}$

(c) $GH = \dfrac{K(s + 10)(s + 20)}{s^2(s + 5)(s + 100)}$

(d) $GH = \dfrac{K(s + 2)(s + 20)^2}{s^2(s + 10)(s + 15)(s + 50)(s + 1000)}$

8-4. For each of the $KZ(s)/P(s)$ functions given below, construct the asymptotic Bode plots, find GM and Φ_m, and describe the stability of the system. Then add a zero, $(s + 1)$, to the Bode plots. Find the new GM and Φ_m values and compare the stability with and without the added zero.

(a) $\dfrac{100}{s^2(s + 100)}$

(b) $\dfrac{107}{(s^2 + 10s + 100)(s + 100)^2}$

(c) $\dfrac{27 \times 10^4}{(s + 3)^3(s + 10)}$

(d) $\dfrac{100(s + 2)}{s^3(s + 20)}$

8-5. For each of the unity feedback systems whose G is given below, construct the asymptotic Bode plots and find the approximate ζ and ω_n values of the dominant roots. Sketch the root locus and see if the results from the Bode plots appear to be reasonable.

(a) $\dfrac{60}{s(s + 10)}$

(b) $\dfrac{100}{s^2 + 10s + 100}$

(c) $\dfrac{200}{s(s + 5)(s + 10)}$

(d) $\dfrac{500(s + 1)}{s(s + 5)^2(s + 10)}$

8-6. (a) Construct Bode plots for a unity feedback system with

$$G = \frac{25A}{s(s^2 + 9s + 25)}$$

Choose the amplifier gain A that will yield $\zeta = 0.5$ for the dominant pair of roots. What is ω_n?

(b) Using the results of (a) write the lowest-order transfer function that will approximately represent the system.

(c) Using the transfer function of (b) construct a Bode plot of the system amplitude ratio (M_{db}).

8-7. Same as Prob. 8-6 but with

$$G = \frac{5 \times 10^3 A(s + 1)}{s(s + 5)(s + 10)(s + 100)}$$

8-8. Same as Prob. 8-6 but with

$$G = \frac{3 \times 10^4 A}{s(s + 5)(s + 20)(s + 100)}$$

8-9. You have received two pieces of equipment with the experimental frequency response data listed below.

ω_f (rad/sec)	AR (db)	Φ (deg)
Equipment A		
0.1	0.01	−4
1	0.73	−45
5	−18.6	−200
10	−34.8	−232
20	−52	−250
50	−76	−262
Equipment B		
0.01	14	−89
0.1	−6	−85
0.5	−19	−69
1	−23	−56
5	−27	−66
10	−31	−104
20	−39	−157
40	−53	−205
100	−74	−242
500	−116	−264

    (a) If a negative and unity feedback loop were closed around each piece, would the resulting system be stable? Why?

    (b) Find an approximate transfer function for each piece of equipment.

8–10. Two unity feedback systems have the open-loop AR_{db} plots shown in Fig. P8–10. The open-loop TFs are minimum phase.

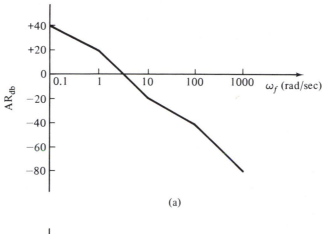

(a)

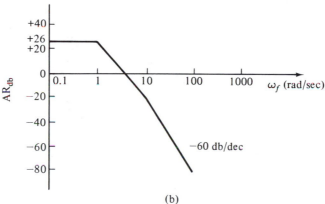

(b)

Fig. P8–10

    (a) Using the Bode plots determine the type of each system and find the error constants. Find ζ and ω_n.

    (b) Find the open-loop transfer function and determine the stability of each system.

    (c) Write the system (closed-loop) transfer function for each system.

8–11. A vibrometer is mounted on the axle of an automobile to measure the roughness of a road and has the transfer function

$$\frac{e_{out \text{ (volts)}}}{x_{in \text{ (inches)}}} = \frac{2.5s^2}{s^2 + 3.5s + 25}$$

    (a) Construct the asymptotic Bode plots.

    (b) Would the vibrometer measure individual potholes and bumps? Explain.

    (c) On what type of roads would this vibrometer be most effective?

8–12. A direct-reading accelerometer has the transfer function

$$\frac{e_{out\ (volts)}}{a_{in\ (ft/sec^2)}} = \frac{0.03}{s^2 + 2s + 1}$$

(a) Sketch the Bode plots.

(b) Would you use this accelerometer to measure accelerations with frequencies greater than 2 Hz? Explain.

9

System Compensation
and Controller Design

9–1 INTRODUCTION

System compensation* is the process of designing a controller that will produce an acceptable transient response while maintaining a desired steady-state accuracy. These two design objectives are conflicting in most systems, since small errors imply high gains but high gains reduce system stability and may even drive the system unstable. *Compensation* may be thought of as the process of increasing the stability of a system without reducing its accuracy below minimum acceptable standards. The three basic methods of compensation are the reduction of gain, the reduction of component time constants, and the addition of poles and zeros. Since gain reduction reduces system accuracy and time constant reduction means larger or more expensive components, neither method is desirable under ordinary circumstances and will not be discussed further.

A feedback control system may be thought of as a filter with two basic design objectives. The first is to pass all command inputs without error and with a minimum of dynamics, and the second is to block all disturbances (unwanted inputs), again with a minimum of dynamics. These are idealized objectives and how closely they are approached is a measure of the "goodness" of a control system and of its controller. A controller itself may also be thought of as filter (or signal conditioner) or as a mechanism for modifying the dynamics of the system so as to meet a set of design and performance criteria.

The generalized control system of Fig. 9–1–1a can be represented, for command inputs only, by the block diagram of Fig. 9–1–1b. *G* is the product of the transfer functions of the plant and of the forward elements of the controller, and *H* is normally part of the controller. In Fig. 9–1–1c the forward and feedback

---

* System compensation is sometimes referred to as *system equalization*.

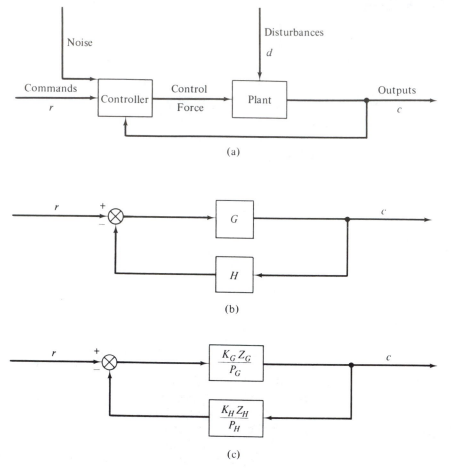

Figure 9–1–1. (a) The generalized feedback control system: (b) and (c) disturbance-free block diagrams.

transfer functions are written in the $KZ(s)/P(s)$ form. The system transfer function can be written as

$$W(s) = \frac{K_G Z_G / P_G}{1 + K_G K_H Z_G Z_H / P_G P_H} = \frac{K_G Z_G P_H}{P_G P_H + K_G K_H Z_G Z_H} \qquad (9\text{--}1\text{--}1)$$

For unity feedback systems the generalized error constant K_e is directly proportional to K_G; the larger the error constant, the smaller the appropriate steady-state error. For non-unity feedback systems, there is generally only one value of K_H that minimizes the steady-state errors, and the influence of K_G is also lessened. Therefore, when adding poles and zeros, we must be aware of any changes in K_G and K_H and consider the possible effects of such changes upon the accuracy and transient behavior of the system.

By examining the right side of Eq. (9–1–1) we can generalize about the effects of adding poles and zeros to the forward and feedback transfer functions. Zeros Z_G added to the forward path will have a stabilizing influence upon the roots of the characteristic equation (see Sec. 6–4), but since they are also closed-loop zeros, they will tend to destabilize the overall transient response (see Sec. 4–5). Fortunately, the stabilizing influence usually prevails. Forward poles P_G affect only the roots and are destabilizing. The feedback zeros Z_H affect only the roots since they are not closed-loop zeros; they are a stabilizing influence. The feedback poles P_H will destabilize the roots, and since they are also closed-loop zeros, they will also destabilize the overall transient response.

These generalizations apply only to the response to command inputs. Disturbances and inputs that enter the system differently than the command inputs will have transfer functions with different numerator functions. Disturbance inputs will be discussed in Sec. 9–6.

Compensation may be described in terms of where it is introduced into the system. It is called *series compensation* if in the forward path (poles and zeros added to *G*), *parallel compensation* if in the feedback path (poles and zeros added to *H*), and *combination* or *series-parallel compensation* if poles and zeros are added to both *G* and *H*. In addition, there can be *inner-loop compensation* that is to be considered a separate category; *stability augmentation* is a special type of inner-loop compensation. There may also be *feedforward compensation* for handling external disturbances that enter at the plant. Compensation is also described as *passive* if passive elements such as *RLC* networks are used, or *active* if active elements such as amplifiers, tachometers, and integrators are used to achieve the desired performance.

The selection of appropriate and suitable compensation is the essence of the design of a control system. There are two basic controller design approaches: synthesis and analysis.

The synthesis approach starts with the desired system performance rather rigidly specified and directly determines the unique compensation required. The inverse root locus method of Sec. 6–8 is one example of the synthesis approach. Synthesis is extensively employed with state variables and modern control theory. Compensation by synthesis is an attractive concept but does have its drawbacks; specification of the desired performance is not a trivial task, and mechanization of the synthesized compensation may be impossible or more complex and expensive than necessary.

The analysis approach on the other hand is a "cut and try," even intuitive, process of selecting suitable compensation. The end result should be an acceptable but hardly unique control system. The basic principles of the analysis approach will be described in the following sections using both root locus and frequency response techniques. No attempt will be made to specify design criteria as such. Keep in mind the general objectives of small steady-state errors (high gain), small peak overshoots (large ζ, small M_p), low rise times (small ζ, large ω_n), low settling times (large $\zeta\omega_n$), and small bandwidths (low ω_n). These broad criteria indicate that the analysis approach relies heavily on the dominant roots concept—the approximation of a system by a simple second-order system.

9–2 PASSIVE SERIES COMPENSATION

To illustrate the principles of passive series compensation we shall use the servomechanism of Fig. 9–2–1, where G_c represents the compensation transfer function. When G_c is equal to unity, we consider the system to be uncompensated. The uncompensated system transfer function for a command input is

$$W(s) = \frac{\dfrac{AK_m}{s(0.1s + 1)(0.05s + 1)}}{1 + \dfrac{AK_m}{s(0.1s + 1)(0.05s + 1)}} \tag{9–2–1}$$

Since this is a type-1 unity feedback system, $E_{p_{ss}}$, the steady-state error due to a position step input, will be zero, but $E_{v_{ss}}$, the steady-state error in position due to a velocity step (ramp) input, will be finite:

$$E_{v_{ss}} = \frac{v_0}{K_v} = \frac{v_0}{AK_m} \tag{9–2–2}$$

where the velocity error constant K_v is obviously equal to AK_m.

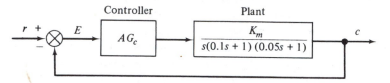

Figure 9–2–1.   A position servomechanism.

Equation (9–2–2) is valid only if the system is stable. To determine stability we sketch the root locus in Fig. 9–2–2 for a variable A using

$$G = K\frac{Z(s)}{P(s)} = \frac{200AK_m}{s(s + 10)(s + 20)} \tag{9–2–3}$$

and see that the system is conditionally stable with respect to A. For each value of $K_v(AK_m)$ there will be a corresponding value of ζ, the damping ratio of the second-order transient mode. As K_v is increased to reduce the steady-state error and improve the accuracy, ζ will be decreased, as indicated on the root locus sketch. When K_v is equal to 30, the system will be marginally stable; larger values of K_v will make the system unstable.

If there is no K_v-ζ combination that will satisfy the performance criteria, we attempt to achieve that desired combination by adding compensation. One simple approach is to add a passive lead network to the controller. The simplest lead network is shown in Fig. 3–3–3 and has the transfer function

$$G_c(s) = \frac{s + 1/\alpha\tau}{s + 1/\tau} \qquad \alpha > 1 \tag{9–2–4}$$

This network adds a zero, which is also a closed-loop zero of this particular transfer function, and a pole to our root locus sketch. The zero is closer to the imaginary axis than the pole; a zero alone would have been more effective but cannot be obtained with a passive network. With this compensation,

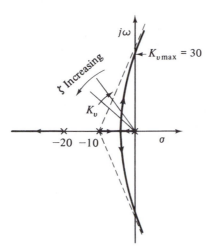

Figure 9–2–2. Root locus sketch of the system of Fig. 9–2–1 without compensation, i.e., $G_c = 1$.

$$G = K\frac{Z(s)}{P(s)} = \frac{200AK_m(s + 1/\alpha\tau)}{s(s + 10)(s + 20)(s + 1/\tau)} \qquad (9\text{--}2\text{--}5)$$

K is unchanged and equal to $200AK_m$; K_v, however, is now equal to AK_m/α. Since $\alpha > 1$, the amplifier gain of the compensated system must be increased to match K_v for the uncompensated system.

It can be shown that the larger α is (the farther apart the added pole and zero), the more effective the compensation is; the maximum value of α is limited by component values and impedance matching. With $\alpha = 10$, a common choice, let us try $\tau = 1/50$ sec. Equation (9–2–5) becomes

$$G = K\frac{Z(s)}{P(s)} = K\frac{(s + 5)}{s(s + 10)(s + 20)(s + 50)} \qquad (9\text{--}2\text{--}6)$$

where K is equal to $200AK_m$ and K_v is equal to $AK_m/10$.

The compensated root locus is sketched in Fig. 9–2–3a. The system is still conditionally stable, but the root locus has been shifted to the left away from the imaginary axis. This shift to the left has increased the upper limit of K_v for stable operation from 30 to 48.5. As a consequence, the damping ratio for a given error constant can be increased, reducing the peak overshoot and number of oscillations. The undamped natural frequency will also be increased somewhat, thus reducing the rise and settling times but increasing the bandwidth. Alternatively the damping ratio can be held constant and the error constant increased, thereby reducing the steady-state error but increasing ω_n.

In addition to shifting the root locus to the left, the compensation has added a real root near the origin, and thus a slowly decaying transient mode is added, as well as a closed-loop zero that is also close to the origin. The effects of this root and closed-loop zero upon the transient response must be considered.

The added root and closed-loop zero may be moved away from the origin by selecting a smaller τ. The root locus for $\tau = 1/150$ sec (which places the added zero to the left of the pole at -10) is sketched in Fig. 9–3–2b. The maximum value of K_v for stability has been increased to 184, and the added real

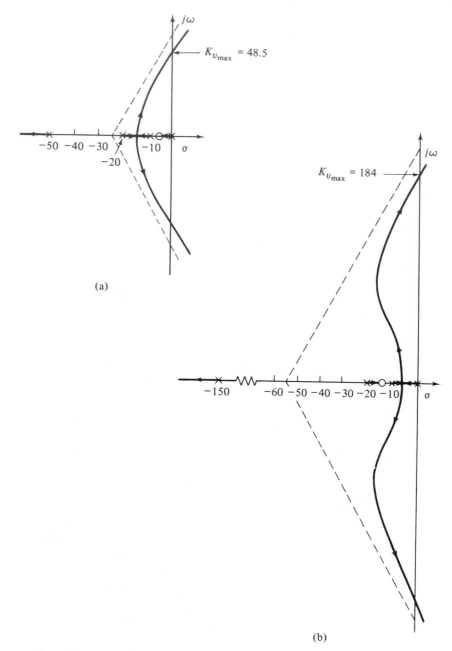

Figure 9–2–3. Root locus sketches of the system of Fig. 9–2–1 with passive lead compensation with $\alpha = 10$: (a) with $\tau = 1/50$ sec; (b) with $\tau = 1/150$ sec.

root, with a faster response time, is farther away from the imaginary axis. Although it may appear that the smaller value of τ is better, the actual choice requires knowledge of the design and performance criteria along with root locus and Bode plots and complete time responses.

Compensation is often easier to visualize and specify in the frequency domain than in the Laplace domain. The frequency transfer function of the lead* network can be written as

$$G_c(j\omega) = \frac{1}{\alpha} \frac{(j\alpha\omega\tau + 1)}{(j\omega\tau + 1)} \qquad \alpha > 1 \qquad (9\text{–}2\text{–}7)$$

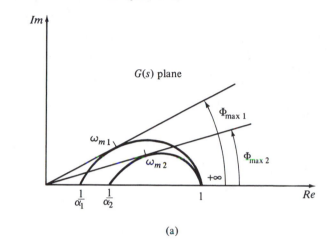

(a)

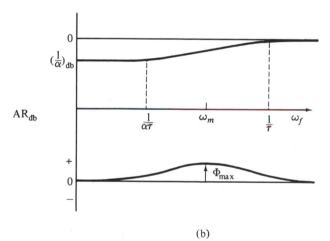

(b)

Figure 9–2–4.   The passive lead network

$$G_c = \frac{1}{\alpha} \left(\frac{j\alpha\omega\tau + 1}{j\omega\tau + 1} \right):$$

(a) polar plot; (b) Bode plots.

The polar plot and Bode plots of this function are shown in Fig. 9–2–4. It is apparent that the maximum phase lead Φ_{max} that can be obtained from the network is directly proportional to α, and that the AR or gain is reduced at low

* A lead network is often called a phase lead network, particularly in the frequency domain.

frequencies. The relationships among $\Phi_{\max}$, α, τ, and ω_m are

$$\Phi_{\max} = \sin^{-1}\frac{\alpha - 1}{\alpha + 1} \tag{9-2-8}$$

and

$$\omega_m = \frac{1}{\tau\sqrt{\alpha}} \tag{9-2-9}$$

The relative simplicity of compensation in the frequency domain is first illustrated in Fig. 9-2-5, where the Nyquist diagram of the uncompensated system is sketched for a large K_v that makes the system unstable. By adding a lead network with a $\omega_m = \omega_{gc}$ and with a sufficiently large $\Phi_{\max}$, we can bend the Nyquist diagram to pass to the right of the -1 point as shown. The gain will have to be judiciously increased so as to maintain the original K_v, as well as to achieve the Φ_m and degree of stability desired.

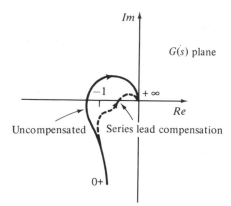

Figure 9-2-5. The Nyquist diagram illustrating series lead compensation.

Bode plots are convenient for determining the characteristics of the compensation network. In Fig. 9-2-6a we have the uncompensated Bode plots of Eq. (9-2-3) for $AK_m = 100$. Here Φ is approximately $-207°$ at $\omega_{gc} = 27$ rad/sec, Φ_m is $-27°$ (less than zero), and thus the system is unstable for this value of AK_m. If we choose α to be equal to 10, Eq. (9-2-8) shows that the $\Phi_{\max}$ value to be added will be approximately $54.9°$. If this phase lead is to be added at $\omega_m = \omega_{gc}$, Eq. (9-2-9) requires $\tau = 0.0117$ sec; the two corner frequencies will be 8.5 and 85 rad/sec. The compensated Bode plots are shown in Fig. 9-2-6b. The system now has a positive phase margin and is stable; however, K_v has been reduced by $1/\alpha$. Increasing A to maintain a desired K_v value will reduce the phase margin; this reduction can be minimized by a wise juggling of α, τ, and ω_m.

A passive lag network may also be used for compensation, but care must be exerted since an unwise choice of τ may make the system unstable. The simple lag network of Fig. 3-3-4 has the transfer function

$$G_c(s) = \frac{1}{\alpha}\frac{s + 1/\tau}{s + 1/\alpha\tau} \qquad \alpha > 1 \tag{9-2-10}$$

If we choose α to be 4 and $\tau = 0.5$ sec and use the uncompensated system of

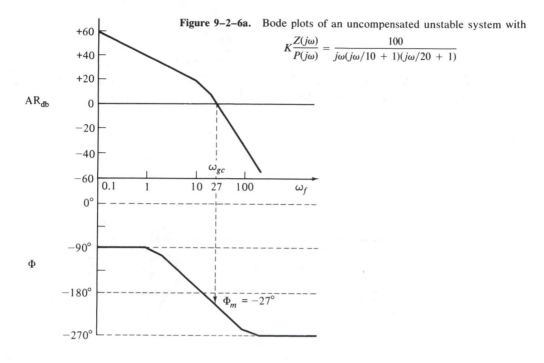

Figure 9-2-6a. Bode plots of an uncompensated unstable system with

$$K\frac{Z(j\omega)}{P(j\omega)} = \frac{100}{j\omega(j\omega/10 + 1)(j\omega/20 + 1)}$$

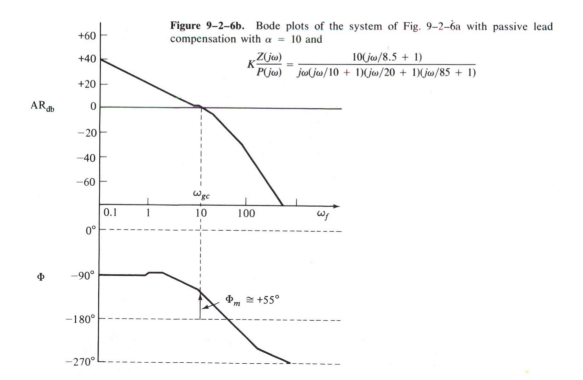

Figure 9-2-6b. Bode plots of the system of Fig. 9-2-6a with passive lead compensation with $\alpha = 10$ and

$$K\frac{Z(j\omega)}{P(j\omega)} = \frac{10(j\omega/8.5 + 1)}{j\omega(j\omega/10 + 1)(j\omega/20 + 1)(j\omega/85 + 1)}$$

Fig. 9–2–1, we have

$$G = K\frac{Z(s)}{P(s)} = \frac{50AK_m(s + 2)}{s(s + 10)(s + 20)(s + 0.5)} \qquad (9\text{–}2\text{–}11)$$

The velocity error constant K_v is unchanged, remaining equal to AK_m. When the lag-compensated root locus sketch of Fig. 9–2–7a is compared with the

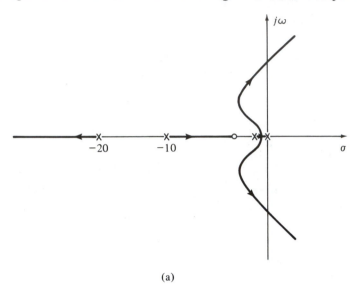

(a)

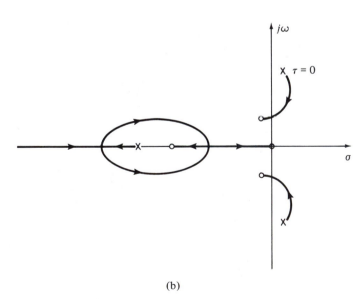

(b)

Figure 9–2–7. Root locus sketches of the system of Fig. 9–2–1 with passive lag compensation with $\alpha = 4$: (a) variable K_v with $\tau = 0.5$ secs; (b) variable τ with $K_v = 50$.

uncompensated root locus sketch of Fig. 9–2–2, the improvement is not obvious but is there; for example, the value of K_v at the imaginary axis is now of the order of 90, versus 30 for the uncompensated system. For a given ζ, K_v can be increased. It can be shown that the undamped natural frequency will be reduced, narrowing the bandwidth but making the system more sluggish. As with lead compensation, another first-order transient mode has been added.

The influence of the magnitude of the lag time constant τ can be illustrated by a root locus sketch for a variable τ. The characteristic equation of the lag-compensated system can be written in the form

$$1 + \frac{\alpha\tau s(s^3 + 30s^2 + 200s + 200AK_m/\alpha)}{(s^3 + 30s^2 + 200s + 200AK_m)} = 0 \qquad (9\text{–}2\text{–}12)$$

Leaving α equal to 4 and setting K_v equal to 50, the root locus of Fig. 9–2–7b shows that the system will be unstable for small values of τ (<0.19). As a rule of thumb, τ should be larger than the largest time constant in the uncompensated system. Increasing τ not only increases the stability of the system for a given K_v but also increases the maximum allowable K_v. For example, increasing τ from 0.5 to 5.0 with an α of 4 increases $K_{v\,max}$ from approximately 90 to approximately 120. A large α is also desirable; increasing α from 4 to 10 with $\tau = 5.0$ increases K_v from 90 to approximately 210.

Bode plots and Nyquist diagrams of the passive lag network would show that lag compensation attenuates (reduces) the gain at high frequencies and requires that the uncompensated system have a substantial phase margin at relatively high frequencies.

Individually applied, both lead and lag compensation have disadvantages. Lead compensation requires additional gain, and lag compensation requires an adequate phase margin and also slows down the system response. The two can be cascaded or combined in a single RC lead-lag (or lag-lead) network as shown in Fig. 9–2–8. Proper selection of the element characteristics will determine the type of network and maximize the advantages and minimize the disadvantages of each. Although the zeros of this network are real, the poles can be real or complex. With a bridged-T network both the zeros and the poles can be

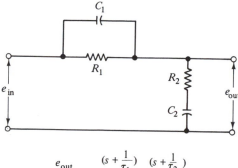

$$\frac{e_{out}}{e_{in}} = \frac{(s + \frac{1}{\tau_1})\ (s + \frac{1}{\tau_2})}{(s + \frac{1}{\alpha\tau_1})\ (s + \frac{\alpha}{\tau_2})}, \alpha > 1$$

Figure 9–2–8. A simple passive lead-lag network.

either real or complex. This is desirable if the network is to be used as a notch filter or to shift poles. A notch filter is very effective in increasing the phase margin in the vicinity of the gain crossover frequency to make the system stable (as portrayed in Fig. 9–2–5) or to increase its robustness. Such frequency response shaping by passive (and active) filters can be extended to include setting the bandwidth and modifying the cutoff rate.

In this section we have used only the simplest of passive networks to illustrate the principles of passive compensation. Complex plants require complex networks. Network synthesis is an important and busy design area in its own right. Compensation networks need not be electrical; they can be mechanical or fluid. Finally, no attempt has been made to describe or discuss the rules and procedures for determining the networks or the parameters that would provide the best compensation.

9–3 ACTIVE COMPENSATION AND MODE CONTROLLERS

The principles of active series compensation will be illustrated by consideration of *mode controllers* frequently used in process control and aerospace applications. We shall use the block diagram of Fig. 9–3–1 with a plant transfer function

$$G_p = \frac{c}{u} = \frac{50}{(s + 10)(s + 20)} \tag{9–3–1}$$

This is a type-0 overdamped plant with no poles at the origin and is typical of many plants and processes. We shall develop the controller transfer function G_c for various active control modes and observe the changes in the response to a unit position step input.

Since the system has unity feedback, the actuating signal ε is the error signal E. We can use the position error constant K_p as a measure of the steady-state accuracy; the larger the K_p, the smaller the error and the greater the accuracy. The system transfer function is

$$W(s) = \frac{G_c G_p}{1 + G_c G_p} \tag{9–3–2}$$

and the error transfer function is

$$\frac{E}{r}(s) = \frac{1}{1 + G_c G_p} \tag{9–3–3}$$

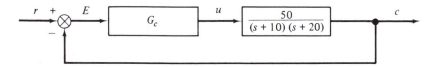

Figure 9–3–1. A control system for process control.

Without any compensation whatsoever G_c will be simply unity and the open-loop transfer function G will be the plant transfer function G_p. The system is type 0; K_p will be 0.25 and $E_{p_{ss}} = 0.8$, which is a large finite error.

The simplest controller is an amplifier with a transfer function

$$G_c = \frac{u}{\varepsilon} = A \qquad (9-3-4)$$

It is referred to as a single-mode controller or a *proportional* (P) *controller* since the control force entering the plant is proportional to the error signal. The open-loop transfer function G will be used to sketch the root locus for a variable controller gain A, where

$$G = \frac{50A}{(s + 10)(s + 20)} \qquad (9-3-5)$$

The root locus in Fig. 9–3–2a shows the system to be a stable second-order system with the damping ratio ζ decreasing as A is increased. Since this is a type-0 system, K_p is found to be equal to $A/4$, and the steady-state position error is

$$E_{p_{ss}} = \frac{4}{4 + A} \qquad (9-3-6)$$

This is a finite error which in process control is called the *offset*. As the gain of the proportional controller is increased, the offset is reduced, but can never be eliminated, and the damping ratio is reduced. The sketches in Fig. 9–3–2b show the effects of increasing A. When A is equal to unity, we have the un-compensated unit step response.

In the preceding section on passive compensation, the uncompensated system had an amplifier. The system, therefore, was actually not uncompensated but had a proportional controller. It is most unusual to see a control system without an amplifier of some sort.

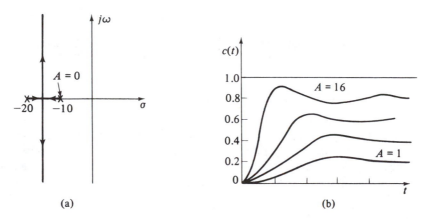

Figure 9–3–2. The system of Fig. 9–3–1 with a proportional controller: (a) root locus for variable A; (b) unit step response for several values of A.

Returning to our illustrative system we shall add an integrator to the controller as schematically shown in Fig. 9–3–3a. We now have a two-mode *proportional-integral* (PI) *controller* with the transfer function

$$G_c = \frac{A(s + 1/\tau_I)}{s} \tag{9–3–7}$$

where $1/\tau_I$ is the gain of the integrator; τ_I has the dimensions of time and is the *integrator time constant*. The open-loop transfer function is now

$$G = \frac{50A(s + 1/\tau_I)}{s(s + 10)(s + 20)} \tag{9–3–8}$$

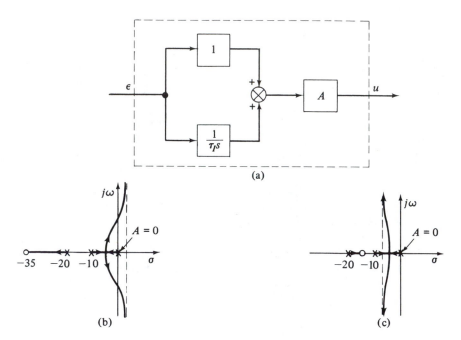

Figure 9–3–3. The system of Fig. 9–3–1 with a proportional-integral controller: (a) controller block diagram; (b) root locus for variable A with $\tau_I = 1/35$; (c) with $\tau_I = 1/15$.

A pole at the origin has been added, making the system type 1; consequently, K_p is infinite and the offset becomes zero.* The added pole, however, reduces the stability of the system, and the unit step response with PI control will in general be more oscillatory, having larger overshoots than with proportional control alone. This destabilizing effect can be reduced by increasing τ_I in order to move the zero closer to the imaginary axis.† The root locus sketches for

---

* In process control the integral mode may be referred to as *reset*.

† This zero is also a closed-loop zero for this transfer function, and its effect on the early portion of the time response must also be considered.

$\tau_I = \frac{1}{35}$ sec and for $\tau_I = \frac{1}{15}$ sec are shown in Fig. 9-3-3b and c. Compare these sketches with each other and with Fig. 9-3-2a.

To reduce the peak overshoot, number of oscillations, and settling time, a third mode, derivative control, is added to produce a *proportional-integral-derivative* (PID) *controller*, sometimes called a three-mode controller. The schematic of a PID controller is shown in Fig. 9-3-4a. Its transfer function is

$$G_c = \frac{\tau_D A \left(s^2 + \dfrac{1}{\tau_I} s + \dfrac{1}{\tau_I \tau_D} \right)}{s} \tag{9-3-9}$$

We see from this equation that we have added another zero, which is also a closed-loop zero. If τ_I and τ_D are chosen to produce a pair of real zeros, the root locus will have the form shown in Fig. 9-3-4b; the system is heavily overdamped and undoubtedly would be too sluggish to be useful. If τ_I and τ_D are decreased to make the pair of zeros a complex conjugate pair, the root locus will have the shape shown in Fig. 9-3-4c. Notice that there is a minimum damping ratio for the dominant pair of roots. Since derivative control senses the rate of change of the actuating signal, it acts to reduce the oscillatory behavior of the output and to speed up the response of the system.

The shape of the transient response with a PID controller is obviously determined by the values selected for the three parameters A, τ_D, and τ_I in Eq.

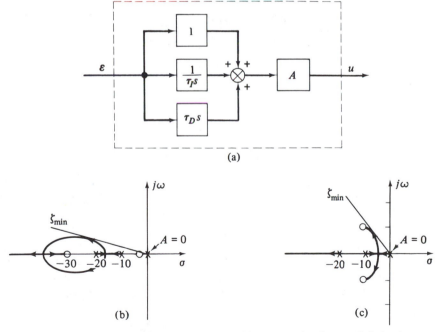

(a)

(b)   (c)

Figure 9-3-4. The system of Fig. 9-3-1 with a proportional-integral-derivative controller: (a) controller block diagram; (b) root locus with real zeros; (c) with complex zeros.

(9–3–9). It should be pointed out that the various active controllers have been represented by convenient schematics and that a common practice is to write the transfer function of a PID controller as

$$Gc = K_p + K_D s + \frac{K_I}{s} \tag{9-3-10}$$

The process of selecting the controller parameters to meet a set of performance criteria is known as *controller tuning*. Several methods often used in process control for first estimates are based on the value of K_p that results in marginal stability with only the proportional mode activated. Among these methods are the Ziegler-Nichols settings, the damped oscillation method, and the reaction curve method. Current practice is to use values obtained from or refined by computer simulations, either analog or numerical.

When an offset, or position step input error, is acceptable,* a *proportional-derivative* (PD) *controller* can be effective. Its schematic is shown in Fig. 9–3–5a, and it has a transfer function

$$\frac{u}{\varepsilon} = A(\tau_D s + 1) \tag{9-3-11}$$

The root locus for the system of Fig. 9–3–1 with a PD controller is shown in Fig. 9–3–5b. The location of the zero must be carefully chosen to avoid making the system too stable and thus too slow to respond.

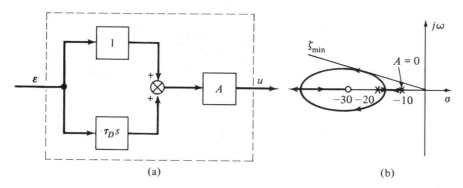

(a)   (b)

Figure 9–3–5.   The system of Fig. 9–3–1 with a proportional-derivative controller: (a) controller block diagram; (b) root locus with $\tau_D = 1/30$.

A comparison of the general shapes of the unit step response for the various active controllers for the same A is shown in Fig. 9–3–6.

The active elements used in controllers are varied; they can be electronic, mechanical, pneumatic, magnetic, and fluidic. Active networks can be used to replace the passive networks of the preceding section, although they are generally

---

* Or there is integration elsewhere in the forward path.

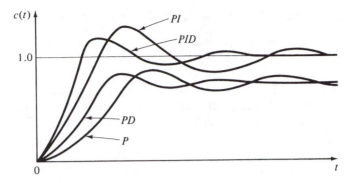

Figure 9–3–6.   Typical unit step responses for various types of active controllers with same amplifier gain A.

more expensive. The proportional-derivative controller adds a zero only and can be considered a pure phase-lead network. A pure phase-lag network is shown in Fig. 9–3–7 where a unity feedback loop is closed around an integrator. A lead-lag or lag-lead network can be formed as shown in Fig. 9–3–8 where the network characteristics are determined by the selection of the two independent time constants.

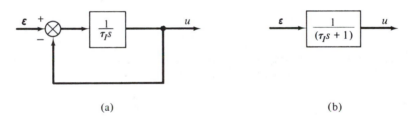

(a)   (b)

Figure 9–3–7.   An active phase-lag network: (a) block diagram; (b) transfer function.

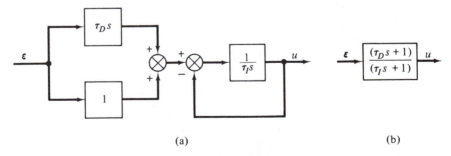

(a)   (b)

Figure 9–3–8.   An active lead-lag or lag-lead network: (a) block diagram; (b) transfer function.

9–4 PARALLEL AND INNER-LOOP COMPENSATION

Parallel compensation is attractive in that the zeros in the feedback loop are *not* closed-loop zeros; however, the poles are. Consequently, parallel compensation accentuates the characteristics of the compensation network being used; a lead network in the feedback path has a stronger stabilizing influence than one in the forward path. Another attraction may be the generally greater accessibility of both the output and the feedback path to the designer. The major disadvantages of parallel compensation are the high sensitivity of the transfer function to changes and variations in the feedback parameters, as shown in Sec. 5–6, and the degradation of the steady-state accuracy of the system by non-unity feedback, as stated in Sec. 5–5.

The addition of the passive network of Sec. 9–2 to the feedback path is straightforward. The root locus sketches will remain unchanged, but the system transient responses will differ because of the changes in the closed-loop zeros, and the error constants will decrease. The same is true for combination or series-parallel compensation in which compensation is introduced into both the forward and feedback paths.

Active parallel compensation in the form of rate or tachometer feedback is a frequently used compensation technique that can be quite effective but requires additional amplification. We have used this type of compensation in an inner loop around the drive motor of our tracking system. Without compensation, the motor-load combination has the transfer function

$$\frac{\theta_{\text{out}}}{e_c} = \frac{k_m}{Js^2} \tag{9-4-1}$$

The simplest tracking system possible using this motor-load combination is shown in Fig. 9–4–1 and has the transfer function

$$\frac{\theta_{\text{out}}}{\theta_{\text{cmd}}} = \frac{AK_m}{Js^2 + Ak_m} \tag{9-4-2}$$

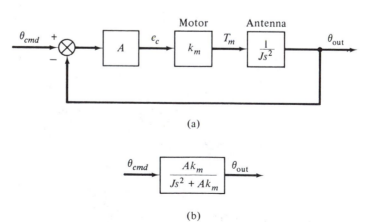

(a)

(b)

Figure 9–4–1. Tracking system with an undamped motor.

This is a marginally stable system, since the pair of complex conjugate roots lies on the imaginary axis for all values of AK_m/J.

If a voltage proportional to the angular velocity of the output shaft is subtracted from the control voltage, we have the two-loop system of Fig. 9–4–2a, which reduces to the unity feedback system of Fig. 9–4–2b. The transfer function of the motor becomes, with feedback,

$$\frac{\theta_{\text{out}}}{e_c} = \frac{k_m}{s(Js + Bk_m)} \tag{9–4–3}$$

The motor has been effectively changed from a constant-acceleration motor to a constant-speed motor with a first-order time lag; the gain and time constant of the motor are established by the value of B. The root locus of the inner loop (the motor) for a variable B is shown in Fig. 9–4–2c. There will always be a root at the origin because of the pole and zero at the origin. If the outer loop with unity feedback were missing, the system would be limitedly stable. This is an example of *stability augmentation* inasmuch as the unstable bare motor-load combination becomes limitedly stable with inner-loop damping and now appears to the system as a limitedly stable constant-speed motor.

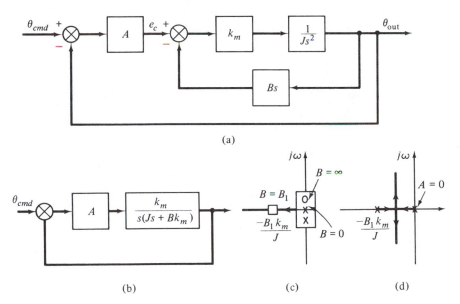

(a)

(b)   (c)   (d)

Figure 9–4–2. The system of Fig. 9–4–1 with inner-loop compensation: (a) block diagram; (b) inner loop reduced; (c) inner-loop root locus for variable B; (d) system root locus for variable A with $B = B_1$.

With the outer loop present in Fig. 9–4–2a, as shown, the system transfer function can be written as

$$\frac{\theta_{\text{out}}}{\theta_{\text{cmd}}} = \frac{Ak_m}{Js^2 + Bk_m s + Ak_m} \tag{9–4–4}$$

The root locus of this stable second-order system of type 1 is sketched in Fig. 9–4–2d for variable A, with B equal to B_1. Notice that the outer-loop pole at $-B_1k_m/J$ is the inner-loop root. This is an example of *inner-loop or minor-loop compensation* in which the characteristics of a component or subsystem are changed to achieve the desired system performance. For this type-1 system K_v is equal to A/B.

If the rate signal is fed back around the amplifier as well, as shown in Fig. 9–4–3a, the compensation becomes *outer-loop or parallel, compensation*. Comparing the reduced block diagram of Fig. 9–4–3b with that of Fig. 9–4–1, we see that we have added a single real zero Z_H to the feedback path. The system transfer function can be written as

$$\frac{\theta_{out}}{\theta_{cmd}} = \frac{Ak_m}{Js^2 + Ak_mBs + Ak_m} \tag{9–4–5}$$

The root locus sketches for variable A and for variable B are shown in Fig. 9–4–3c and d. Comparing Fig. 9–4–3c with Fig. 9–4–2d, we see that outer-loop compensation results in a more stable system; this effect on the damping ratio can be seen by comparing the coefficient of the second term in the characteristic function of Eq. (9–4–5) with the corresponding coefficient in Eq. (9–4–4). However, K_v is now equal only to $1/B$; the amplifier gain A no longer affects the accuracy.

In this example, inner-loop compensation has increased the stability of both the motor and the system by moving a root of the motor, which is also a pole

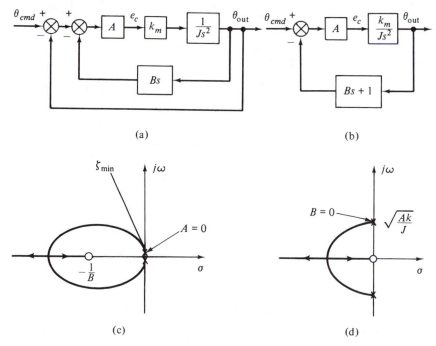

(a)   (b)

(c)   (d)

Figure 9–4–3. The system of Fig. 9–4–1 with outer-loop compensation: (a) and (b) block diagrams; (c) system root locus for variable A; (d) for variable B.

of the system, away from the origin of the *s* plane. Outer-loop compensation leaves the stability of the motor unchanged, increases the stability of the system by adding a real zero to the system feedback loop, and greatly reduces the accuracy. If the outer-loop were broken for any reason, the first system would degrade to a limitedly stable condition; the second system would become unstable. Furthermore, in using outer-loop compensation, care must be taken to avoid reducing the type of the system (and thus the accuracy) by introducing non-unity feedback. For these reasons, outer-loop compensation is not generally recommended.

*9–5 POLE CANCELLATION AND SHIFTING

When a plant has a pair of complex conjugate poles in the left half plane near the imaginary axis, it might be tempting to cancel them with a corresponding pair of controller zeros from either a second-order passive network or a PID controller. Resist the temptation and do *not* cancel.

First of all, there is really no such thing as pole cancellation. Placing a zero on top of a pole creates a *stationary root* that is independent of changes in the parameter of interest. This stationary root is still part of the system and is canceled only when there is an identical closed-loop zero in the appropriate transfer function. Such apparent cancellation occurs frequently in multivariable plants when a plant zero cancels a plant pole in one or more of the input-output transfer functions [see Eq. (2–6–4a)]; most physical plants are multivariable in that there are usually one or more inputs affecting one or more outputs. Generally, stationary roots that are canceled for command inputs are not canceled for disturbance inputs. When canceled, a stationary root is not observable; when not canceled, it is not controllable.

Second, it is not physically possible to achieve exact cancellation because of modeling uncertainties and approximations and because of uncertainties in and drifting of system hardware parameters. At first glance, whether cancellation is exact or inexact seems to have little effect on the transient response of the system. However, if the added zeros are above and/or to the right of the poles, the roots will be shifted toward the imaginary axis as the gain is increased, and the stability will be decreased. If the poles are near or close to the imaginary axis, the system might even become unstable.

If inexact cancellation of complex poles is to be used as a compensation technique, the complex zeros should be placed relatively close to but sufficiently far to the left of and below the poles so that drift of either the poles or zeros will not produce an unfavorable juxtaposition, i.e., a decrease in stability as the gain is increased. Since the branches connecting the poles and zeros are short, the complex root itself will be close to both the poles and zeros. Consequently, the amplitudes of the transient mode associated with the root should be small, but only when these added open-loop zeros are also closed-loop zeros (when the root and closed-loop zero form a dipole).

As the compensation zeros are moved farther to the left (and usually down), the inexact cancellation technique blends into the *pole shifting* technique, which concentrates on improving the characteristics of the transient mode by moving the root associated with the complex pole away from the imaginary axis and closer to the real axis rather than attempting to supress the undesirable pole(s). The techniques of the preceding sections in this chapter are examples of pole shifting.

Let us illustrate the differences between cancellation and pole shifting by example. Consider a unity feedback system with the type-1 forward transfer function

$$G(s) = \frac{K(s + 0.6)}{s(s + 10)(s^2 + 0.3s + 4)} \qquad (9\text{-}5\text{-}1)$$

where the quadratic term represents a complex plant pole with $\zeta = 0.075$, $\omega_n = 2$ rad/sec, and $t_s = 20$ sec. The root locus for positive values of K, as sketched in Fig. 9–5–1a, shows not only that the system is conditionally

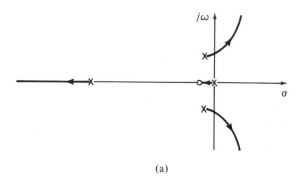

(a)

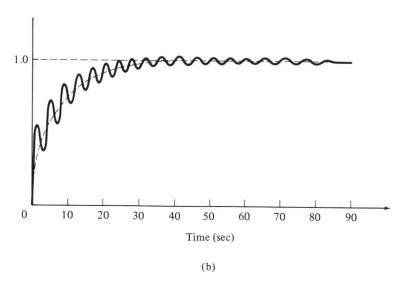

Time (sec)

(b)

Figure 9–5–1. The uncompensated type-1 system with unity feedback and the forward transfer function of Eq. (9–5–1): (a) root locus; (b) unit step time response.

stable but also that the stability decreases rapidly as K is increased. Furthermore, the maximum value of K for stability is only 14.56, which implies a large velocity error. Setting K equal to 10 yields two real roots at -10.1 and -0.12, with a response time of approximately 25 sec for the smaller root and a complex conjugate root with ζ on the order of 0.01 and a settling time of approximately 75 sec. The unit step response of Fig. 9–5–1b shows the lightly damped second-order mode superimposed upon a dominant first-order mode from the smaller real root. This is the uncompensated system.

Let us assume that the pole at the origin in Eq. (9–5–1) was introduced by the integral mode in a PID controller and that compensation will be accomplished by suitable placement of the two zeros of the controller. With exact cancellation, the open-loop transfer function becomes

$$G(s) = \frac{K(s + 0.6)}{s(s + 10)} \qquad (9\text{–}5\text{–}2)$$

and the system becomes overdamped, as shown in the root locus sketch of Fig. 9–5–2a. The system is stable for all positive values of K. As K is increased, the smaller real root approaches the value of the plant zero, decreasing the time constant (and response time) and the influence of the slower mode. Larger values of K also reduce the velocity error.

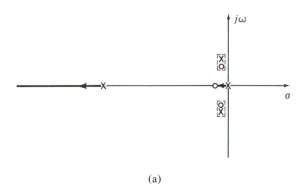

(a)

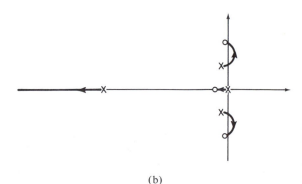

(b)

Figure 9–5–2. Compensated root locus of the system of Fig. 9–5–1: (a) with exact pole cancellation; (b) with undesirable inexact cancellation.

As a first try for inexact cancellation, the compensated open-loop transfer function is

$$G(s) = \frac{K(s + 0.6)(s^2 + 0.15s + 16)}{s(s + 10)(s^2 + 0.3s + 4)} \qquad (9\text{--}5\text{--}3)$$

which places the zeros to the right of and above the complex poles. The root locus sketch of Fig. 9–2–5b shows the system to be conditionally stable and not obviously better than the uncompensated system.

The zeros for the second try at inexact cancellation are placed to the left of and below the complex poles so that the open-loop transfer function becomes

$$G(s) = \frac{K(s + 0.6)(s^2 + 1.25s + 1.36)}{s(s + 10)(s^2 + 0.3s + 4)} \qquad (9\text{--}5\text{--}4)$$

The root locus sketch of Fig. 9–5–3a shows the system to be stable for all positive K. Setting $K = 100$ yields a real root at -108 that can be ignored, a real root at -0.435 which has a first-order response time of approximately 7 sec, and a complex root with $\zeta = 0.23$, $\omega_n = 1.3$ rad/sec, and t_s approximately

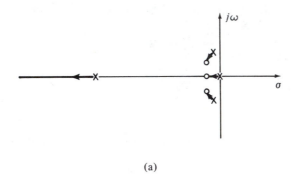

(a)

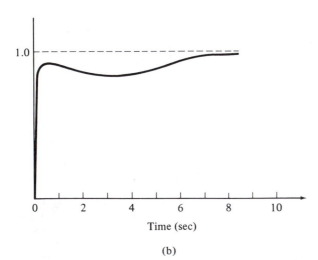

Time (sec)

(b)

Figure 9–5–3. Acceptable inexact pole cancellation of system of Fig. 9–5–1: (a) root locus; (b) unit step time response.

5 sec. The unit step response is sketched in Fig. 9-5-3b and should be compared with the uncompensated step response of Fig. 9-5-1b; notice the difference in the time scales.

Now let us shift the poles by arbitrarily locating the zeros so that

$$G(s) = \frac{K(s + 0.6)(s^2 + 10s + 26)}{s(s + 10)(s^2 + 0.3s + 4)} \tag{9-5-5}$$

We see from the root locus sketch of Fig. 9-5-4a that the system is still stable for all K, but that as K is increased the complex roots move farther away from the imaginary axis than with inexact cancellation. With $K = 100$, there will be a real root at -100, a real root at -0.588 (with a response time of approximately 5 sec), and a complex conjugate root with $\zeta = 0.97$, $\omega_n = 5.2$ rad/sec, and an approximate settling time of 0.6 sec. The response of this system will resemble that of a first-order system whose response time is inversely proportional to K. With this type of compensation there are no "inexactly canceled" roots that might cause trouble with other input-output combinations.

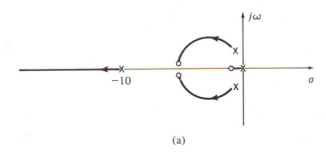

(a)

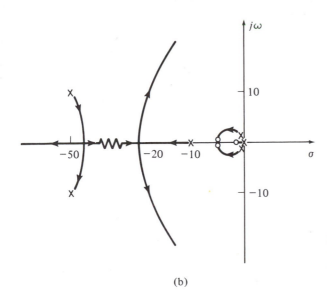

(b)

Figure 9-5-4. Pole shifting compensation of system of Fig. 9-5-1: (a) with zeros only; (b) with zeros and accompanying poles.

Passive zeros can also be used with these compensation techniques. The accompanying poles are sufficiently far away from the imaginary axis to leave the shape essentially unchanged for lower values of the gain but, by introducing two additional excess poles, will make the system conditionally stable (see Fig. 9–5–4b) so that large values of gain can introduce an additional pair of complex conjugate roots with decreasing damping as K is further increased. Consequently, care must be taken to avoid having this pair of roots become a dominant pair. Furthermore, when using either active or passive zeros for compensation, as is done here, the designer must consider the saturation limits of amplifiers, motors, and other components and sensors in selecting the actual gain to be used.

Pole placement (pole assignment) is another compensation design technique in which the desired positions of the system roots (closed-loop poles or eigenvalues) are specified and then the controller is found. The inverse root locus of Sec. 6–8 is a graphical approach to pole placement controller design, and state-variable feedback with the controller in the feedback path is a purely mathematical approach. Both are synthesis techniques with physical and mathematical limitations, and both result in a controller that is unique only for the specified root locations and that may not be physically realizable or practical.

9–6 DISTURBANCE MINIMIZATION

System disturbances can be internal or external. Internal disturbances are noisy signals from or changes in the parameters of the plant and components; their effects are reduced by feedback. External disturbances can be noise entering with commands or physical disturbances entering at the plant itself. Noise, which is usually of higher frequency than commands or signals, may well be sufficiently attenuated by the frequency characteristics and bandwidth of the system; if not, then filters may be needed for signal conditioning. We shall consider only external physical disturbances, which may be outside temperature changes in the case of a process control system or gusts in the case of an aircraft flight control system.

Figure 9–1–1a shows a plant with two inputs and one output. The plant itself is redrawn in Fig. 9–6–1a to emphasize the fact that there are two distinct

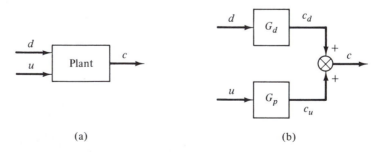

(a)   (b)

Figure 9–6–1. Plant with two inputs and one output: (a) single functional block; (b) block diagram representation.

inputs. Therefore, two transfer functions are required to relate the output of the plant to each of these inputs, as shown in Fig. 9–6–1b, where c_d represents the output for a disturbance input and c_u the output for a control force input.

   With unity feedback, the system with both inputs can be represented by the single block diagram of Fig. 9–6–2a or by the two individual diagrams of Fig. 9–6–2b. Note in Fig. 9–6–2c that G_d is outside the feedback loop and does not affect the stability of the system. For the time being then, let us treat c_d as the undesirable input rather than the disturbance itself. With the component transfer functions written in the $KZ(s)/P(s)$ form, the transformed output can be written as

$$c(s) = \frac{K_c K_p Z_c Z_p\, r(s)}{P_c P_P + K_c K_P Z_c Z_P} + \frac{P_c P_P\, c_d(s)}{P_c P_P + K_c K_P Z_c Z_P} \qquad (9\text{–}6\text{–}1)$$

where each term on the right has the same denominator (the closed-loop char-

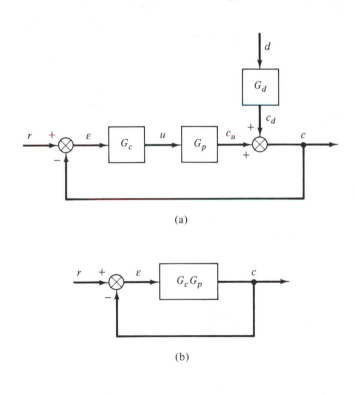

(a)

(b)

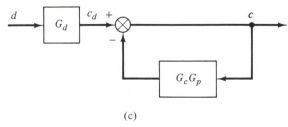

(c)

Figure 9–6–2. Block diagrams of a system with both a command and a disturbance input: (a) with both inputs; (b) with command input only; (c) with disturbance input only.

acteristic function) but a different numerator. The objectives of the controller are to make the first term unity and the second term zero. Equation (9–6–1) shows that, in a general sense, both objectives can be accomplished by setting $P_c P_P$ equal to zero and approached by making $K_c K_P Z_c Z_P \gg P_c P_P$. The former indicates the desirability of having pure integrations in the loop, in either the plant or the controller, and the latter calls for a high loop gain $K_c K_P$. Remember that increasing the type of a system (adding integrations) and increasing the gain often adversely affect the stability and transient response.

Let us return to a unity feedback system with the forward transfer function of Eq. (9–5–5). The disturbance transfer function with $K = 100$ is

$$\frac{c(s)}{c_d(s)} = \frac{1}{1 + G} = \frac{s(s + 10)(s^2 + 0.3s + 4)}{(s + 0.59)(s + 100)(s^2 + 10s + 27)} \qquad (9\text{–}6\text{–}2)$$

Since the system is stable, the presence of the s in the numerator means that for a step in c_d the steady-state value of c (the error) will be zero, as can be verified by the final value theorem. For a unit ramp in c_d, the steady-state value of c will be finite and equal to 0.025 (if the system were type 2, the error would be zero) and is explicitly independent of the value of the loop gain. In Eq. (9–6–2), notice that the poles of the forward transfer function are the closed-loop zeros of this particular transfer function and that the numerator and denominator are of the same order. Consequently, the initial time response will exhibit the effects of the numerator dynamics, and the frequency response will be flat at the higher frequencies. Neither is particularly desirable and, since external disturbances are often random and periodic in nature, the lack of a high cutoff rate may not be acceptable. However, d and not c_d is the real disturbance input. With $G_d(s)$ in the form $K_d Z_d(s)/P_d(s)$, the second term of Eq. (9–6–1) can be written as

$$c(s) = \frac{K_d Z_d P_c P_P \, d(s)}{P_d(P_c P_P + K_c K_P Z_c Z_P)} \qquad (9\text{–}6\text{–}3)$$

Although G_d is dependent on the nature of the plant and of the disturbance and is not easy to define, its excess poles (which may well include one or more of the plant poles) will reduce the order of the numerator of Eq. (9–6–3) and steepen the cutoff rate. Furthermore, it will modify the shape and form of the disturbance input. A step in d will not produce a step in c_d; it is more likely that c_d will be ramplike in nature or a decaying oscillation.

This is passive disturbance minimization or control in that no action is taken by the feedback loop or controller until c_d actually appears and has been sensed. *Feedforward compensation* is a form of active control in which a signal from a measurable disturbance is used to generate an appropriate control force to counteract or mitigate the effects of the disturbance; in other words, to keep c_d from developing or to reduce its magnitude. Figure 9–6–3 indicates how such a compensation scheme might function, where G_a represents an accessible control force generator such as an aircraft elevator actuator for counteracting vertical acceleration arising from a gust or a burner for counteracting an external temperature change. If G_f could be set equal to $G_d/G_a G_c$, which is usually physically un-

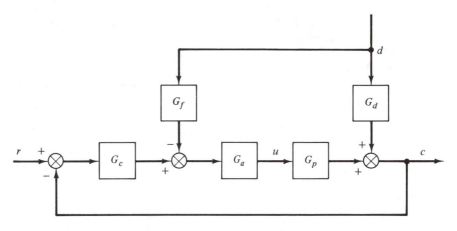

Figure 9-6-3.   Feedforward compensation block diagram.

realizable, the system would be isolated from any effects of d. In practice, simpler forms of G_d, even a simple gain, can be surprisingly effective in minimizing the effects of external disturbances that can be measured with a reasonable degree of accuracy.

Feedforward compensation can also be used in feedback control systems to "decouple" the input-output variables of multivariable plants so that one input controls and affects only one output. An example would be an aircraft flight control system in which the elevator affects only the flight path angle and the throttle affects only the airspeed.

*9-7 THE TRANSPORT LAG

Transport lag* is not a compensation technique; on the contrary, it is a strong destabilizing influence. It is discussed in this chapter for two basic reasons. The first reason is that transport lags, unless inherent in the plant, are usually neglected during the initial analysis and design efforts. The second reason is that there does not appear to be any better place in the book to discuss it.

Transport lag is a time delay in the transmission of a signal or an input; it is a period during which the system is unaware of a change. In manual systems the reaction time of the human operator is a transport lag. The time it takes a driver to realize that a pedestrian has darted out in front of him and then to apply the brakes is a transport lag. If the reaction time of the driver is zero and he or she still hits the pedestrian, obviously he or she will hit him if his or her reaction time is finite. If the driver does not hit the pedestrian when we assume the reaction time is zero, we should then consider the effect of a finite reaction time.

---

* Transport lag is sometimes called *transportation lag* or *dead time*.

Transport lag occurs in physical processes and in automatic control systems when the point of measurement or input is at some distance from the point of control. Consider the portion of a liquid-level control system shown in Fig. 9–7–1a, where the flow rate into the tank is controlled by a valve located a distance d upstream. A change in the tank level will call for a change in the valve setting and thus in the input flow rate. The increase in the flow rate at the valve will not be felt at the tank until T units of time later, where $T = d/v$, v being the velocity of the fluid; this is a transport lag.

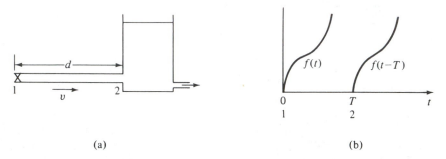

Figure 9–7–1. A transport lag: (a) in a flow rate; (b) effect upon an $f(t)$.

Figure 9–7–1 shows that $f(t)$ appearing at point 1, the valve, at a time arbitrarily set equal to zero will appear at point 2, the tank, T units of time later. This is a shift of T units in the time domain. The second shifting theorem of Sec. 2–3 states that the Laplace transform of $f(t - T)$ at point 2 is equal to e^{-Ts} times the Laplace transform of $f(t)$ appearing at point 1. Therefore, the *transfer function of a transport lag is e^{-Ts}*.

Let us assume the presence of a transport lag in the forward path of a control system, as shown in Fig. 9–7–2. The system transfer function becomes

$$W(s) = \frac{c}{r}(s) = \frac{Ge^{-Ts}}{1 + GHe^{-Ts}} \tag{9-7-1}$$

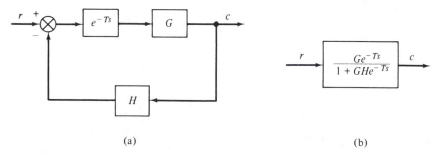

(a)   (b)

Figure 9–7–2. A transport lag in a system: (a) block diagram; (b) transfer function.

The characteristic equation with transport lag is

$$1 + GHe^{-Ts} = 0 \tag{9–7–2}$$

or

$$GH = -e^{+Ts} \tag{9–7–3}$$

No longer is the open-loop transfer function or any $KZ(s)/P(s)$ function equal to -1 as before, but rather it is equal to e^{+Ts}. Since $s = \sigma + j\omega$

$$e^{+Ts} = c^{(\sigma + j\omega)T} = e^{\sigma T} \cdot e^{j\omega T} \tag{9–7–4}$$

From Eq. (9–7–4) we see that the magnitude of the complex transfer function is $e^{\sigma T}$ and that its associated angle is ωT.

If we wish to extend the root locus method to systems with transport lag, we need new criteria to be derived from Eq. (9–7–3). The magnitude criterion is

$$|GH| = \left| K \frac{Z(s)}{P(s)} \right| = e^{\sigma T} \tag{9–7–5}$$

and the angle criterion is

$$\Sigma \phi_i = (2k + 1)180° + \omega T \tag{9–7–6}$$

The magnitude criterion changes with changes in the real part of s. Furthermore, the angle criterion is no longer unique; it will change as the imaginary part of s changes. Although the root locus can be plotted exactly using a Chu plot, it is a somewhat tedious exercise and of doubtful value.

Two not very good approximations of the transport lag will simplify the root locus sketch and will suffice to show the nature of the destabilizing influence of the transport lag. Consider a simple second-order system with transport lag in the forward path and with the transfer function

$$W(s) = \frac{Ke^{-Ts}}{s^2 + 2s + Ke^{-Ts}} \tag{9–7–7}$$

For a variable K,

$$G = K \frac{Z(s)}{P(s)} = \frac{Ke^{-Ts}}{s(s + 2)} \tag{9–7–8}$$

If T were equal to zero, there would be no transport lag and the system would be stable for all values of K, as indicated by the root locus of Fig. 9–7–3a.

The first approximation will be to write e^{-Ts} as $1/e^{+Ts}$ and expand in an exponential series, saving only the first two terms. Then

$$e^{-Ts} \cong \frac{1}{1 + Ts} = \frac{1}{T(s + 1/T)} \tag{9–7–9}$$

We have approximated the transport lag by a real pole in the left-half plane. Equation (9–7–8) becomes

$$G = K \frac{Z(s)}{P(s)} = \frac{K/T}{s(s + 2)(s + 1/T)} \tag{9–7–10}$$

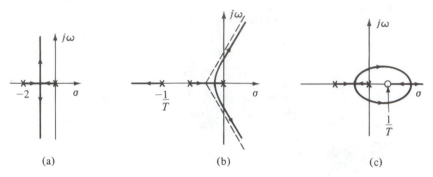

Figure 9–7–3. Root locus sketches of a simple second-order system: (a) without transport lag; (b) with $e^{-Ts} \cong 1/(Ts + 1)$; (c) with $e^{-Ts} \cong (1 - Ts)$.

The root locus will be that of a conditionally stable system as shown in Fig. 9–7–3b; as T increases, the stability of the system decreases.

For the second approximation we shall expand e^{-Ts} directly in an exponential series, again retaining only the first two terms so that

$$e^{-Ts} \cong 1 - Ts \qquad (9\text{--}7\text{--}11)$$

and

$$G = K\frac{Z(s)}{P(s)} = \frac{K(1 - Ts)}{s(s + 2)} = \frac{-KT(s - 1/T)}{s(s + 2)} \qquad (9\text{--}7\text{--}12)$$

This time we have approximated the transport lag by a real zero in the right-half plane; $KZ(s)/P(s)$ is now a nonminimum phase function. The minus sign means that we will have to use the zero-angle root locus rules of Sec. 6–7 for positive values of K. The root locus will have the shape shown in Fig. 9–7–3c, that of a conditionally stable system whose stability will decrease as T increases.

If greater realism is desired, Padé approximations could be used to linearize the transport lag, particularly if computer solutions are to be used to plot the root locus. Although such root locus plots may be of interest and possible value, frequency domain techniques are both exact and simpler.

In the frequency domain, the transport lag can be handled exactly and easily. Since $s = j\omega$,

$$e^{-Ts} = e^{-j\omega T} \qquad (9\text{--}7\text{--}13)$$

where the AR is equal to unity ($AR_{db} = 0$) and Φ is equal to $-\omega T$. Since the phase angle increases in magnitude without limit as ω increases, the transport lag is definitely a nonminimum phase function. From Eq. (9–7–8) we obtain the frequency function

$$G(j\omega) = K\frac{Z(j\omega)}{P(j\omega)} = \frac{Ke^{-j\omega T}}{2(j\omega)(j\omega/2 + 1)} \qquad (9\text{--}7\text{--}14)$$

The value of AR has not been changed by the presence of the transport lag, but Φ becomes

$$\Phi(\omega) = -90° - \tan^{-1}\frac{\omega}{2} - \omega T \qquad (9\text{–}7\text{–}15)$$

With $T = 0$ the function is minimum phase, the phase angle approaches $-180°$ as ω goes to infinity, and the system is stable since the Nyquist curve in Fig. 9–7–4 never crosses the negative real axis. As soon as T becomes finite, however, the phase angle will become infinite as ω becomes infinite, and the Nyquist curve will cross the negative real axis, many times in fact. The system is now conditionally stable with respect to T. As indicated in Fig. 9–7–4a, increasing T with K held constant will eventually cause the curve to encircle the -1 point, indicating that the system has become unstable. With T constant and finite, the system can be driven toward instability by increasing K, as indicated in Fig. 9–7–4b. Since we know the system to be conditionally stable with respect to K and T, we can use the phase margin Φ_m as a stability criterion and as a measure of both the degree of stabilty and the performance of the closed-loop system.

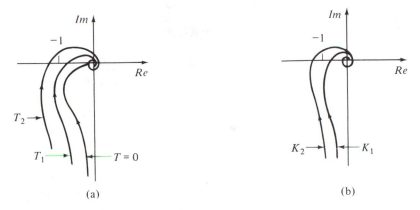

Figure 9–7–4. Nyquist sketches of a simple second-order system: (a) with K constant and variable T; (b) with T constant and variable K.

The Bode plots of the open-loop frequency function of Eq. (9–7–8) are shown in Fig. 9–7–5 for $K = 20$ and for $T = 0$, 0.1, and 1.0 sec. It can be seen that for a given K, increasing T will eventually cause the phase margin to become negative and the system to be unstable; the larger the value of K, the faster the system will become unstable.

Although transport lags can be compensated, it is to the control system designer's advantage to eliminate them completely or keep them as small as possible by proper design.

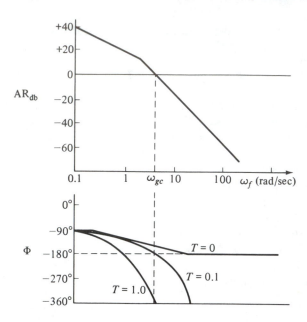

Figure 9–7–5. Bode plots of Eq. (9–5–8) for $K = 20$ and several values of the delay time T.

PROBLEMS

NOTE: In these problems do not hesitate to use any of the techniques available, namely, root locus sketches, Nyquist diagrams, Bode plots, and the Routh-Hurwitz criterion, even if not specified or required.

9–1. A unity feedback system has

$$G = \frac{50A}{s(s + 5)(s + 10)}$$

(a) Construct asymptotic Bode plots for an arbitrary value of A.

(b) From these Bode plots, determine the amplifier gain A required for $GM = 5$ and find the corresponding error constants and steady-state errors.

(c) From the Bode plots of (a) determine the amplifier gain A for a phase margin of $+35°$. Determine ζ, ω_n, M_p, and ω_p for the dominant pair of roots.

(d) Use the Routh-Hurwitz criterion to find the exact value of A required to satisfy (b) and determine the percentage error from using the asymptotic Bode plots.

(e) Using the value of A from (c), write the exact characteristic equation and find the corresponding values to compare with those of (c).

9–2. Same as Prob. 9–1, but with

$$G = \frac{150A(s + 1)}{s^2(s + 10)(s + 15)}$$

9–3. For the system of Fig. P9–3, the design objectives are $\zeta \geqslant 0.5$ and $K_v \geqslant 5$. The plant transfer function is

$$G_p = \frac{10}{(s + 5)(s^2 + 1.9s + 10)}$$

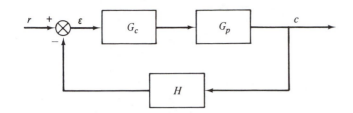

Figure P9–3

(a) With $H = 1$ and $G_c = A$ (proportional control), can this system meet the specifications? What is the system type?

(b) Using proportional control, $G_c = A$, and scalar parallel (outer-loop) compensation, $H = b_0$, select the best value for b_0. Can this system meet the specifications? If not, what is the system type and has its performance been improved?

9–4. With the G_p of Prob. 9–3, let $H = 1$ and use passive lead compensation with $G_c = A(s + 1)/(s + 10)$. Can this system meet the specifications?

9–5. For the G_p of Prob. 9–3, use frequency response techniques to define the most effective passive lead network for the set of given specifications. Be sure to sketch the Nyquist diagram with and without the lead network.

9–6. With the G_p of Prob. 9–3, let $H = 1$ and use a PID controller with $G_c = A(s + 0.5)(s + 2)/s$. Can this system meet the specifications?

9–7. The plant transfer function of the system of Fig. P9–3 is

$$G_p = \frac{150}{s(s + 10)(s + 15)}$$

(a) Same as Prob. 9–3 but with the above G_p.
(b) Same as Prob. 9–4 but with the above G_p.
(c) Same as Prob. 9–6 but with the above G_p.

9–8. With the G_p of Prob. 9–7 and $H = 1$, evaluate the performance in light of the specifications in Prob. 9–3 using a series passive lag network:
(a) $G_c = (s + 2)/4(s + 0.5)$
(b) $G_c = (s + 2)/10(s + 0.2)$
(c) Compare the results of (a) and (b).

9–9. With the G_p of Prob. 9–7 and $H = 1$, try to define a more effective passive lag network, using frequency domain techniques, than that given in Prob. 9–8.

9–10. With the G_p of Prob. 9–7 and $G_c = 1$, try meeting the specifications with non-scalar parallel compensation:
(a) $H = 0.2s + 1$
(b) $H = 0.2s + 0.4$
(c) Discuss the effects of (a) and (b).

9–11. The system of Fig. P9–3 has a type-0 plant, $G_p = 1/(s^2 + 4s + 5)$, and the only design requirement at this time is that the steady-state error for a constant-velocity input be zero, i.e., a type-2 system.
(a) With $H = 1$ and $G_c = A/s^2$ (to constitute a type-2 system), sketch the root locus and determine the stability. Is this system acceptable?
(b) Add rate feedback so that $H = s + 1$; i.e., add a zero to the feedback path. Determine the stability of the system. Is it still a type-2 system and does it satisfy the design requirement?
(c) Reset H to 1, for unity feedback and add the zero to the controller so that

$G_c = A(s + 1)/s^2$. Now determine the stability of the system and whether it is still a type-2 system.

   (d) Move the zero in (c) closer to the imaginary axis so that $G_c = A(s + 0.5)/s^2$. Has the performance been improved?

9–12. Let the plant of Fig. P9–3 be a type-2 plant, such as the inertially loaded motor of Fig. 3–5–1 plus a first-order time lag, with $G_p = 5/[s^2(s + 10)]$.

   (a) With proportional control, $G_c = A$, and unity feedback, sketch the root locus, the Nyquist diagram, and the Bode plots and determine the stability of the system.

   (b) Add rate feedback to the outer loop so that $H = s + 1$ and determine the stability of the system. Has the type been changed and what are the values of K_v and $E_{v_{ss}}$?

   (c) Maintain unity feedback, $H = 1$, and use proportional-derivative (PD) control so that $G_c = A(s + 1)$. Now determine the stability of the system, the system type, and the values of K_v and $E_{v_{ss}}$.

   (d) If the outer loop in (c) is broken, write the new transfer function $c(s)/r(s)$ and find the time response to a unit step in r.

9–13. For the plant and system in Prob. 9–12, introduce inner-loop rate feedback around the motor-load combination, as shown in Fig. P9–13.

   (a) Sketch the inner-loop root locus and find the value of B that yields a ζ of 0.5 for the inner-loop roots. Write the inner-loop transfer function $c(s)/e_c(s)$. What is the type of this "augmented" plant?

   (b) With $H = 1$ and a PD controller with $G_c = A(s + 1)$, what is the system type? Determine the stability of the system and the minimum value of $E_{v_{ss}}$.

   (c) Select a PID controller that will result in a $E_{v_{ss}}$ of zero and a minimum value of ζ of 0.5 for the dominant pair of roots.

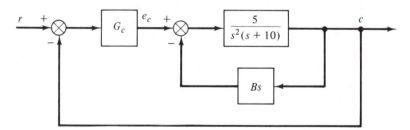

Figure P9–13

9–14. In the system of Fig. P9–14, the plant transfer function is

$$G_p = \frac{3}{(s - 3)(s + 5)}$$

   (a) Describe the stability of the plant.

   (b) With integral control only, $G_c = A/s$, sketch the root locus and determine the stability of the plant.

   (c) Add a zero to the controller that will "cancel" the plant pole in the right-half plane so that $G_c = A(s - 3)/s$. Sketch the root locus and determine the stability of the system.

   (d) With the controller in (c) and with $d = 0$, write the system transfer function

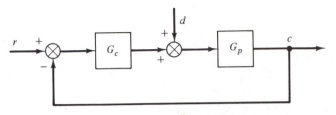

Figure P9–14

c(s)/r(s), retaining the stationary root. Will that root be canceled? Find the steady-state value of c for a unit step in r.

(e) With the controller in (c) and with $c = 0$, write the disturbance transfer function $c(s)/d(s)$, retaining the stationary root. Has that root been canceled? Find the steady-state value of c for a unit step in d.

9–15. For the plant and controller in Prob. 9–14, the controller zero drifts so that $G_c = A(s - 2)/s$.

(a) Sketch the root locus and determine the stability of the system.

(b) Locate the roots for $A = 10$ and write the system transfer function $c(s)/r(s)$. Is this the transfer function of a stable system? Compare it with the transfer function in Prob. 9–14d.

9–16. With respect to the plant in Prob. 9–14, rather than attempting to cancel the pole in the right half of the s plane, let us try to shift it into the left half of the s plane with the PID controller, $G_c = A(s + 2)^2/s$.

(a) Sketch the root locus and determine the conditions for stability.

(b) Write the system transfer function $c(s)/r(s)$ and find the steady-state value of c for a unit step in r. For a unit ramp in r.

(c) Write the disturbance transfer function $c(s)/d(s)$ and find the steady-state value of c for a unit step in d. For a unit ramp in d.

9–17. The system of Fig. P9–14 has a type-0 plant, $G_p = 10/(s + 10)(s + 20)$, and an integral controller, $G_c = A/s$.

(a) Sketch the root locus and determine the stability and type of the system.

(b) Determine the steady-state error for a unit step in c and for a unit step in d.

(c) Determine the steady-state error for a unit ramp in c and for a unit ramp in d.

9–18. Do Prob. 9–17 for a type-1 plant, $G_p = 10/[s(s + 10)(s + 20)]$. Does the location of the integrator affect the steady-state error for a command input? For a disturbance input?

9–19. A unity feedback system with transport lag has a forward transfer function $G = 5Ae^{-s}/(s + 5)(s + 10)$; i.e., $T = 1$ sec.

(a) Sketch on the same figure the root locus without transport lag ($T = 0$) and with transport lag (use an approximation for e^{-s}). Discuss system stability in each case.

(b) Construct the Bode plots without and with transport lag. Find the critical value of A that yields $\zeta = 0$ in each case.

9–20. Same as Prob. 9–15 but with $G = 20Ae^{-0.5s}/s(s + 2)(s + 10)$.

10

Nonlinear Control Systems

10–1 INTRODUCTION

Although nonlinear systems are not simple to analyze and are not part of classical control theory, to ignore them would be a serious omission. Inherent nonlinearities may result in an unstable system or one whose performance is unacceptable even though a linear analysis indicates otherwise. Furthermore, intentional nonlinearities in the form of nonlinear controllers can produce a control system that has better performance and/or is cheaper than one with a linear controller.

The treatment of nonlinearities in this chapter attempts only to be illustrative and most certainly is not exhaustive. The phase plane is emphasized because it is relatively easy to visualize and interpret, because it is used at times to present the results of linear analyses, and because it provides a transition to the discussion of the state space and of state variables in the next chapter.

10–2 THE PHASE PLANE

The *phase plane* is a two-dimensional representation of the time response of a system to nonzero initial conditions in which the first time derivative or velocity of a variable (usually the output) is plotted against the variable itself. Phase-plane analysis is limited to systems that are either simple second-order systems or can be approximated by a simple second-order system (a pair of dominant roots) and to inputs that can be represented by initial conditions. The phase plane may be used with either linear or nonlinear systems.

To illustrate the linear phase-plane solution we shall use our old friend the second-order tracking system, which has the transfer function

$$\frac{\theta_{out}}{\theta_{cmd}} = \frac{AK_m}{\tau_m s^2 + s + AK_m} = \frac{\omega_n^2}{s^2 + 2\zeta\omega_n s + \omega_n^2} \qquad (10\text{--}2\text{--}1)$$

With θ_{cmd} set equal to zero we can recover the unforced differential equation, which in parametric form is

$$\ddot{\theta}_{out} + 2\zeta\omega_n\dot{\theta}_{out} + \theta_{out} = 0 \qquad (10\text{--}2\text{--}2)$$

If we substitute x for θ_{out} and v for $\dot{\theta}_{out}$, then $\ddot{\theta}_{out}$ becomes

$$\ddot{\theta}_{out} = \frac{dv}{dt} = \frac{dv}{dx}\frac{dx}{dt} = v\frac{dv}{dx} \qquad (10\text{--}2\text{--}3)$$

With these relationships Eq. (10–2–2) can be written as

$$v\frac{dv}{dx} + 2\zeta\omega_n v + \omega_n^2 x = 0 \qquad (10\text{--}2\text{--}4)$$

Since time has been explicitly eliminated from this equation, it can be solved directly for v in terms of x and the initial conditions. When v is plotted against x for a specified set of initial conditions, the resulting curve is called a *phase trajectory*. If many such trajectories are plotted, the result is a *phase portrait*.

Phase trajectories and portraits may be plotted for both linear and nonlinear systems. Since the equations of most nonlinear systems cannot be solved directly, graphical techniques are commonly used; they are often used with linear systems. The various graphical methods available include the Lienard construction, Pell's method, the delta method, and the method of isoclines. We shall restrict ourselves to the *method of isoclines*.

An *isocline* is a line or curve in the phase plane along which the slope of the phase trajectory is constant. If we denote the trajectory slope dv/dx by the symbol λ, Eq. (10–2–4) can be written as

$$\lambda v + 2\zeta\omega_n v + \omega_n^2 x = 0 \qquad (10\text{--}2\text{--}5)$$

Solving for λ and v yields

$$\lambda = -\frac{\omega_n^2 x}{v} - 2\zeta\omega_n \qquad (10\text{--}2\text{--}6)$$

and

$$v = -\frac{\omega_n^2 x}{\lambda + 2\zeta\omega_n} \qquad (10\text{--}2\text{--}7)$$

From these two equations the isoclines for various values of λ can be obtained.

Although it is common practice to further nondimensionalize the equations of the isoclines, we shall select specific values of ζ and ω_n so as to relate the phase-plane solution to our previous knowledge of the behavior of simple second-order systems. With $\zeta = 0.5$ and $\omega_n = 2$ rad/sec, Eqs. (10–2–6) and (10–2–7) become

$$\lambda = -\frac{4x}{v} - 2 \qquad (10\text{--}2\text{--}8)$$

and

$$v = -\frac{4x}{\lambda + 2} \qquad (10\text{--}2\text{--}9)$$

The isoclines are calculated from these equations for various values of λ. Equation (10–2–9) is the equation of the isoclines and is a straight line passing through the origin of the phase plane. For example, when $\lambda = 0$, $v = -2x$; when $x = 0$, $\lambda = -2$ along the v axis; and when $v = 0$, $\lambda = \infty$ along the x axis. Additional isoclines are calculated and plotted as shown in Fig. 10–2–1a; the number of

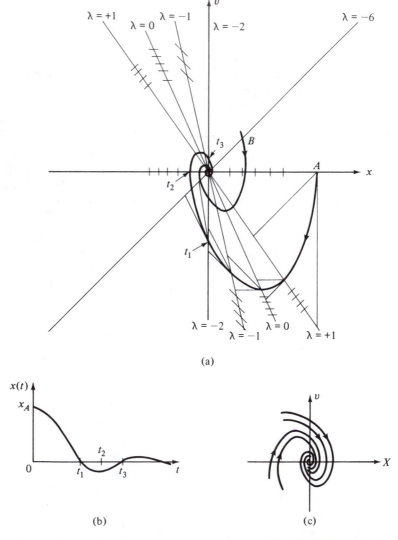

(a)

(b)   (c)

Figure 10–2–1. The linear second-order system of Eq. (10–2–1): (a) phase-plane trajectories; (b) phase trajectory A versus time; (c) phase portrait.

isoclines used is a function of the accuracy desired. Small slope lines are often drawn on each isocline indicating the slope that each trajectory must have when it crosses that particular isocline. It can be shown that motion along a trajectory as time increases is always in a clockwise direction.

To sketch a phase trajectory, select the desired initial conditions, such as those at point A which represent a position step input. Now draw two straight lines through point A, one with a slope equal to that specified for A and the other with a slope equal to that shown for the crossing of the next isocline, in this case $+1$. On this next isocline the midpoint of the segment between the two straight lines from A is taken as a point on the trajectory, and the procedure is repeated. When all the midpoints are plotted, the trajectory is drawn through these midpoints.

Each point on the phase trajectory corresponds to a specific instant of time, which can be determined and marked if so desired. One method for determining time marks is to use the definition of v as dx/dt to write

$$dt = \frac{1}{v}dx \tag{10-2-10}$$

so that

$$t_n - t_{n-1} = \int_{x_{n-1}}^{x_n} \frac{1}{v}dx \tag{10-2-11}$$

Plot $1/v$ against x and integrate graphically to obtain the time interval. Another method is completely graphical and is called the method of *isochrones*.

If the phase trajectory of Fig. 10-2-1a were plotted against time as is done in Fig. 10-2-1b, the response would be that of a simple second-order system to a negative step in position. Some key points of identical times are indicated in both plots to tie the two together.

If phase trajectories for different combinations of initial conditions for this underdamped system were plotted, the result would be the phase portrait of Fig. 10-2-1c. All the trajectories spiral monotonically inward and terminate at the phase-plane origin, which represents the equilibrium position of the radar system and is a singular point. A *singular point* is a point in the phase plane at which $\dot{v} = v = 0$. The slope of the trajectory at a singular point is indeterminate since

$$\lambda = \frac{dv}{dx} = \frac{dv/dt}{dx/dt} = \frac{\dot{v}}{v} = \frac{0}{0} \tag{10-2-12}$$

Singular points are classified in terms of the behavior of the trajectories in their vicinity. Since a singular point is an equilibrium point, its stability can be used to determine the stability of the system, whether linear or nonlinear.

For linear systems the nature of a singular point is a function of the location of the roots of the characteristic equation in the s plane. The system of Fig. 10-2-1 has a pair of complex conjugate roots on the left-half plane. Consequently, the system is stable and the singular point is called a *stable focus*. The relationship between this particular location of the roots and the phase portrait is shown in Fig. 10-2-2a.

If the roots were complex conjugates but were in the right-half plane, the system would be unstable. The singular point would now be an *unstable focus* as shown in the phase portrait of Fig. 10–2–2b. An ideal second-order physical system cannot theoretically be unstable. However, when a higher-order system is approximated by a pair of dominant roots, the resulting second-order system can have unstable transient modes. As an example, examine the root locus sketch of the third-order tracking radar in Fig. 6–4–2 and consider the movement and location of the dominant pair of roots as K is increased beyond K_c.

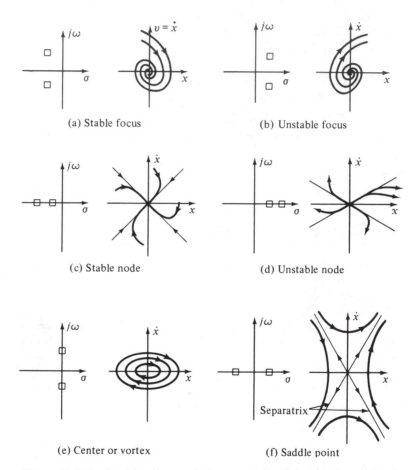

Figure 10–2–2. Root locations and phase portraits of the singular points of a linear second-order system.

If the roots are real and both are on the same side of the imaginary axis, there will be isoclines whose slope will be equal to the slope of the trajectory for that isocline. These are barrier isoclines that cannot be crossed by any phase trajectory. If the roots are equal (the magnitude of ζ is unity), there will be one barrier isocline. If the roots are unequal (the magnitude of ζ is greater than unity), there will be two barrier isoclines. If the roots are both in the left-half plane (ζ is positive), upon reaching a barrier isocline a trajectory will move directly to the singular point, which is now called a *stable node*. If both roots are in the right-half plane (ζ is negative), a trajectory will move directly away from the singular point, which is called an *unstable node*. These two types of singular points are illustrated in Fig. 10–2–2c and d.

   If ζ is zero, the system is marginally stable and the phase portrait will be a family of concentric curves about the singular point, which is now called a *center* or *vortex*. If the roots are real but one is negative and the other is positive, the system is unstable and the singular point is called a *saddle point*. The center and saddle point are shown in Fig. 10–2–2e and f.

   In the next section phase-plane analysis will be used to examine the response of systems with nonlinear controllers to initial-condition inputs.

10–3 NONLINEAR CONTROLLERS

In systems with linear controllers the control force applied to the plant is proportional in some manner to the actuating signal, which in turn is a measure of the instantaneous error of the system. If the system has unity feedback, the control force would be proportional to the instantaneous error itself. In the tracking radar example, the torque applied by the drive motor at any instant of time is the control force and is proportional to the error at that instant of time. The relationships between the motor torque and the error signal and between the error signal and time for a step input are shown in Fig. 10–3–1. Implicit in the definition of a linear control system is the assumption that the amplifier and motor will not saturate. This assumption means that for linear operation T_{max} must be greater than the torque called for by the maximum error signal expected. From Fig. 10–3–1 we see that the motor is loafing most of the time.

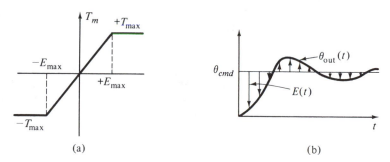

(a)   (b)

Figure 10–3–1.   Two linear controller relationships: (a) motor torque and the error signal; (b) the error signal and time.

   If an ideal relay is placed in the controller, the motor can be operated at its maximum torque whenever there is any error signal whatsoever, as shown in Fig. 10–3–2 for a motor with tachometer feedback and with a pure inertia load. Systems with this type of nonlinear control are variously referred to as

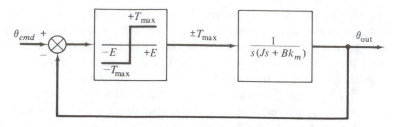

Figure 10–3–2.   A second-order tracking servo with a relay controller.

relay, on-off, bang-bang, or *contactor control systems.* Some of the advantages and disadvantages of such systems will be mentioned as the discussion progresses.

It is no longer possible, however, to write a transfer function for the complete system. From the transfer function of the linear portion we can write the corresponding differential equation

$$J\ddot{\theta}_{out} + Bk_m\dot{\theta}_{out} = \pm T_{max} \tag{10–3–1}$$

or

$$\ddot{\theta}_{out} + \frac{1}{\tau_m}\dot{\theta}_{out} = \pm a \tag{10–3–2}$$

where τ_m is the time constant of the motor and a (the control force) is the maximum acceleration of the output shaft with no feedback damping. With the substitution of x for θ_{out}, v for $\dot{\theta}_{out}$, and λ for dv/dx, Eq. (10–3–2) becomes

$$\lambda v + \frac{1}{\tau_m}v = \pm a \tag{10–3–3}$$

Solving for λ and v yields the expressions

$$\lambda = \frac{\pm a}{v} - \frac{1}{\tau_m} \tag{10–3–4}$$

and

$$v = \frac{\pm a}{\lambda + 1/\tau_m} \tag{10–3–5}$$

To avoid the necessity of specifying the motor and load characteristics, these equations are further nondimensionalized by letting $v' = v/a\tau_m$, $x' = x/a\tau_m^2$, and $\lambda' = \tau_m\lambda$. The slope and isocline equations can now be written as

$$\lambda' = \pm\frac{1}{v'} - 1 \tag{10–3–6}$$

and

$$v' = \frac{\pm 1}{\lambda' + 1} \tag{10–3–7}$$

Since θ_{cmd} is always equal to zero when using the phase-plane method, the instantaneous error will be the negative of the instantaneous value of θ_{out}. When θ_{out} is greater than zero, the error will be negative and the motor will develop a maximum negative torque to drive the system toward the equilibrium point. Similarly, when θ_{out} is less than zero, the applied torque will be positive. Consequently, the phase plane is divided into two regions separated by a switching line which coincides with the v' axis for this ideal relay. In the region to the right of the switching line the negative sign is used in Eqs. (10–3–6) and (10–3–7), and the positive sign is used for the region to the left of the switching line. The isoclines for these two regions are drawn in Fig. 10–3–3. Notice the two barrier isoclines at $v' = -1$ for the $-T_{max}$ region and at $v' = +1$ for the $+T_{max}$ region. These isoclines represent the maximum output velocity attainable for any given motor-load combination.

One complete trajectory and one partial trajectory are sketched in Fig. 10–3–3. Point B represents either the same motor as at A but with a larger initial displacement, or the same displacement and a smaller motor. When the trajectory starts at B, the maximum velocity is attained prior to reaching the switching line. The same system with a linear proportional controller and with the same peak overshoot and developing the same initial torque would have the trajectory shown by the dashed curve. The linear system would require approximately 50 percent longer to reach the first peak overshoot, and the second overshoot will be somewhat larger than that of the nonlinear system. The nonlinear system will reach the vicinity of the equilibrium point faster than the linear system, but with a much larger number of oscillations.

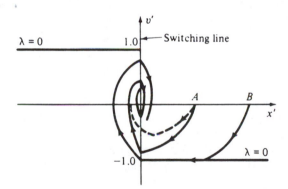

Figure 10–3–3. Phase trajectories of the tracking servo with an ideal switching relay.

Instantaneous switching is difficult to achieve in a relay and is hard on both the relay contacts and the motor. A dead zone not only increases the life of the relay and motor but also provides a means for controlling and even optimizing the system response. The characteristics of a switching relay with a symmetrical dead zone* are shown in Fig. 10–3–4a. The dead zone has the effect of introducing two switching lines and creating a third region, in which the applied torque is zero. This third region is the dead zone itself. Since the acceleration in the dead zone is zero, Eq. (10–3–4) shows that λ will be equal

---

* Other dead zone configurations are possible and are used.

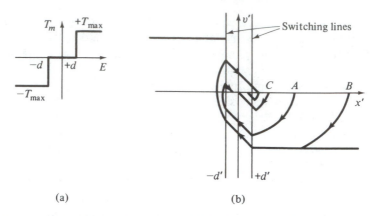

(a)   (b)

Figure 10–3–4. Relay with a symmetrical dead zone: (a) relay characteristics;
(b) some phase-plane trajectories.

to $-1/\tau_m$ and λ' will be equal to -1 throughout this zone. The isoclines outside
the dead zone will remain unchanged.

With $d' = d/a\tau_m^2$, some typical trajectories for step inputs are sketched in
Fig. 10–3–4b. When a trajectory reaches the dead zone, the motor is turned
off and the system coasts within the dead zone. Whenever the velocity goes to
zero in the dead zone, the trajectory ends, leading to a steady-state error with
maximum values of $\pm d$. Compare the trajectory starting at A with the corresponding
trajectory in Fig. 10–3–3. We see that the dead zone has reduced the number
of oscillations as well as the time required to reach a steady-state value. Since
the response of a system is strongly influenced by the type and width of the
dead zone, relay controllers are used in certain classes of time-optimal control
systems.

With hysteresis effects considered, a relay with a dead zone might have
the characteristics shown in Fig. 10–3–5a. Now there will be four switching
lines, as shown in Fig. 10–3–5b along with several typical trajectories.

With linear systems the stability and dynamic behavior are independent of
the magnitude of the input and the initial conditions. For example, the percent

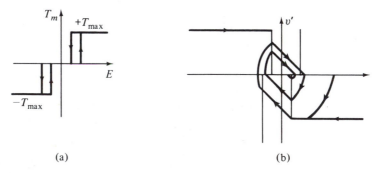

(a)   (b)

Figure 10–3–5. Relay with dead zone and hysteresis: (a) relay characteristics;
(b) phase-plane trajectories.

overshoot, number of oscillations, and settling time of the step response of a linear system are determined by the system characteristics alone. This is not true with nonlinear systems, as can be seen by examining the trajectories of Fig. 10-3-4b and 10-3-5b. When the magnitude of the input step, represented by the initial conditions, is small, there is no overshoot or oscillation and the settling time is small. As the input magnitude is increased, overshoots and oscillations appear and the settling time increases. The percent overshoot, the number of oscillations, and the settling time are all strongly dependent on the magnitude of the input. A fundamental characteristic of nonlinear systems is that the stability of the system is dependent on both the magnitude of the input and the initial conditions, as well as on the system characteristics.

Relay controllers generate a control force u that is discontinuous; it is either on or off, depending upon the value of the output (or instantaneous error). Other types of nonlinear controllers may be used that generate a continuous control force that is a nonlinear function of the output. One such nonlinear control force might be a two-mode proportional plus derivative control with

$$u = ke + \frac{e}{|e|} \tag{10-3-8}$$

where the amount of derivative control is inversely proportional to the magnitude of the instantaneous error and manifests itself as nonlinear damping; the damping is low for large errors and high for small errors. The response to large step inputs has been improved at the cost of extreme sluggishness for small step inputs.

Inherent nonlinearities can also be examined by use of the phase plane. For example, the straight-line isoclines of a linear system with coulomb friction will not pass through the origin but will be shifted in opposite directions in the upper and lower halves of the phase plane; consequently, coulomb friction introduces steady-state position errors.

Saturation and threshold can be handled by using pseudo switching lines to divide the phase plane into the regions shown in Fig. 10-3-6. Gear trains resemble relays with dead zones and hysteresis, and so on. The specific nature

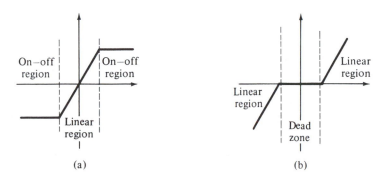

                 (a)   (b)

Figure 10-3-6. Phase-plane division by pseudo switching lines: (a) saturation; (b) a threshold.

of the nonlinearity determines how the regions of the phase plane are to be laid out.

The response to ramp, unit impulse, and ramp plus step inputs can be analyzed by the proper choice of initial conditions. We shall conclude this section with the statement that the analog computer with a two-dimensional plotter can be a very useful tool in obtaining phase-plane trajectories.

*10–4 THE DESCRIBING FUNCTION

The output of a nonlinear system subjected to a disturbance or a step input may go into a steady-state oscillation about an equilibrium point and still be considered stable. Such an oscillation is called a *limit cycle* and is a periodic though not sinusoidal oscillation whose amplitude and frequency are dependent only upon the characteristics of the system. Limit cycles are not to be confused with the periodic oscillations about a center or vortex that are associated with undamped or conservative systems and have a frequency and amplitude dependent upon the initial conditions as well as the system characteristics.

Limit cycles may be stable or unstable. A *stable limit cycle* in the phase plane is a closed curve which is approached asymptotically by all trajectories in its vicinity. The phase portrait of a typical stable limit cycle is shown in Fig. 10–4–1a. If a trajectory starts anywhere within the limit cycle, it will spiral outward until it reaches the limit cycle itself. A stable limit cycle is often tolerable if the amplitude is sufficiently small and may even be desirable for certain systems.

An *unstable limit cycle* is a closed curve in the phase plane from which all trajectories diverge. A typical phase portrait is shown in Fig. 10–4–1b. A trajectory starting within the limit cycle will move into the equilibrium point; this is a region of stable operation. A trajectory starting outside the limit cycle will spiral outward. Although an unstable limit cycle can never be achieved

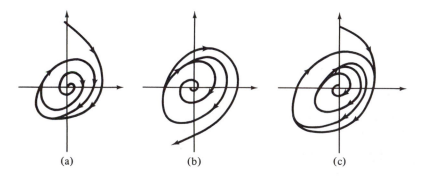

     (a)   (b)   (c)

Figure 10–4–1.   The limit cycle: (a) stable; (b) unstable; (c) combination.

physically, it does divide the phase plane into regions of different dynamic behavior. Consider the combination shown in Fig. 10-4-1c of an unstable limit cycle within a stable limit cycle. Trajectories originating within the unstable limit cycle will terminate at the singular point; those starting outside the unstable limit cycle will end up in the stable limit cycle. Other combinations, of course, are possible.

A phase-plane analysis may hint at or even reveal the existence and characteristics of a limit cycle. The trajectories of Fig. 10-3-3, if examined closely, indicate the possibility of a stable limit cycle about the equilibrium point known as relay chatter, since the amplitude of the limit cycle approaches zero and the frequency is extremely high. Chatter is very hard on relay contacts. For other systems and nonlinearities the presence of a limit cycle might be more obvious. In any case, limiting the phase plane to second-order systems forces the neglect of higher-order dynamic effects, for instance the dynamics of the relay and of the amplifier. These dynamics in conjunction with a nonlinearity might easily lead to a limit cycle not discernible by the phase-plane solution. The describing function method* is a technique for detecting limit cycles, examining their stability, and determining approximately their amplitude and frequency.

A *describing function* is an attempt to approximate a nonlinearity by a variable gain and phase angle. If the input to a nonlinear element is sinusoidal, $X \sin \omega t$, the output $y(t)$ will be periodic and consist of a fundamental component plus a series of harmonics of higher order. This periodic output can be expressed in a Fourier series as

$$y(t) = Y_1 \sin (\omega t + \Phi) + \text{harmonics} \qquad (10\text{-}4\text{-}1)$$

where Y_1 and Φ are the amplitude and phase angle of the fundamental component and are both functions of the amplitude of the sinusoidal input. The constant term in the Fourier series has been omitted with the assumption that the nonlinearity is symmetric and the average value of $y(t)$, therefore, is zero. If the amplitudes of the harmonics are small with respect to that of the fundamental, which is generally the case, the harmonics can be neglected. The nonlinear element can now be represented by a describing function N, where

$$N(X,j\omega) = \frac{Y_1}{X} e^{j\Phi} \qquad (10\text{-}4\text{-}2)$$

If the nonlinearity is single-valued, such as an ideal relay, a dead zone, or a saturation, the describing function will be real with no phase angle per se and can be represented by a variable gain that is a function of the input amplitude only. If the nonlinearity is double-valued, as with hysteresis or backlash, or contains an energy storage element, the describing function will be complex, having both a magnitude and a phase angle. Describing functions have been derived or experimentally determined for most nonlinearities and can be found in works devoted to nonlinear systems. We shall not derive any describing

---

* This method is also known as *harmonic linearization.*

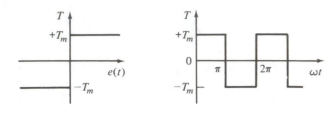

$$\text{(a) } N = \frac{4T_m}{\pi E}$$

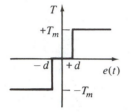

$$\text{(b) } N = 0 \text{ for } E < d; N = \frac{4T_m}{\pi E} \sqrt{1 - (\frac{d}{E})^2} \text{ for } E > d$$

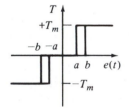

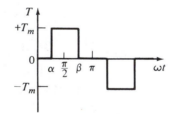

$$\text{(c) } N = \frac{4T_m}{\pi E} \sin(\frac{\beta - \alpha}{2})e^{j\Phi}; \Phi = \frac{\pi}{2} + (\frac{\alpha + \beta}{2}); \begin{array}{l} \alpha = \sin^{-1}\frac{b}{E} \\ \beta = \pi - \sin^{-1}\frac{a}{E} \end{array}$$

Figure 10–4–2. Relay characteristics output wave form for $e(t) = E \sin \omega t$, and describing function: (a) ideal relay; (b) relay with dead zone; (c) relay with dead zone and hysteresis.

functions but shall instead use those given in Fig. 10–4–2 for three relay controllers in the limit-cycle analysis of the third-order system of Fig. 10–4–3.

If the system of Fig. 10–4–3 goes into a limit cycle in response to a command or disturbance, the input into the relay will be periodic. Although not a pure

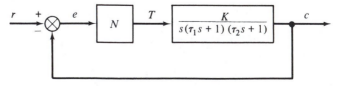

Figure 10–4–3. A relay controlled third-order plant.

sinusoid, this relay input will not differ greatly, particularly with respect to the amplitude and frequency, since the output of the relay must pass through the linear portion of the system before entering the relay again. The linear elements of this system and of most useful control systems function as low-pass filters to attenuate the higher-order harmonics. This is our justification for using the describing function in a limit-cycle analysis.

For a linear system the closed-loop frequency transfer function can be written as

$$W(j\omega) = \frac{G(j\omega)}{1 + G(j\omega)H(j\omega)} \tag{10–4–3}$$

If the Nyquist diagram of GH passes through the -1 point, i.e., $1 + GH = 0$, then the system is marginally stable and a step input will result in a sustained sinusoidal oscillation of the output.* With caution we write for our nonlinear system with unity feedback a pseudo frequency transfer function

$$W(E,j\omega) = \frac{N(E,j\omega)G(j\omega)}{1 + N(E,j\omega)G(j\omega)} \tag{10–4–4}$$

where E is the amplitude of the signal entering the nonlinear element. Extending the concept of marginal stability for a linear system to the limit cycle of a nonlinear system, we can establish as a requirement for the existence of a limit cycle that

$$1 + N(E,j\omega)G(j\omega) = 0 \tag{10–4–5}$$

or that

$$G(j\omega) = -\frac{1}{N(E,j\omega)} \tag{10–4–6}$$

In essence $-1/N$ is considered to be analogous to the -1 point used in the linear analysis. We can now draw the Nyquist diagram of G and the locus of $-1/N$ on the same plot and see if the two curves intersect and, if so, for what set of conditions.

For the ideal relay with an error signal input,

$$-\frac{1}{N} = \frac{-\pi E}{4T_m} \tag{10–4–7}$$

where E is the amplitude of the periodic error signal. This function plots in the complex plane as the entire negative real axis and is shown in Fig. 10–4–4a along with Nyquist diagrams of the third-order plant of Fig. 10–4–3 for two values of K. When $K = K_1$, the two curves intersect at a point A, indicating the presence of a limit cycle. The amplitude and frequency of the limit cycle will be the amplitude ratio and frequency of G at the point of intersection.

We should like to know whether the limit cycle is stable or unstable. Although there is an analytic test based on a truncated Taylor's series, we shall use a heuristic test based on our extension of the significance of the -1 point to the $-1/N$ locus. If a system in a limit cycle is subjected to a small disturbance

* This is not a limit cycle.

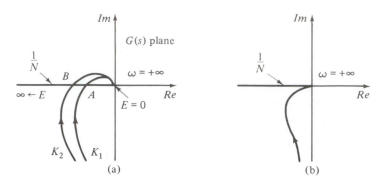

Figure 10–4–4. An ideal relay controller: (a) with a third-order plant; (b) with a second-order plant.

that changes the amplitude of the input to the nonlinear element, the $-1/N$ point will move. If the move is in such a direction that the Nyquist diagram encircles the new $-1/N$ point, we say that the system is unstable and that the amplitude of the periodic oscillation will increase. Conversely, if a disturbance shifts the $-1/N$ point so that the Nyquist diagram does not encircle it, we say that the system is stable and that the amplitude of the oscillation will decrease. Applying this argument to the limit cycle at point A, a slight disturbance that increases E will shift the $-1/N$ point to the left so that it is not encircled by the $G(j\omega)$ plot. The system, therefore, is stable; E will decrease, and the $-1/N$ point will move back to A. If the slight disturbance decreases E, the operating point will shift to the right and be enclosed by the Nyquist diagram. The system is now unstable; the amplitude E will increase, returning the $-1/N$ point to A. Therefore, point A is a stable equilibrium point and the limit cycle is also stable. Increasing K to K_2 moves the point of intersection to point B, which represents a stable limit cycle with the same frequency ω_1 but with a larger amplitude than the limit cycle of point A.

The Nyquist diagram of a second-order system is sketched along with the $-1/N$ locus in Fig. 10–4–4b. The only possible intersection is at the origin, indicating a stable limit cycle of zero amplitude and infinite frequency. This is the relay chatter indicated by the phase trajectories of Fig. 10–3–3. Any time lag(s) in the system components, no matter how small, will cause the Nyquist diagram to cross the $-1/N$ locus. Consequently, control by an ideal relay will always result in a stable limit cycle.

If we introduce a symmetrical dead zone into the relay controller, then $-1/N$ can be written as

$$-\frac{1}{N} = -\frac{\pi E}{4T_m\sqrt{1 - (d/E)^2}} \qquad (10\text{–}4\text{–}8)$$

For a given value of d, the $-1/N$ locus is a function of the ratio E/d, as shown in Fig. 10–4–5. As E/d is increased, the locus moves toward the origin until E/d becomes equal to $\sqrt{2}$; this is the closest point of approach. As E/d increases

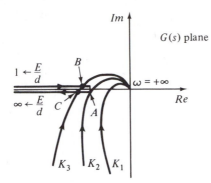

Figure 10–4–5. Control of a third-order plant by a relay with a symmetrical dead zone.

further, the locus moves outward toward $-\infty$. Since the $-1/N$ locus no longer extends to the origin, as was the case with the ideal relay, it is now possible to have the system respond without a limit cycle. When K is equal to K_1, there is no intersection of the two curves and no limit cycle exists.

If K is increased to K_2* so that the Nyquist diagram just touches the $-1/N$ locus at point A, there will be a limit cycle. If the system is operating at point A and is disturbed so as to increase E, the $-1/N$ point will be moved to the left of the Nyquist curve. This is a stable mode of operation; the amplitude E will decrease and move the $-1/N$ point back to A. If, however, the disturbance decreases E, the operating point will again be moved to the left and the system will be stable. The amplitude will continue to decrease, moving the operating point further to the left until the oscillations die out completely. Thus, point A represents an unstable limit cycle.

If K is further increased to K_3, there will be two intersections at points B and C. It can be seen that point B represents an unstable limit cycle; any disturbance that increases the amplitude E will move the $-1/N$ point to point C, which represents a stable limit cycle.

When a relay with both a dead zone and hysteresis is used (see Fig. 10–4–2c), the describing function is no longer real. It now has a phase angle that is a function of the input amplitude. For a specified ratio of a/b, say $a/b = 0.2$, the $-1/N$ locus would have the general shape shown in Fig. 10–4–6. When the Nyquist diagram is sketched for three increasing values of K, we have, respectively, no limit cycle, two limit cycles (unstable at A, stable at B), and one stable limit cycle at C. Since the $-1/N$ locus with hysteresis is below the negative real axis, it is now possible for a second-order system to have limit cycles. Notice that an a/b ratio of unity reduces the $-1/N$ locus to that of the relay with a dead zone only. If d is then set equal to zero, the $-1/N$ locus further reduces to that of the ideal relay.

Gain-phase plots can be used for limit-cycle analysis. They are particularly useful when the describing function is complex and/or when the designer is using Nichols charts to examine the overall system stability. Bode plots can

---

* Or if the width of the dead zone is decreased and K is held constant.

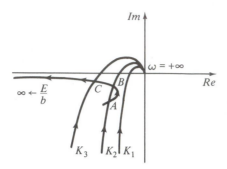

Figure 10–4–6. Control of a third-order plant by a relay with dead zone and hysteresis.

also be used directly but are somewhat more difficult to interpret; as before, Bode plots are a good source of data for Nyquist or gain-phase plots. Although the root locus is used by some authors in conjunction with the describing function, it is not recommended since the describing function is *not* a transfer function and is valid only for sinusoidal inputs.

The describing function method is not limited to second-order systems as is the phase plane. In fact, the higher the order, the more accurate the limit-cycle analysis. This method, however, provides no explicit information about the transient behavior or the overall stability of the system. Other techniques should be used to examine these facets of a nonlinear system, reserving the describing function method for what it does best, limit-cycle analysis.

Although the examples in this section are minimum phase functions and conditionally stable in the simple sense, these are not limiting cases. With nonminimum phase functions and nonsimple conditionally stable systems, the principles of the describing function method still hold; just exercise care and judgment in applying them.

In closing we should be aware that there are other definitions of the describing function that can be used in place of the conventional one used in this section. They are more complicated but can give more accurate results for certain system-nonlinearity combinations.

PROBLEMS

10–1. When $A = A_c$ the system of Fig. P10–1 can be approximated by the transfer function

$$W(s) = \frac{c(s)}{r(s)} = \frac{3}{s^2 + 3}$$

(a) Write the unforced (homogeneous) differential equation.
(b) With $x = c$ and $v = \dot{c}$ write the equation for the phase trajectories.
(c) Using the method of isoclines sketch the trajectory originating at $c(0+) = 1$ and $\dot{c}(0+) = 0$. Sketch the corresponding time response as in Fig. 10–2–1b.
(d) Same as (c) but with $c(0+) = 0.5$ and $\dot{c}(0+) = 0.5$ on the same figure as (c).
(e) Locate and classify the singular point.

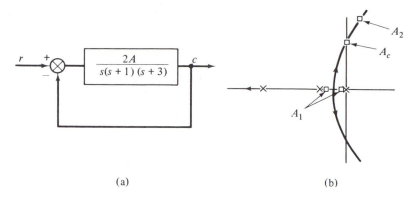

(a)   (b)

Figure P10–1

10–2. Same as Prob. 10–1 but with $A = A_1$ and the system approximated by

$$W(s) = \frac{c(s)}{r(s)} = \frac{0.12}{(s + 0.6)(s + 0.2)}$$

10–3. Same as Prob. 10–1 but with $A = A_2$ and the system approximated by

$$W(s) = \frac{c(s)}{r(s)} = \frac{5}{s^2 - 1.34s + 5}$$

10–4. The relay control system of Fig. P10–4 has the following component values: $e_{max} = \pm 10$ V; $K_m = 5$ deg/V-sec; and $\tau_m = 0.5$ sec. The switching relay has no dead zone; i.e., $\pm d = 0$.
  (a) Write the isocline equations.
  (b) Sketch the trajectory for an initial value of $\theta_{out} = +50°$. What is the steady-state error?
  (c) Same as (b) but for $\theta_{out}(0+) = 150°$.

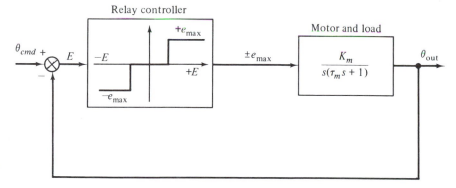

Figure P10–4

10–5. Same as Prob. 10–4 but with a symmetrical dead zone in the relay. Let $d = \pm 10°$. What are the effects of the dead zone upon the performance of the system?

10–6. Same as Prob. 10–5 but with an asymmetrical dead zone. Let $d = +10°$ and $-5°$.

10–7. Same as Prob. 10–5 but with $\tau_m = 0.1$ sec.

10–8. For the system of Fig. P10–4 the actual transfer function of the motor-load combination contains another first-order time lag so that

$$\frac{\theta_{\text{out}}}{e_{\text{max}}} = \frac{\pm 5}{s(0.5s + 1)(0.1s + 1)}$$

With no dead zone, $d = 0$, determine the frequency, amplitude, and stability of any limit cycles that might be present for $\theta_{\text{out}}(0+) = +50°$ and $\theta_{\text{out}}(0+) = +150°$.

10–9. Same as Prob. 10–8 but with a symmetrical dead zone with $d = \pm 10°$.

10–10. The amplifier of the system of Fig. P10–10 has a gain of 5 but saturates at a voltage level of ± 20 V for any error signal in excess of $\pm 4°$. As a first approximation neglect the time lag with $\tau = 0.01$ sec and sketch the phase trajectories for $\theta_{\text{out}}(0+) = +10°$ and $+50°$.

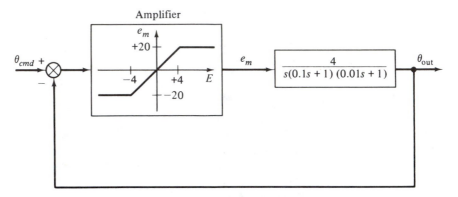

Figure P10–10

10–11. For Prob. 10–10 the $-1/N$ locus of the describing function for the saturating amplifier is as shown in Fig. P10–11. Using the complete transfer function perform a limit-cycle analysis.

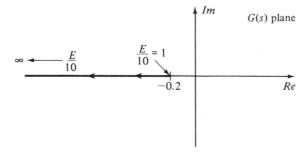

Figure P10–11

11

An Introduction

to Modern Control Theory

11–1 INTRODUCTION

Classical control theory is based on input-output relationships, principally the transfer function. When differential equations are encountered, they are linearized and subjected to whatever constraints are necessary to establish useful input-output relationships. Modern control theory, on the other hand, is based on direct use of the differential equations* themselves. Transfer functions, when available or experimentally determined, are used only to obtain the corresponding differential equations.

Although its techniques are powerful and relatively simple, classical control theory does have limitations and shortcomings that multiply as plants and control systems become more complex—thus the advent of modern control theory. For example, as the number of inputs and outputs increases, the number of transfer functions needed to describe the plant or system increases dramatically. A plant with 10 inputs and 10 outputs requires 100 transfer functions with classical theory, and only a single matrix vector equation with modern theory. Only impulse, step, and sinusoidal inputs are easily handled by classical theory; modern theory is indifferent to the type of input. Initial conditions must be added and treated separately in the classical approach, whereas they are automatically included in the modern approach.

The modern control theory representation of a plant or system does not obscure the internal behavior as do the input-output relationships of classical control theory, nor is it limited to stationary systems. Whereas classical control theory is limited to linear controllers, usually closed loop, modern control theory

---

* Difference equations are used in the case of sampled-data systems.

is not, leading to the development of optimal controllers, which are generally nonlinear. Finally, modern control theory makes provisions for both the inclusion of detailed and varied performance criteria and the direct design of controllers by synthesis, in contrast to the trial-and-error design and performance evaluation techniques of classical control theory.

In principle modern control theory has no inherent limitations. There are, however, many practical limitations and they can be severe. Representation of complex multivariable plants and systems may be simple, but solutions do not always exist. When they do, they are generally involved and require numerical solutions by a digital computer. Algebraic manipulations of system parameters (other than in the steady-state equations) and analytic solutions are virtually nonexistent; parameters must be given numerical values prior to solution. In modern control theory we put numbers in and get numbers out. Determining the effects of parameter and equilibrium condition changes requires many computer runs, and the resulting data are often difficult to analyze and correlate. Finally, it may be difficult, if not impossible, to write the differential equations for a complex plant or subsystem; the experimentally determined transfer function can be most helpful in some of these situations.

This brief comparison was made to show that modern control theory should not yet be considered a full replacement for classical control theory but rather as an extension and as a complement. In analyzing a plant and designing a control system, the designer should, as always, exploit the advantages of each approach.

Although the treatment and discussion in this and the succeeding chapter are sketchy, the objectives are also modest. The first objective is to make the reader aware of other types of control systems and of other analytical and design approaches and to attempt to relate them to classical systems and techniques. The second objective is to acquaint the reader with basic features and some of the vocabulary. The final and most important objective is to whet the enthusiasm of the reader—to generate an interest in further study of control theory and its applications.

11–2 STATE-VARIABLE REPRESENTATION

The dynamic equation of a plant represented by the simple mechanical system of Fig. 3–2–2 is

$$m\ddot{x} + B\dot{x} + kx = f_a(t) \tag{11–2–1}$$

where f_a is the applied force. If x is replaced by x_1 and $\dot{x}$ by x_2, the second-order differential equation can be written as a set of two first-order equations:

$$\dot{x}_1 = x_2$$

$$\dot{x}_2 = -\frac{k}{m}x_1 - \frac{B}{m}x_2 + \frac{f_a}{m} \tag{11–2–2}$$

The variables x_1 and x_2 are called the *state variables* since they define the state of the plant. The *state* of the plant is defined as the smallest set of state variables which must be known at some time $t = t_0$ in order to predict uniquely the future behavior of the plant for any specified input. Although the state of a plant is unique at any instant of time, a set of state variables is not; in fact, the proper choice of a set of state variables may greatly simplify the solution of the control problem. The individual state variables, however, must be linearly independent; there must not be any set of constants α_i, other than zero, that satisfies the equation

$$\alpha_1 x_1 + \alpha_2 x_2 + \cdots + \alpha_n x_n = 0 \qquad (11\text{--}2\text{--}3)$$

These two points of nonuniqueness and independence can be illustrated by use of the block diagram of Fig. 11–2–1, which represents the plant of Eq. (11–2–1) and in which f_d and f_s are the forces exerted on the mass by the dashpot and spring, respectively. The state variables $x_1 = x$ and $x_2 = \dot{x}$ are linearly independent and define the state of the plant. If we now let $x_1 = x$ but $x_2 = f_s$, then

$$\alpha_1 x_1 + \alpha_2 x_2 = \alpha_1 x + \alpha_2 kx = 0 \qquad (11\text{--}2\text{--}4)$$

Since this equation can be satisfied with $\alpha_1 = k$ and $\alpha_2 = 1$, x and f_s do not form a valid set of state variables. However, if we let $x_1 = x$ and $x_2 = f_d$, then

$$\alpha_1 x_1 + \alpha_2 x_2 = \alpha_1 x + \alpha_2 B\dot{x} = 0 \qquad (11\text{--}2\text{--}5)$$

These two variables are linearly independent and can be used as a valid set of state variables.

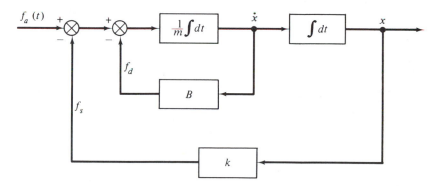

Figure 11–2–1. Block diagram of the linear and stationary plant of Eq. (11–2–1).

If n state variables are needed to define the state of a plant, they can be considered the n components of a *state vector* x. *State space* is then defined as an n-dimensional space with $x_1, x_2, \ldots, x_n$ as the coordinates. The state of the plant at any instant of time will be a point in the n-dimensional state space. A series of such points form a *state trajectory*. At this point the reader may wish to review Secs. 10–2 and 10–3 in terms of state variables and a two-dimensional state space.

We shall limit our discussion of state-variable representation to lumped parameter plants with continuous variables so as to work with ordinary differential equations. A nonlinear, nonstationary version of the simple mechanical system is shown in Fig. 11–2–2. With x and $\dot{x}$ as the two state variables x_1 and x_2, the

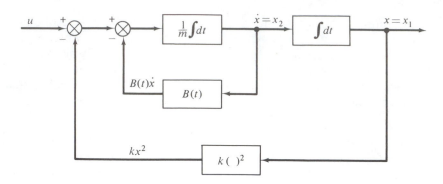

Figure 11-2-2.   Block diagram of the nonlinear and nonstationary plant of Eq. (11-2-6).

two first-order equations describing the plant are

$$\dot{x}_1 = x_2$$

$$\dot{x}_2 = -\frac{k}{m}x_1^2 - \frac{B(t)}{m}x_2 + \frac{u(t)}{m} \qquad (11\text{-}2\text{-}6)$$

where $u(t)$ represents the input or control force. Extending the two-dimensional representation, an nth-order plant can be represented by a set of n first-order equations in the form

$$\dot{x}_1 = f_1(x_1, \ldots, x_n; u_1, \ldots, u_m; t)$$

$$. \qquad (11\text{-}2\text{-}7)$$

$$\dot{x}_n = f_n(x_1, \ldots, x_n; u_1, \ldots, u_m; t)$$

To simplify the representation, this set of equations can be written as the single vector equation

$$\dot{\mathbf{x}}(t) = \mathbf{f}(\mathbf{x}, \mathbf{u}, t) \qquad (11\text{-}2\text{-}8)$$

where $\mathbf{x}$, $\mathbf{f}$, and $\mathbf{u}$ are the *state vector, function vector,* and *control vector,* respectively.

A state variable is not necessarily an output variable, or vice versa. Separate *output variables* may appear when we are interested in the behavior of variables other than the state variables or when the state variables are either not physical quantities or not physically measurable. In any event we can always write a generalized vector equation relating the *output vector* $\mathbf{c}$ to the state and control vectors as

$$\mathbf{c}(t) = \mathbf{g}(\mathbf{x}, \mathbf{u}, t) \qquad (11\text{-}2\text{-}9)$$

If the plant model is linear or linearized, the vector equations can be written as vector matrix equations in normal form as

$$\dot{\mathbf{x}} = \mathbf{A}(t)\mathbf{x} + \mathbf{B}(t)\mathbf{u} \qquad (11\text{-}2\text{-}10)$$
$$\scriptstyle (n \times 1) \ (n \times n) \ (n \times 1) \ (n \times m) \ (m \times 1)$$

$$\mathbf{c} = \mathbf{C}(t)\mathbf{x} + \mathbf{D}(t)\mathbf{u} \qquad (11\text{-}2\text{-}11)$$
$$\scriptstyle (p \times 1) \ (p \times n) \ (n \times 1) \ (p \times m) \ (m \times 1)$$

where $\mathbf{A}(t)$ is known as the *plant matrix* and $\mathbf{B}(t)$ as the *control matrix*. The unnamed matrices $\mathbf{C}(t)$ and $\mathbf{D}(t)$ relate the output variables to the state and control variables. The terms in parentheses below each vector and matrix indicate its size.* The stationary plant of Eq. (11–2–2) would have the following vectors and matrices

$$\mathbf{x} = \underset{(2 \times 1)}{\begin{bmatrix} x_1 \\ x_2 \end{bmatrix}} \quad \underset{(1 \times 1)}{\mathbf{u} = f_a} \quad \underset{(1 \times 1)}{\mathbf{c} = x_1}$$

(11–2–12)

$$\mathbf{A} = \underset{(2 \times 2)}{\begin{bmatrix} 0 & 1 \\ -k/m & -B/m \end{bmatrix}} \quad \mathbf{B} = \underset{(2 \times 1)}{\begin{bmatrix} 0 \\ 1/m \end{bmatrix}} \quad \underset{(1 \times 2)}{\mathbf{C} = [1 \; 0]} \quad \mathbf{D} = \underset{(2 \times 1)}{\begin{bmatrix} 0 \\ 0 \end{bmatrix}}$$

If transfer functions have been obtained experimentally or are otherwise available, we need to be able to translate them into state-variable models. Consider a single-variable plant with the transfer function

$$G_p(s) = \frac{c}{u}(s) = \frac{6}{s^3 + 8s^2 + 19s + 12}$$

(11–2–13)

The corresponding third-order differential equation is

$$\dddot{c} + 8\ddot{c} + 19\dot{c} + 12c = 6u$$

(11–2–14)

A simple choice of state variables would be the output itself and its first two derivatives. The state vector will then be

$$\mathbf{x} = \begin{bmatrix} x_1 \\ x_2 \\ x_3 \end{bmatrix} = \begin{bmatrix} c \\ \dot{c} \\ \ddot{c} \end{bmatrix}$$

(11–2–15)

Equation (11–2–14) can now be written as a set of three first-order equations:

$$\begin{aligned} \dot{x}_1 &= x_2 \\ \dot{x}_2 &= x_3 \\ \dot{x}_3 &= -12x_1 - 19x_2 - 8x_3 + 6u \end{aligned}$$

(11–2–16)

In vector matrix form, this set can also be written as

$$\begin{aligned} \dot{\mathbf{x}} &= \mathbf{A}\mathbf{x} + \mathbf{B}\mathbf{u} \\ \mathbf{c} &= \mathbf{C}\mathbf{x} \end{aligned}$$

(11–2–17)

$$\mathbf{A} = \begin{bmatrix} 0 & 1 & 0 \\ 0 & 0 & 1 \\ -12 & -19 & -8 \end{bmatrix} \quad \mathbf{B} = \begin{bmatrix} 0 \\ 0 \\ 6 \end{bmatrix}$$

$$\mathbf{C} = [1 \quad 0 \quad 0] \qquad\qquad \mathbf{u} = u$$

(11–2–18)

A different state model of the same plant can be obtained by expanding

---

* Appendix C briefly discusses matrices and determinants.

the transfer function in partial fractions as

$$\frac{c}{u}(s) = \frac{1}{s + 1} - \frac{3}{s + 3} + \frac{2}{s + 4} \tag{11-2-19}$$

so that

$$c(s) = \frac{u(s)}{s + 1} - \frac{3u(s)}{s + 3} + \frac{2u(s)}{s + 4} \tag{11-2-20}$$

We now let

$$x_1(s) = \frac{u(s)}{s + 1} \qquad sx_1 + x_1 = u$$

$$x_2(s) = -\frac{3u(s)}{s + 3} \qquad sx_2 + 3x_2 = -3u \tag{11-2-21}$$

$$x_3(s) = \frac{2u(s)}{s + 4} \qquad sx_3 + 4x_3 = 2u$$

Inverting these last relationships yields

$$\dot{x}_1 = -x_1 + u$$

$$\dot{x}_2 = -3x_2 - 3u \tag{11-2-22}$$

$$\dot{x}_3 = -4x_3 + 2u$$

Equations (11-2-20) and (11-2-21) will provide the relationship

$$c(t) = x_1 + x_2 + x_3 \tag{11-2-23}$$

We now have another state model in the form of Eq. (11-2-17) but with

$$\mathbf{A} = \begin{bmatrix} -1 & 0 & 0 \\ 0 & -3 & 0 \\ 0 & 0 & -4 \end{bmatrix} \qquad \mathbf{B} = \begin{bmatrix} 1 \\ -3 \\ 2 \end{bmatrix} \tag{11-2-24}$$

$$\mathbf{C} = [1 \quad 1 \quad 1]$$

Although the state models are different and have different block diagrams, they do represent the same plant. The two models are related by a nonsingular transformation matrix $\mathbf{P}$ that can be determined if so desired. Both plant matrices $\mathbf{A}$ will have identical characteristic equations and eigenvalues. In the second model the plant matrix is a diagonal matrix whose elements are the eigenvalues of the plant. This special form is called *canonical* and is an extremely useful form; for one thing, it mathematically decouples the state variables from each other. Notice that in the second model the state variables are not physical quantities as they are in the first model.

Let us now add a zero to the plant transfer function so that Eq. (11-2-13) becomes

$$\frac{c}{u}(s) = \frac{2(s + 4)}{s^3 + 8s^2 + 19s + 12} \tag{11-2-25}$$

Although this particular zero coincides with a pole in the denominator, do *not* cancel. If you do, you will lose a transient mode and have a model that does not accurately represent the plant. The differential equation corresponding to Eq. (11–2–25) is

$$\dddot{c} + 8\ddot{c} + 19\dot{c} + 12c = 2\dot{u} + 8u \qquad (11\text{–}2\text{–}26)$$

If we define the state variables in accordance with Eq. (11–2–15), the set of first-order equations will be

$$\dot{x}_1 = x_2$$
$$\dot{x}_2 = x_3 \qquad (11\text{–}2\text{–}27)$$
$$\dot{x}_3 = -12x_1 - 19x_2 - 8x_3 + 2\dot{u} + 8u$$

The right side of the last equation contains a $\dot{u}$ term for which there is no provision in the form of Eq. (11–2–17). Of the various techniques available for eliminating this anomaly we shall define a new state vector

$$x_1 = c$$
$$x_2 = \dot{c} \qquad (11\text{–}2\text{–}28)$$
$$x_3 = \ddot{c} + ku$$

where k is a constant. Substituting these new state variables into Eq. (11–2–26) produces

$$\dot{x}_3 - k\dot{u} + 8x_3 - 8ku + 19x_2 + 12x_1 = 2\dot{u} + 8u \qquad (11\text{–}2\text{–}29)$$

If k is set equal to -2, the $\dot{u}$ terms drop out, leaving

$$\dot{x}_1 = x_2$$
$$\dot{x}_2 = x_3 + 2u \qquad (11\text{–}2\text{–}30)$$
$$\dot{x}_3 = -12x_1 - 19x_2 - 8x_3 - 8u$$

or in matrix vector form

$$\begin{bmatrix} \dot{x}_1 \\ \dot{x}_2 \\ \dot{x}_3 \end{bmatrix} = \begin{bmatrix} 0 & 1 & 0 \\ 0 & 0 & 1 \\ -12 & -19 & -8 \end{bmatrix} \begin{bmatrix} x_1 \\ x_2 \\ x_3 \end{bmatrix} + \begin{bmatrix} 0 \\ 2 \\ -8 \end{bmatrix} u \qquad (11\text{–}2\text{–}31)$$

Although these examples have been limited to single-variable plants, the techniques for multivariable plants would be the same, as would be the form of the vector matrix equations. If we wish to include disturbance inputs into the plant, we only need to introduce a disturbance vector $\mathbf{z}$ and a disturbance matrix $\mathbf{F}$. Equation (11–2–17) would be written as

$$\dot{\mathbf{x}} = \mathbf{Ax} + \mathbf{Bu} + \mathbf{Fz} \qquad (11\text{–}2\text{–}32)$$

or with some loss of detail as

$$\dot{\mathbf{x}} = \mathbf{Ax} + \mathbf{B'u'} \qquad (11\text{–}2\text{–}33)$$

where

$$\mathbf{B'} = [\mathbf{B}, \mathbf{F}] \qquad \mathbf{u'} = \begin{bmatrix} \mathbf{u} \\ \mathbf{z} \end{bmatrix} \qquad (11\text{–}2\text{–}34)$$

11–3 LINEAR SOLUTIONS

A linear stationary plant* can be represented by the vector matrix equation

$$\dot{\mathbf{x}}(t) = \mathbf{A}\mathbf{x}(t) + \mathbf{B}\mathbf{u}(t) \qquad (11\text{–}3\text{–}1)$$

with

$$\mathbf{x}(0) = \mathbf{x}_0$$

where $\mathbf{A}$ and $\mathbf{B}$ are constant matrices and $\mathbf{x}_0$ is the initial-state vector representing the initial conditions (the initial time is taken to be zero). The complete solution to this equation consists of a homogeneous part and a particular part. The former is independent of the input and depends only upon the initial-state vector $\mathbf{x}_0$; it is variously referred to as the *transient, unforced, undriven,* or *free* solution. The latter part of the complete solution is independent of the initial state, but depends upon the control vector and the time interval during which control is applied; this part is generally referred to as the *forced* or *driven* solution.

We shall obtain the unforced solution first. With $\mathbf{u}$ set equal to zero, Eq. (11–3–1) becomes

$$\dot{\mathbf{x}} = \mathbf{A}\mathbf{x} \qquad \text{with} \qquad \mathbf{x}(0) = \mathbf{x}_0 \qquad (11\text{–}3\text{–}2)$$

Assume a solution in the form of a Taylor's series expansion of $\mathbf{x}(t)$ about t equal to zero, so that

$$\mathbf{x}(t) = \mathbf{x}_0 + \dot{\mathbf{x}}_0 t + \ddot{\mathbf{x}}_0\frac{t^2}{2!} + \dddot{\mathbf{x}}_0\frac{t^3}{3!} + \cdots \qquad (11\text{–}3\text{–}3)$$

where

$$\dot{\mathbf{x}}_0 = \dot{\mathbf{x}}(t)\big|_{t=0} = \mathbf{A}\mathbf{x}(t)\big|_{t=0} = \mathbf{A}\mathbf{x}_0$$

$$\ddot{\mathbf{x}}_0 = \frac{d\dot{\mathbf{x}}(t)}{dt}\bigg|_{t=0} = \mathbf{A}\cdot\dot{\mathbf{x}}(t)\big|_{t=0} = \mathbf{A}\cdot\mathbf{A}\mathbf{x}_0 = \mathbf{A}^2\mathbf{x}_0$$

$$\cdot$$
$$\cdot \qquad\qquad (11\text{–}3\text{–}4)$$
$$\cdot$$

$$\frac{d\mathbf{x}_0^j}{dt^j} = \frac{d\mathbf{x}^j(t)}{dt^j}\bigg|_{t=0} = \mathbf{A}^j\mathbf{x}_0$$

Notice that we used Eq. (11–3–2) in evaluating the derivatives. With the evaluations of Eq. (11–3–4), the coefficients in Eq. (11–3–3), which are constants, can be written in terms of $\mathbf{A}$ and $\mathbf{x}_0$ so that

$$\mathbf{x}(t) = \left(\mathbf{I} + \mathbf{A}t + \mathbf{A}^2\frac{t^2}{2!} + \mathbf{A}^3\frac{t^3}{3!} + \cdots\right)\mathbf{x}_0 \qquad (11\text{–}3\text{–}5)$$

where $\mathbf{I}$ is the identity matrix. The infinite coefficient series within the parentheses defines what is known as the *matrix exponential function.* Equation (11–3–5),

---

* This type of plant is still the easiest to analyze and solve.

which is the unforced solution, can be written as

$$\mathbf{x}(t) = e^{\mathbf{A}t}\mathbf{x}_0 = \boldsymbol{\Phi}(t)\mathbf{x}_0 \tag{11–3–6}$$

where $\boldsymbol{\Phi}(t)$ is the accepted symbol for the matrix exponential. This matrix exponential function can be shown to be convergent for all square matrices and is called the *fundamental* or *transition* matrix. It can be thought of as a transformation matrix in that Eq. (11–3–6) represents the linear transformation of the plant from the initial state $\mathbf{x}_0$ to a new state $\mathbf{x}(t)$ where $t > 0$.

The transition matrix $\boldsymbol{\Phi}(t)$ can be evaluated from the series expansion shown in Eq. (11–3–5). This is a task well suited to the digital computer. However, if the plant matrix $\mathbf{A}$ is in parametric form, numerical values must be substituted prior to the computer solution.

This transition matrix possesses most of the properties of a scalar exponential function, one being that

$$\frac{d}{dt}(e^{\mathbf{A}t}) = \mathbf{A}e^{\mathbf{A}t} \tag{11–3–7}$$

or

$$\dot{\boldsymbol{\Phi}}(t) = \mathbf{A}\boldsymbol{\Phi}(t) \tag{11 3–8}$$

We shall make use of Eq. (11–3–8) in obtaining the forced solution.

Let us assume a forced solution $\mathbf{x}_p$ of the form

$$\mathbf{x}_p(t) = \boldsymbol{\Phi}(t)\mathbf{g}(t) \tag{11–3–9}$$

and substitute it in Eq. (11–3–1) to obtain

$$\dot{\boldsymbol{\Phi}}\mathbf{g} + \boldsymbol{\Phi}\dot{\mathbf{g}} = \mathbf{A}\boldsymbol{\Phi}\mathbf{g} + \mathbf{B}\mathbf{u} \tag{11–3–10}$$

With Eq. (11–3–8), Eq. (11–3–10) becomes

$$\mathbf{A}\boldsymbol{\Phi}\mathbf{g} + \boldsymbol{\Phi}\dot{\mathbf{g}} = \mathbf{A}\boldsymbol{\Phi}\mathbf{g} + \mathbf{B}\mathbf{u} \tag{11–3–11}$$

Canceling like terms, we have

$$\boldsymbol{\Phi}\frac{d\mathbf{g}}{dt} = \mathbf{B}\mathbf{u} \tag{11–3–12}$$

This equation can be integrated to find

$$\mathbf{g}(t) = \int_0^t \boldsymbol{\Phi}^{-1}(\tau)\mathbf{B}\mathbf{u}(\tau)\,d\tau \tag{11–3–13}$$

where

$$\boldsymbol{\Phi}^{-1}(\tau) = e^{-\mathbf{A}\tau} \tag{11–3–14}$$

Substituting Eq. (11–3–13) into Eq. (11–3–9) gives us the forced solution

$$\mathbf{x}_p(t) = \boldsymbol{\Phi}(t)\int_0^t \boldsymbol{\Phi}^{-1}(\tau)\mathbf{B}\mathbf{u}(\tau)\,d\tau \tag{11–3–15}$$

which can also be written as

$$\mathbf{x}_p(t) = \int_0^t \boldsymbol{\Phi}(t-\tau)\mathbf{B}\mathbf{u}(\tau)\,d\tau \tag{11–3–16}$$

where
$$\Phi(t - \tau) = \Phi(t)\Phi^{-1}(\tau) = e^{A(t-\tau)} \qquad (11-3-17)$$
The complete solution to Eq. (11–3–1) is the sum of the two individual solutions or

$$\mathbf{x}(t) = \Phi(t)\mathbf{x}_0 + \int_0^t \Phi(t - \tau)\mathbf{B}\mathbf{u}(\tau)\, d\tau \qquad (11-3-18)$$

If the initial time is taken to be t_0 rather than zero, the transition matrix becomes
$$\Phi(t - t_0) = e^{A(t-t_0)} \qquad (11-3-19)$$
and $\mathbf{x}_0$ becomes $\mathbf{x}(t_0)$. The complete solution (the time response) for any time interval $(t - t_0)$ is therefore

$$\mathbf{x}(t) = \Phi(t - t_0)\mathbf{x}_0 + \int_{t_0}^t \Phi(t - \tau)\mathbf{B}\mathbf{u}(\tau)\, d\tau \qquad (11-3-20)$$

If the linear plant is nonstationary, the governing vector equation becomes
$$\dot{\mathbf{x}}(t) = \mathbf{A}(t)\mathbf{x}(t) + \mathbf{B}(t)\mathbf{u}(t) \qquad (11-3-21)$$
The complete solution to this equation can be written as

$$\mathbf{x}(t) = \Phi(t, t_0)\mathbf{x}_0 + \int_{t_0}^t \Phi(t, \tau)\mathbf{B}(\tau)\mathbf{u}(\tau)\, d\tau \qquad (11-3-22)$$

where the transition matrix must satisfy the two conditions that
$$\dot{\Phi}(t, t_0) = \mathbf{A}(t)\Phi(t, t_0) \qquad (11-3-23a)$$
and
$$\Phi(t_0, t_0) = \mathbf{I} \qquad (11-3-23b)$$
The transition matrix $\Phi(t, t_0)$ can be evaluated by the numerical integration of Eq. (11–3–23a).

In reference to the linear stationary plant, Eq. (11–3–1) can also be solved by the use of the Laplace transformation. Since $\mathbf{A}$ and $\mathbf{B}$ are constant matrices, the Laplace transform of Eq. (11–3–1) is
$$s\mathbf{x}(s) - \mathbf{x}_0 = \mathbf{A}\mathbf{x}(s) + \mathbf{B}\mathbf{u}(s) \qquad (11-3-24)$$
where $\mathbf{x}(s)$ and $\mathbf{u}(s)$ are the Laplace transforms of the two vectors and $\mathbf{x}_0$ is the initial-state vector. Equation (11–3–24) is now an algebraic vector equation and can be rearranged and written as
$$(s\mathbf{I} - \mathbf{A})\mathbf{x}(s) = \mathbf{x}_0 + \mathbf{B}\mathbf{u}(s) \qquad (11-3-25)$$
Premultiplying both sides by the inverse of $(s\mathbf{I} - \mathbf{A})$ yields the expression for the Laplace transform of the state vector:
$$\mathbf{x}(s) = (s\mathbf{I} - \mathbf{A})^{-1}\mathbf{x}_0 + (s\mathbf{I} - \mathbf{A})^{-1}\mathbf{B}\mathbf{u}(s) \qquad (11-3-26)$$
Consequently, the state vector $\mathbf{x}(t)$ representing the time response, or complete solution, is
$$\mathbf{x}(t) = L^{-1}[(s\mathbf{I} - \mathbf{A})^{-1}]\mathbf{x}_0 + L^{-1}[(s\mathbf{I} - \mathbf{A})^{-1}\mathbf{B}\mathbf{u}(s)] \qquad (11-3-27)$$
with the inversion normally accomplished by a digital computer.

Comparing the first term of Eq. (11–3–27) with the first term of Eq. (11–3–18) provides an explicit expression for the time-invariant transition matrix

$$\mathbf{\Phi}(t) = e^{\mathbf{A}t} = L^{-1}[(s\mathbf{I} - \mathbf{A})^{-1}] \qquad (11-3-28)$$

If the initial-state vector is zero and the output vector is defined by*

$$\mathbf{c}(t) = \mathbf{C}\mathbf{x}(t) \qquad (11-3-29)$$

then with Eq. (11–3–26) the transform of the output vector can be written as

$$\mathbf{c}(s) = \mathbf{C}(s\mathbf{I} - \mathbf{A})^{-1}\mathbf{B}\mathbf{u}(s) \qquad (11-3-30)$$

Recalling the definition of a transfer function as the ratio of the transformed output to the transformed input, we can express Eq. (11–3–30) as

$$\mathbf{c}(s) = \mathbf{G}_p(s)\mathbf{u}(s) \qquad (11-3-31)$$

where $\mathbf{G}_p(s)$ is defined as the *transfer function matrix* of the plant. The block diagrams of the plant with state variables and with a transfer function matrix are shown in Fig. 11–3–1. The simplicity of the transfer function matrix representation is accomplished by obscuring the internal behavior of the plant.

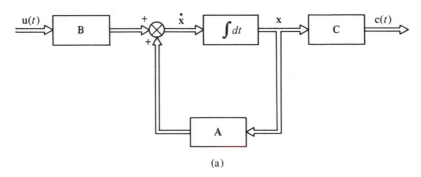

(a)

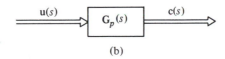

(b)

Figure 11–3–1. A linear stationary plant: (a) state-variable representation; (b) transfer function matrix.

With the definition for the inverse of a matrix we can write the plant transfer function matrix as

$$\mathbf{G}_p(s) = \frac{\mathbf{c}}{\mathbf{u}}(s) = \frac{\mathbf{C}\ \text{adj}\ (s\mathbf{I} - \mathbf{A})}{|s\mathbf{I} - \mathbf{A}|} \qquad (11-3-32)$$

The characteristic equation of the transfer function matrix is

$$|s\mathbf{I} - \mathbf{A}| = 0 \qquad (11-3-33)$$

This scalar equation is the characteristic equation of classical control theory,

* **D** is set equal to zero in Eq. (11–2–11) for simplicity.

and its roots are the *eigenvalues* of the plant matrix $\mathbf{A}$. Conversely, the eigenvalues of $\mathbf{A}$ are the roots of the characteristic equation of the plant, and we can use them to determine the stability of the plant.

Plant representation is as typical of modern control theory as system representation is of classical control theory. The modern control theory designer seeks a control vector that will transfer a given plant from one state to another along an acceptable or prescribed trajectory and then is concerned with the design of a practical controller to generate such a control vector. The control scheme may be open loop or closed loop, linear or nonlinear—thus the emphasis on plant rather than system representation. With classical control theory the input-output relationship is of primary importance, and the control force per se receives little attention.

It is possible, of course, to represent a closed-loop system by vector equations. The simplest representation would be to replace the control vector $\mathbf{u}(t)$ by a command input vector $\mathbf{r}(t)$. Then a linear stationary system could be described by the vector equations

$$\dot{\mathbf{x}}(t) = \mathbf{A}\mathbf{x}(t) + \mathbf{B}\mathbf{r}(t) \qquad (11\text{--}3\text{--}34)$$
$$\mathbf{c}(t) = \mathbf{C}\mathbf{x}(t)$$

where $\mathbf{A}$ is now the system matrix and its eigenvalues are the roots of the characteristic equation of the system. The state vector $\mathbf{x}(t)$ describes the state of the system, and $\mathbf{B}$ is the input matrix.

If we wish to retain knowledge of the plant and other internal behavior, we can add a third equation describing the control vector to the original plant and output equations. A closed-loop system can then be represented by

$$\dot{\mathbf{x}}(t) = \mathbf{A}\mathbf{x}(t) + \mathbf{B}\mathbf{u}(t)$$
$$\mathbf{c}(t) = \mathbf{C}\mathbf{x}(t) \qquad (11\text{--}3\text{--}35)$$
$$\mathbf{u}(t) = \mathbf{K}\mathbf{E}(t) = \mathbf{K}\mathbf{r}(t) - \mathbf{K}\mathbf{K}_{fb}\mathbf{c}(t)$$

The block diagram of these equations is shown in Fig. 11–3–2 along with the

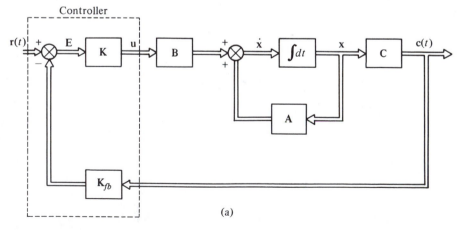

(a)

Figure 11–3–2. A closed-loop control system with output feedback: (a) state-variable representation; (b) transfer matrices.

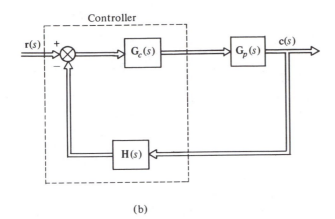

(b)   **Figure 11–3–2.** (Cont.)

transfer matrix representation of the same system. In this figure the output variables are fed back to close the loop; state-variable feedback is also possible, at least mathematically if not always physically. Signal flow graphs can be used instead of block diagrams, and in fact are preferred by many designers.

State-variable representation is deceptively simple; the solutions, however, are usually complex, involving large computational effort. The transfer matrix solution of a single-variable plant or system, for example, requires considerably more effort than the transfer function solution.

11–4 CONTROLLABILITY, OBSERVABILITY, AND STABILITY

A plant (or system) is said to be *completely state-controllable* if it is possible to find an unconstrained control vector $\mathbf{u}(t)$ that will transfer any initial state $\mathbf{x}(t_0)$ to any other state $\mathbf{x}(t)$ in a finite time interval. Since complete state controllability does not necessarily mean complete control of the output, and vice versa, *complete output controllability* is separately defined in the same manner. A plant is said to be *completely observable* if the state $\mathbf{x}(t)$ can be determined from a knowledge of the output $\mathbf{c}(t)$ over a finite time interval.

The dual concepts of controllability and observability are fundamental to the control of multivariable plants, particularly with regard to optimal control. Complete controllability ensures the existence of an unconstrained control vector and thus the existence of a possible controller. It does not, however, tell how to design the controller, nor does it guarantee either a realistic control vector or a practical controller. Complete observability ensures a knowledge of the state or internal behavior of the plant from a knowledge of the output. It does not, however, guarantee that the output variables are physically measurable.

The significance of these two concepts can be illustrated by consideration of a generalized *n*th-order plant, which will have *n* state variables and thus *n* transient (dynamic) modes. The number of control variables will be designated by *m*, and the number of output variables by *p*. In a practical control system we expect *m* and *p* to be less than *n* and would like them both to be small in number. If the plant is not completely controllable, there will be modes (state variables) that cannot be controlled in any way by one or more of the control variables; these modes are decoupled from the control vector. If the plant is not completely observable, there will be modes whose behavior cannot be determined; these modes are decoupled from the output vector.

A plant can be divided into four subsystems, as shown in Fig. 11–4–1. Since only the first subsystem A, which is both controllable and observable, has an input-output relationship, it is the only subsystem that can be represented by a transfer function or a transfer function matrix. Conversely, a transfer function or matrix representation of this plant reveals nothing about the dynamic behavior of subsystems B and D and provides no control over the behavior of subsystems C and D. If, for example, the modes comprising subsystem B reacted violently to any of the control variables, the output variables would give no indication of such behavior. Undesirable transients in subsystem C would affect the output, but nothing could be done to modify them. This plant can be made completely controllable by appropriately adding control variables. The task of making the plant completely observable, however, is more difficult and will not be discussed further.

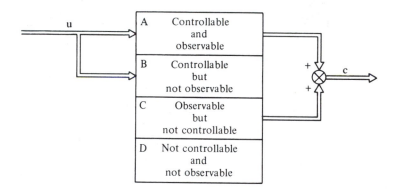

Figure 11–4–1. A plant divided into subsystems on the basis of controllability and observability.

There are various tests and rules developed by matrix algebra for determining the controllability and observability of a linear plant, either stationary or non-stationary. One such rule states that a stationary plant is completely controllable and observable if there are no cancellations in the appropriate transfer functions or matrices. For instance, the generalized plant of Fig. 11–4–2a is completely state-controllable if there are no common factors in the numerator and denominator of $\mathbf{x}(s)/\mathbf{u}(s)$ and is completely observable if there are no common factors in $\mathbf{c}(s)/\mathbf{u}(s)$.

Let us apply these tests to the specific subsystem of Fig. 11–4–2b. The transfer functions relating the state variables of the plant and the output variable

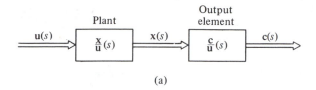

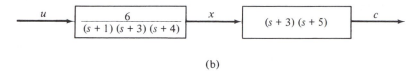

(b)

Figure 11-4-2. A subsystem comprising a plant and output element: (a) general; (b) specific.

to the single control variable are

$$\frac{x}{u}(s) = \frac{6}{(s + 1)(s + 3)(s + 4)} \tag{11-4-1}$$

and

$$\frac{c}{u}(s) = \frac{6(s + 3)(s + 5)}{(s + 1)(s + 3)(s + 4)} \tag{11-4-2}$$

There are no common factors in Eq. (11-4-1), indicating that the plant is completely state-controllable. However, there is a common factor $(s + 3)$ in the transfer function of Eq. (11-4-2), indicating that the plant is not completely observable and that the subsystem is not state-controllable.

These conclusions can be verified by deriving state models of the plant and subsystem from the transfer functions. To determine state controllability we use the transfer function of Eq. (11-4-1), which is identical to that of Eq. (11-2-13), and the techniques of Sec. 11-2 to write the vector matrix equation of the plant in canonical form as

$$\begin{bmatrix} \dot{x}_1 \\ \dot{x}_2 \\ \dot{x}_3 \end{bmatrix} = \begin{bmatrix} -1 & 0 & 0 \\ 0 & -3 & 0 \\ 0 & 0 & -4 \end{bmatrix} \begin{bmatrix} x_1 \\ x_2 \\ x_3 \end{bmatrix} + \begin{bmatrix} 1 \\ -3 \\ 2 \end{bmatrix} [u] \tag{11-4-3}$$

The corresponding set of differential equations is

$$\dot{x}_1 + x_1 = u$$

$$\dot{x}_2 + 3x_2 = -3u \tag{11-4-4}$$

$$\dot{x}_3 + 4x_3 = 2u$$

where x_1, x_2, and x_3 are the state variables of the plant alone. We see from these equations that each variable is controlled by the single control variable u. If there had been any zeros in the control matrix **B**, the plant would not have been completely state-controllable. Therefore, examination of the control matrix for any zero values when the plant matrix $\mathbf{A}$ is in canonical form* is another test for state controllability. A more direct but purely mathematical test for controllability is to form a specific matrix from $\mathbf{A}$ and $\mathbf{B}$ and examine its rank. This test is known as *Kalman's test*.

To verify observability we make use of the transfer function that relates the output to the state variables:

$$\frac{c}{x}(s) = (s + 3)(s + 5) \tag{11–4–5}$$

The transform of the output variable can be written as

$$c(s) = (s^2 + 8s + 15)x(s) \tag{11–4–6}$$

where

$$x(s) = x_1(s) + x_2(s) + x_3(s) \tag{11–4–7}$$

The equations of Eq. (11–4–4) are transformed to obtain expressions for $x_1(s)$, $x_2(s)$, and $x_3(s)$, which are substituted into Eq. (11–4–6) which is then inverted to obtain the differential equation for $c(t)$. This equation can be written in the form

$$c(t) = [8 \quad 0 \quad -1] \begin{bmatrix} x_1 \\ x_2 \\ x_3 \end{bmatrix} \tag{11–4–8}$$

or simply as

$$c(t) = 8x_1(t) - x_3(t) \tag{11–4–9}$$

From Eq. (11–4–9) we see that the behavior of x_2 does not affect the output; a state variable has been lost, as is also indicated by the zero in Eq. (11–4–8).

To examine state controllability of the subsystem we shall represent the transfer function of Eq. (11–4–2) by the vector equations

$$\dot{\mathbf{z}} = \mathbf{A}\mathbf{z} + \mathbf{B}\mathbf{u} \tag{11–4–10}$$

$$\mathbf{c} = \mathbf{C}\mathbf{z}$$

where $\mathbf{z}$ is the state vector of the subsystem composed of the plant and output element. Retaining the common factor in Eq. (11–4–2), $c(s)$ can be expanded in partial fractions as

$$c(s) = \frac{8u(s)}{s + 1} + \frac{0}{s + 3} - \frac{2u(s)}{s + 4} \tag{11–4–11}$$

If we let

$$z_1(s) = \frac{8u(s)}{s + 1} \qquad sz_1 + z_1 = 8u$$

$$z_2(s) = \frac{0}{s + 3} \qquad sz_2 + 3z_2 = 0 \tag{11–4–12}$$

$$z_3(s) = \frac{-2u(s)}{s + 4} \qquad sz_3 + 4z_3 = -2u$$

---

* For multiple eigenvalues the Jordan canonical form is required; zeros in the control matrix that correspond to the last row of each Jordan block denote noncontrollability.

we have the equations

$$\begin{bmatrix} z_1 \\ z_2 \\ z_3 \end{bmatrix} = \begin{bmatrix} -1 & 0 & 0 \\ 0 & -3 & 0 \\ 0 & 0 & -4 \end{bmatrix} \begin{bmatrix} z_1 \\ z_2 \\ z_3 \end{bmatrix} + \begin{bmatrix} 8 \\ 0 \\ -2 \end{bmatrix} [u] \qquad (11\text{–}4\text{–}13)$$

and

$$c = \begin{bmatrix} 1 & 1 & 1 \end{bmatrix} \begin{bmatrix} z_1 \\ z_2 \\ z_3 \end{bmatrix} \qquad (11\text{–}4\text{–}14)$$

With $\mathbf{A}$ in canonical form the zero in the control matrix $\mathbf{B}$ means that z_2 cannot be controlled and that the subsystem is not state-controllable; the absence of zeros in Eq. (11–4–14) means that the subsystem, however, is observable.

Leaving the concepts of controllability and observability, we need to reexamine the concepts and definitions of stability with regard to continuous-variable systems in general. The stability of stationary linear systems is relatively straightforward in that it is a property of the system characteristics only, being independent of the initial state and of the magnitude and type of inputs. There is one finite equilibrium (singular) state, and we call the system stable if it returns to that state if disturbed. Stability is determined by the location of the eigenvalues (roots of the characteristic equation), and there are various techniques for locating the eigenvalues.

For nonstationary linear systems and particularly for nonlinear systems, stability is no longer dependent only upon the system properties but is also dependent upon the initial state and the type and magnitude of any input. Furthermore, there may well be more than one equilibrium state. To discuss stability for these systems, additional definitions and criteria are necessary. We shall limit ourselves to autonomous systems* since stability theory for arbitrary inputs is still undeveloped.

A system is said to be *stable* if trajectories leaving an initial state return to and remain within a specified region surrounding an equilibrium state. This general definition of stability is often referred to as *stability in the sense of Liapunov* and permits limit cycles and vortices. If the trajectories of a system that is stable in the sense of Liapunov eventually converge to the equilibrium state, the system is said to be *asymptotically stable*. If the system is stable only for initial states within a bounded region of state space, it is said to be *locally stable* or *stable in the small*. If it is stable for all initial states within the entire state space, it is said to be *globally stable* or *stable in the large*.

We should like our control systems to have asymptotic stability, preferably global; if not global, then the region of asymptotic stability should be large enough to include any anticipated disturbances. The stability of classical control theory is asymptotic. It may appear at first glance to be global, but in reality it is local since no system is truly linear. Only local asymptotic stability with respect to the established equilibrium state can be guaranteed for linear analyses.

---

* Essentially this refers to systems with only initial-state inputs.

There are three basic methods for determining the stability of nonlinear autonomous systems. One method is to approximate the actual system by a second-order system, plot many trajectories in the phase plane, and examine the resulting phase portrait for regions of stability and instability. The describing function method can be used in conjunction with the phase plane to search for and identify limit cycles. Another method is known as the *first*, or *indirect*, *method of Liapunov*. It consists of linearizing the nonlinear vector equations about each equilibrium state by means of the Jacobian matrix and then examining the corresponding eigenvalues for local stability only. The two methods just mentioned are sometimes lumped together as Liapunov's first method.

The third technique is the *second*, or *direct, method of Liapunov*, so called because it does not require solution of the differential equations. It is applicable to all types of differential equations of any order, provides some answers to global as well as local stability, and is widely used.

In using the second method of Liapunov, the equilibrium state being investigated is translated to the origin of the state space so that the autonomous system can be represented by the equation

$$\dot{\mathbf{x}} = \mathbf{f}(\mathbf{x}) \tag{11–4–15}$$

with the equilibrium state $\mathbf{x}_{eq} = 0$. The *asymptotic stability theorem of Liapunov* is the essence of this direct method. This theorem states that the system of Eq. (11–4–15) is asymptotically stable within the closed region R surrounding the origin if there exists a positive definite scalar function $V(\mathbf{x})$ which will vanish with time along all trajectories originating within the region R. If the region R includes all of the state space, the system is globally stable; if not, the system is locally stable within the finite region R. The scalar function $V(\mathbf{x})$ is known as a *Liapunov function*. It must be continuous within the region R, as must its first partial derivatives. The requirement that it be positive definite means that $V(\mathbf{x})$ must be greater than zero for all nonzero values of the state variables and that $V(0)$ be equal to zero. In order for $V(\mathbf{x})$ to vanish along all trajectories starting within R, $dV(\mathbf{x})/dt$ must be less than zero, that is, be negative definite. If $\dot{V}(\mathbf{x}) \leq 0$, it is negative semidefinite and the system is guaranteed to be stable only in the sense of Liapunov; if $\dot{V}(\mathbf{x}) < 0$ along a trajectory, the system is asymptotically stable. Finally, if $\dot{V}(\mathbf{x})$ is indefinite, nothing has been proved with respect to the stability of the system, and we must try different $V(\mathbf{x})$ functions until the system has been shown to be either stable or unstable. Incidentally, for stable systems the size of the region of guaranteed stability can vary with the choice of the Liapunov function.

The major shortcoming of the second method of Liapunov is the problem of finding or constructing a suitable Liapunov function. The simplest positive definite function is the quadratic form

$$Q(\mathbf{x}) = \sum_{i=1}^{n} \sum_{j=1}^{n} a_{ij} x_i x_j = \mathbf{x}^T \mathbf{A} \mathbf{x} \tag{11–4–16}$$

Not all quadratic forms, however, are positive definite, so be careful. The quadratic form of Eq. (11–4–16) often produces a satisfactory Liapunov function for linear systems but seldom does for nonlinear systems. Three techniques for generating Liapunov functions for nonlinear systems are Krasovskii's method which uses the Jacobian matrix of the system, the variable gradient method of Scholtz and Gibson which uses the gradient vector grad V, and Lur'e's method which is applicable to a class of systems with one nonlinear element and uses the canonical form.

12

Digital and Other Control Systems

12–1 DIGITAL CONTROL SYSTEMS

Digital control and *digital control systems* are terms that have different meanings to different people. Rather than be purists and semanticists, let us use the terms interchangeably but be more specific in describing a digital control system.

The simplest digital control system is one in which an analog controller is replaced by a digital computer but the remaining components and sensors are all analog, similar to the situation in Sec. 10–3 where linear controllers were replaced by relay controllers. Such a system is sketched in Fig. 12–1–1. Notice the addition of two blocks, one labeled D/A and the other A/D. These are digital-to-analog and analog-to-digital converters, respectively, required because a digital computer can only handle digital (discrete) inputs and uses difference equations in its algorithms. If all the sensors and components are digital, which is not yet the usual case, converters are not necessary and the functional block diagrams of analog and digital control systems are indistinguishable.

The trend is toward all-digital control systems in which the controller and the instrumentation are digital. Among the many reasons for doing so are the following:

1. Flexibility associated with programming that improves control performance, robustness, measurement of variables, and back calculation and estimation of unmeasurable variables
2. Higher resolution and increased accuracy in measurements
3. Capacity for storing data and control laws
4. Improved troubleshooting and maintenance procedures

5. Reduced cost, volume, and weight

6. Ability to handle multivariable inputs and outputs.

The ability to program is a definite advantage in using a computer as a controller. It simplifies, for example, the determination of changes in operating conditions and the appropriate scheduling of controller gains.

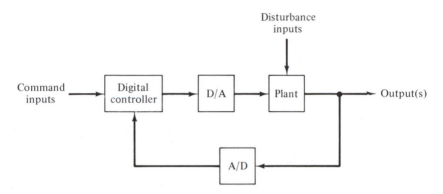

Figure 12–1–1.   Feedback control system with a digital controller.

Digital systems can be analyzed and controllers designed in several ways. One way is to treat them as though they were continuous-variable systems and then approximate the controller by a digital filter. A second is to represent the system as a sampled-data system and extend the transfer function concept and techniques into the sampled-data regime. A third, and probably more direct, approach is to use state variables and difference equations and design the controller by mathematical techniques.

12–2 SAMPLED-DATA (DISCRETE) SYSTEMS

A sampled-data system (or discrete system) can be defined as a system with one or more variables in the form of pulsed data or numerically coded data. These pulsed variables are called *sampled signals* and provide information only at discrete instants of time. If sampled signals occur as the consequence of using a particular class of components or subsystems, we speak of *inherent sampling*. Some examples are the use of scanning radar, the time-sharing of a telemetry data channel, and digital computation and/or control. If sampled signals are deliberately introduced, for example, to increase the preciseness and sensitivity of control or to permit the storage of numerically coded data, we speak of *intentional sampling*. The distinction is not important and is somewhat academic. Sampled-data control systems form an ever-growing class of control systems, particularly digitally controlled systems and processes.

It is necessary to relate a sampled signal to the continuous signal it represents. If the continuous signal is $f(t)$, the symbol for the sampled signal is $f^*(t)$, which is read "f star of t." The mathematical representation of the relationship between a sampled signal and its corresponding continuous signal is known as the *sampler*. The exact form of the sampler is determined by the choice of the sampling scheme being used.

The elements of a sampling scheme are the frequency of sampling and the method of transmitting information. The frequency of sampling can be uniform, periodic, cyclic, multirate, skip, and even random. Information can be conveyed by the height of the pulse, known as *pulse-amplitude modulation* (PAM), or by the width or duration of the pulse, described respectively as *pulse-width modulation* (PWM) or *pulse-duration modulation* (PDM).

The most commonly used sampling scheme is the combination of uniform-rate sampling and pulse-amplitude modulation. The symbol for this scheme is the switch shown in Fig. 12–2–1, where T, the sampling period, is the time between samples or pulses. The relationship between the continuous and sampled signals is also shown in Fig. 12–2–1. The choice of the sampling frequency is important. There is no upper limit since an infinite frequency obviously leads to the continuous signal itself. As the frequency is decreased (the sampling period increased), the representation of the continuous signal becomes poorer until no information at all is transmitted. Shannon's sampling theorem defines the mathematical lower frequency limit for information retrieval. In addition, lowering the frequency adversely affects the stability of closed-loop systems; most systems will become unstable before Shannon's frequency limit is reached.

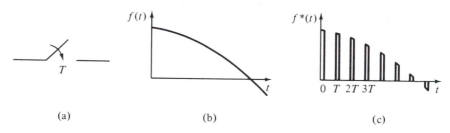

(a)   (b)   (c)

Figure 12–2–1. A uniform-rate, pulse-amplitude-modulation sampling scheme: (a) symbol; (b) a continuous signal; (c) the sampled signal.

A sampled signal is not suitable for controlling a drive component such as a motor because it contains very little energy and is intermittent. Furthermore, we may want output signals in a continuous form. In either case we face the task of reconstructing the original continuous signal from the sampled signal. Reconstruction is essentially an extrapolation and smoothing process; perfect reconstruction is difficult, if not impossible, is expensive, and is detrimental to the stability of the system. Continuous signal recovery is accomplished by filters or demodulators. One important class of such devices is referred to as *data holds* or data extrapolators. The simplest data hold is the zero-order hold, sometimes called a clamp or clamping device. A *zero-order hold* uses the value of the last sample to represent the continuous signal at any instant of time as shown in Fig. 12–2–2b. The *first-order hold* of Fig. 12–2–2c uses the last two samples to determine the instantaneous values of the continuous signal. Higher-order holds are possible; an *n*th-order hold would use the last $(n + 1)$ samples. As the order increases, so does the complexity of the device; further, the increased holding time degrades the closed-loop stability. Fortunately, the zero-order hold with a reasonably high sampling frequency gives an acceptable signal reconstruction along with an acceptable system performance. This combination is often encountered in practice. The *exponential hold* of Fig. 12–2–2d is of passing interest in that it can be obtained with a simple *RC* low-pass network.

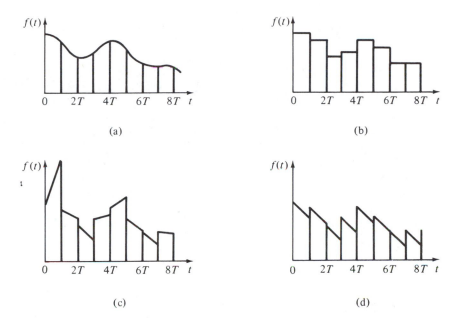

Figure 12–2–2.   Use of data holds to reconstruct a continuous signal: (a) continuous signal and sampling period; (b) zero-order hold; (c) first-order hold; (d) exponential hold.

Let us compare the two systems shown in Fig. 12–2–3. The two systems differ only in that all the signals of the first are continuous, whereas the second has a sampled error signal with uniform-rate sampling and pulse-amplitude modulation. In addition, a zero-order hold has been added to the discrete system in order that the input into the integrator representing the plant will be continuous. The equations describing both systems will be linear and first-order; however, that of the first system will be a differential equation and that of the second a difference equation.

The continuous system is stable for all positive values of K; the unit step response will smoothly approach unity with no overshoot or steady-state error and will have a response time of $3/K$. The sampled-data system, on the other

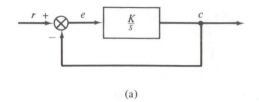

(a)

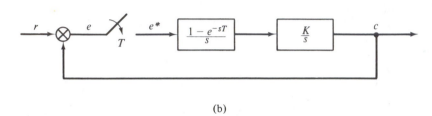

(b)

Figure 12–2–3. Two linear first-order systems: (a) continuous variables; (b) sampled error signal and zero-order hold.

hand, is conditionally stable with respect to KT, the product of the loop gain and the sampling period. The nature of the unit step response is strongly influenced by the value of KT, as can be seen in Fig. 12–2–4. When $KT < 1$, the response resembles that of the continuous system. When $KT = 1$, i.e., $T = \tau$, we have what is known as the *deadbeat response*. As KT is further increased, the response becomes oscillatory and finally unstable. When stable, the discrete system also has a zero steady-state error. It should be noted that although the output of the

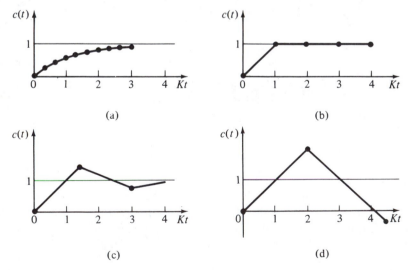

Figure 12–2–4. The unit step response of the system of Fig. 12–2–3b for: (a) $KT = 0.33$; (b) $KT = 1.0$, deadbeat response; (c) $KT = 1.5$; (d) $KT = 2.2$.

sampled-data system is continuous, it is known or specified only at the sampling times.

In order to analyze a sampled-data system, a mathematical model is required. The sampling scheme of Fig. 12–2–1 converts the continuous signal $f(t)$ into a train of amplitude-modulated pulses given the symbol $f^*(t)$. If the duration of the pulses is much shorter than the sampling period and the characteristic times of the continuous signal, the train of pulses may be approximated by a train of impulses, where the area of the impulse at any time $t = nT$ is $f(nT)$. This train of impulses can now be written as the infinite series

$$f^*(t) = f(0)I(t) + f(T)I(t - T) + \cdots \qquad (12\text{–}2\text{–}1)$$
$$+ f(nT)I(t - nT) + \cdots$$

or

$$f^*(t) = \sum_{n=0}^{\infty} f(nT)I(t - nT) \qquad (12\text{–}2\text{–}2)$$

Applying the second shifting theorem, the Laplace transform of Eq. (12–2–2) is

$$f^*(s) = L[f^*(t)] = \sum_{n=0}^{\infty} f(nT)e^{-nTs} \qquad (12\text{–}2\text{–}3)$$

Unfortunately, Eq. (12–2–3) is a nonalgebraic equation and thus difficult to use and manipulate. If, however, we introduce a change in variable by setting

$$z = e^{Ts} \qquad (12\text{–}2\text{–}4)$$

or

$$s = \frac{1}{T}\ln z \qquad (12\text{–}2\text{–}5)$$

the Laplace transform of $f^*(t)$ becomes the z *transform* of $f(t)$ and is given the symbol $f(z)$. The relationships among the time, Laplace, and z domains can be expressed by

$$f(z) = Z[f(t)] = L[f^*(t)]_{s=1/T \ln z} \qquad (12\text{–}2\text{–}6)$$

The z transforms of the continuous signal represented by the sampled signal $f^*(t)$ can be obtained by finding the Laplace transform $f^*(s)$ and then substituting z for $\exp(Ts)$. Thus, the z transform of the sampled signal of Eq. (12–2–2) is found from Eq. (12–2–3) to be

$$f(z) = \sum_{n=0}^{\infty} f(nT)z^{-n} \qquad (12\text{–}2\text{–}7)$$

If $f(t)$ is a unit step $u(t)$, then $f(nT) = 1$ for all values of n. From Eq. (12–2–3),

$$f^*(s) = L[u^*(t)] = 1 + e^{-Ts} + e^{-2Ts} + \cdots \qquad (12\text{–}2\text{–}8)$$

With the substitution of Eq. (12–2–4), the z transform is

$$f(z) = Z[u(t)] = 1 + \frac{1}{z} + \frac{1}{z^2} + \cdots = \frac{z}{z - 1} \qquad (12\text{–}2\text{–}9)$$

The z transform of any continuous function that has a Laplace transform can be found by considering the function to be the input to a fictitious sampler that has the same sampling period as any physical samplers in the system. Consequently, the z transform can be obtained directly from Eq. (12–2–7). It is now possible to construct a table of z transforms. A few examples are shown in Table 12–2–1 along with the corresponding Laplace transforms. Such a table in conjunction with methods such as partial fractions and residues is useful both in directly obtaining z transforms from known Laplace transforms and in inverting z transforms to find the response in the time domain. As might be expected, there are theorems that are helpful in the z-transformation and its application.

The z transform can be used to develop z transfer functions for sampled-data systems that can, in turn, be used in the same manner as the transfer functions of classical control theory. Consider the linear element of Fig. 12–2–5. The Laplace transform of the continuous output can be written as

$$c(s) = G(s)e^*(s) \qquad (12\text{--}2\text{--}10)$$

It can be shown that the introduction of a fictitious sampler has the effect of starring both $c(s)$ and $G(s)$ so that

$$c^*(s) = G^*(s)e^*(s) \qquad (12\text{--}2\text{--}11)$$

The function $G^*(s)$, which defines a starred input-output relationship, is called the *pulse transfer function* of the element. We can now find the z transform of both sides of Eq. (12–2–11) to be

$$c(z) = G(z)e(z) \qquad (12\text{--}2\text{--}12)$$

and $G(z)$ becomes the z transfer function of this linear sampled-data element.

The location of physical samplers with respect to the elements of the system affects the resulting z transfer function. When the sampler precedes two elements in series, as in Fig. 12–2–6a,

$$c(s) = G_1(s)G_2(s)e^*(s) \qquad (12\text{--}2\text{--}13)$$

$$c^*(s) = [G_1(s)G_2(s)]^*e^*(s) \qquad (12\text{--}2\text{--}14)$$

TABLE 12–2–1.   SOME z TRANSFORMS

$f(t)$	$f(z)$	$f(s)$
Unit impulse $I(t)$	1	1
Unit step $u(t)$	$\dfrac{z}{z-1}$	$\dfrac{1}{s}$
Unit ramp t	$\dfrac{Tz}{(z-1)^2}$	$\dfrac{1}{s^2}$
e^{-at}	$\dfrac{z}{z-e^{-at}}$	$\dfrac{1}{s+a}$
$\sin \omega t$	$\dfrac{z \sin \omega T}{z^2 - 2z \cos \omega T + 1}$	$\dfrac{\omega}{s^2 + \omega^2}$

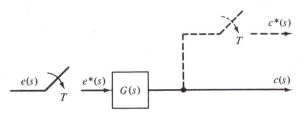

Figure 12–2–5. A linear element with a sampled input and a fictitious sampler.

and

$$c(z) = G_1G_2(z)e^*(z) \qquad (12\text{–}2\text{–}15)$$

In Eq. (12–2–15) the z transfer function $G_1G_2(z)$ is the z transform of the product $G_1(s)G_2(s)$. If, however, the sampler is between the two elements, as in Fig. 12–2–6b,

$$c(s) = G_2(s)m^*(s) \qquad (12\text{–}2\text{–}16)$$

but

$$m(s) = G_1(s)e(s) \qquad (12\text{–}2\text{–}17)$$

Starring both sides of Eq. (12–2–17) yields

$$m^*(s) = G_1^*(s)e^*(s) \qquad (12\text{–}2\text{–}18)$$

which substituted into Eq. (12–2–16) produces

$$c(s) = G_1^*(s)G_2(s)e^*(s) \qquad (12\text{–}2\text{–}19)$$

Starring, or taking the pulse transform of this equation, gives

$$c^*(s) = G_1^*(s)G_2^*(s)e^*(s) \qquad (12\text{–}2\text{–}20)$$

and

$$c(z) = G_1(z)G_2(z)e(z) \qquad (12\text{–}2\text{–}21)$$

In this case the z transfer function is the product of the z transfer functions of the individual elements. Note that

$$G_1(z)G_2(z) \neq G_1G_2(z) \qquad (12\text{–}2\text{–}22)$$

The z transfer functions can be written for closed-loop systems such as the

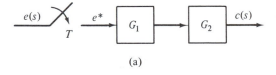

(a)

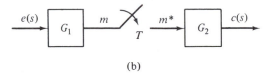

(b)

Figure 12–2–6. Two linear elements and a sampler: (a) not between the elements; (b) between the elements.

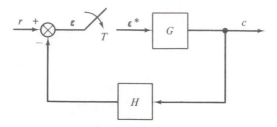

Figure 12–2–7. A closed-loop system with a sampled actuating signal.

one with one sampler represented by the block diagram* of Fig. 12–2–7. Applying the techniques used in obtaining the z transfer functions of the open-loop elements results in the closed-loop z transfer function

$$W(z) = \frac{c}{r}(z) = \frac{G(z)}{1 + GH(z)} \qquad (12\text{–}2\text{–}23)$$

where $1 + GH(z) = 0$ is the characteristic equation of the system. If the system were continuous, the closed-loop transfer function would be

$$W(s) = \frac{c}{r}(s) = \frac{G(s)}{1 + G(s)H(s)} \qquad (12\text{–}2\text{–}24)$$

and the stability of the system would be determined by the location in the s plane of the roots of the characteristic equation. Similarly the location in the z plane of the roots of the sampled-data characteristic equation determines the stability of the sampled-data system. The z-transformation maps the left half of the s plane into the inside of a unit circle in the z plane as indicated in Fig. 12–2–8. Therefore, the criterion for sampled-data stability is that *all* the roots of the characteristic equation of the z transfer function lie within the unit circle in the z plane.

With this stability criterion, all the techniques for examining the stability and the nature of the transient response of stationary linear systems with continuous variables can be extended to sampled-data systems that are linear and stationary. As an example, the rules for constructing the root locus in the z plane are exactly the same as those used for the root locus in the s plane. The Routh-Hurwitz criterion can be used after a bilinear transformation. Such a transformation is recommended to simplify the construction and interpretation of Nyquist and Bode plots. The Schur-Cohn criterion and Jury's criterion are tests similar to the Routh-Hurwitz criterion and tell whether or not all the roots lie within the unit circle in the z plane.

The z-transform method does have a fundamental shortcoming in that it provides information about the system response only at the sampling times, even though the output signal is a continuous function. As a consequence, the inverse z transform is not unique in that two or more continuous functions can have the

---

* Signal flow graphs are often used with pulse transfer functions to represent a system.

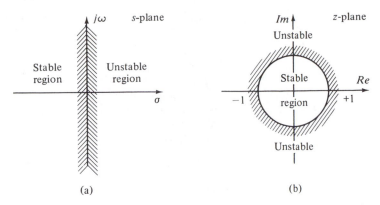

Figure 12–2–8. System stability vis-à-vis the location of the roots of the characteristic equation: (a) in the s plane; (b) in the z plane.

same z transform. Furthermore, the inverse z transform furnishes no information as to the behavior of the output between samplings and, even worse, may hide a divergent (unstable) oscillation. The modified z transform and submultiple sampling methods are two techniques that can be used to provide more information about the system behavior between sampling times.

The w plane can be used as an alternative to the z plane for discrete system analysis. A z transfer function can be converted into a w-domain transfer function by means of an inverse bilinear transformation. The advantages of the w domain are that most of the techniques and rules developed for the analysis and design of continuous-variable systems can be carried directly over into the discrete world. For example, stability boundaries for the root locus are associated with the left-half plane, as in the s plane, rather than the unit circle of the z plane, and the Bode plots are similar to those for continuous-variable systems. The w transfer function is of the same form as the Laplace transfer function, with w approximating s at all frequencies and approaching s as the sampling period approaches zero.

Sampled-data systems need not be linear and stationary nor restricted to the analytical and design techniques taken from classical control theory. They can and do appear in all the forms assumed by continuous-variable systems. They can be linear, nonlinear, stationary, nonstationary, deterministic, stochastic, optimal, adaptive, etc. The techniques and methods for handling any particular class of continuous systems can be extended to its sampled-data counterpart. There is, as a partial illustration of this statement, at least one book (Saucedo and Schiring) that treats stationary linear systems, both continuous and sampled-data, in a unified manner with both classical and modern control theory techniques. The extension and application of continuous system methods to sampled-data systems are not, however, necessarily simple.

The increasing use and potential of sampled-data systems warrants further study by the contol system engineer and their inclusion in his or her repertoire.

12–3 OPTIMAL CONTROL SYSTEMS

The potential and basic features of modern control theory were discussed in Chap. 11, but there was no mention of controllers per se or of their design. Classical control system design can be considered the process of trying various controller configurations until acceptable input-output relationships are attained; the control force itself is accepted as the output of the controller under investigation. Modern control system design, on the other hand, is concerned with finding a mathematical control force that is acceptable and then looking for a controller configuration that will generate such a control force.

If the control force is such that the behavior of the plant output is the best with respect to some preselected performance criteria, it is called an *optimal control*. Some typical optimal control problems are those of minimum time, terminal control, minimum control effort, minimum energy, tracking, and regulation. Each of these problems will usually contain elements of one or more of other problems. For example, we may be interested in the minimum time required to transfer from an initial to a final state (minimum time) with the minimum expenditure of fuel (minimum control effort).

The design sequence for an optimal control system is deceptively simple, having five basic steps:

1. Modeling of the plant
2. Establishment of constraints
3. Selection of the performance index
4. Minimization of the performance index
5. Determination of the controller configuration.

Although an *optimal control system* is defined in terms of the fourth step as one that *minimizes the performance index*, we should not assume that the other steps are of lesser importance; furthermore, they are the steps that tax the judgment and physical understanding of the designer.

For continuous, deterministic, and lumped parameter plants the end results of the first step are the state and output equations

$$\dot{\mathbf{x}}(t) = \mathbf{f}(\mathbf{x}, \mathbf{u}, t) \qquad (12\text{–}3\text{–}1)$$

$$\mathbf{c}(t) = \mathbf{g}(\mathbf{x}, \mathbf{u}, t) \qquad (12\text{–}3\text{–}2)$$

Obviously, these equations must adequately describe the plant. Plant modeling is not a trivial task, nor is the selection of the best state, control, and output variables. Complete controllability in the mathematical sense* is a necessary but not sufficient condition for the existence of an optimal control. In addition, if control is to be feedback (closed-loop), the plant must be completely observable. Remember that observability does not guarantee physical measurability.

The constraints of the second step are the physical constraints imposed on the state and control variables as well as any other physical constraints that

---

* Most physical systems are controllable, but their mathematical models may not be.

might affect the performance of the plant. The lack of proper constraints leads to physically unrealistic and even ridiculous solutions. State constraints may be *equality constraints* whereby the initial and/or final states are specified or *inequality constraints* restricting the range of permissible values of specific state variables. Control and other constraints are generally inequality constraints; e.g. the maximum acceleration of the plant or the fuel used must be less than a specified value. State trajectories and controls that satisfy all the constraints are called *admissible trajectories* and *admissible controls* and are candidates for further investigation. Those trajectories and controls that do not satisfy the constraints are termed *inadmissible* and are rejected.

The formulation of the performance index* (PI) may well be the most critical and difficult step of all. The *performance index* is an attempt to express quantitatively the deviations in plant performance from an ideal performance. The performance index is written in one form as the functional

$$J = J_1[\mathbf{x}(t_f)] + \int_{t_0}^{t_f} J_2[\mathbf{x}, \mathbf{u}, t] \, dt \qquad (12\text{--}3\text{--}3)$$

where t_0 and t_f are the initial and final times; t_f may or may not be specified. J_1 is evaluated at the final state and is not necessarily specified. J_2, the cost or loss function, is evaluated over the entire control interval $(t_f - t_0)$. Weighting factors are used to assign relative importance to various terms in J_2 that describe the deviations from the ideal performance. Each admissible control will yield a single value for a given performance index. This value of J can be used as a figure of merit in comparing competing controls; the smaller the value of J, the better the control—mathematically that is. The admissible control that produces an admissible trajectory and minimizes the value of the performance index is called the *optimal control* and given the symbol $\mathbf{u}^*$. The trajectory is called the *optimal trajectory* and symbolized by $\mathbf{x}^*$.

If an optimal control, which is a function of time, is specified only for a particular initial state

$$\mathbf{u}^*(t) = \mathbf{f}_1[\mathbf{x}(t_0), t] \qquad (12\text{--}3\text{--}4)$$

then the control is open-loop. If, however, the optimal control is a function of both time and the state

$$\mathbf{u}^*(t) = \mathbf{f}_2[\mathbf{x}(t), t] \qquad (12\text{--}3\text{--}5)$$

then the control is closed loop with state-variable feedback and we call $\mathbf{u}^*$ the *optimal control law*. If, for example,

$$\mathbf{u}^*(t) = \mathbf{E}\mathbf{x}(t) \qquad (12\text{--}3\text{--}6)$$

where $\mathbf{E}$ is a constant matrix, the optimal control law is a stationary linear feedback of the state variables.

An optimal control need not exist for a given performance index nor be unique if it does exist. In addition, changing the performance index will result

---

* Also variously described as the *performance measure, loss functional, cost functional,* and *index of performance.*

in a different optimal control and trajectory. The designer must be capable of interpreting and choosing from several optimal controls in terms of physical and practical considerations.

With the exception of some special cases, minimization of the performance index to obtain an optimal control does not yield analytical solutions and the computational effort is high. The two basic methods for minimization are dynamic programming and a calculus of variations approach known as Pontryagin's minimum principle.*

Dynamic programming is a multistage decision process that searches directly for the minimum of the performance index, which is written as a recurrence equation. The distinguishing characteristic of dynamic programming is the use of the principle of optimality to reduce the area of search sufficiently to make direct search feasible. The reduction of the search area by the principle of optimality and the state and control constraints is illustrated in two dimensions in Fig. 12–3–1. As applied to the optimal control problem the *principle of optimality* states that a control that is optimal over a complete interval must be optimal over every subinterval. The computational procedure is to start at a final state and work backward in stages to an initial state, finding the optimal control for each stage in turn. If the optimal control is found for N stages, it contains the optimal control for any lesser number of stages ending at the same final state. This is known as the *principle of embedding*.

Dynamic programming yields an optimal control law (closed-loop control) but not in analytical form. If analytical approximation is not possible, the tabulated control values must be stored and be accessible when needed. Dynamic programming basically uses difference equations†; they may be an approximation of the differential equations of a continuous plant or may represent an actual sampled-data plant. The major disadvantage of dynamic programming is the requirement for rapid-access storage with a large capacity that increases rapidly as the dimensionality of the plant increases.

With the exception of some linear plants, the variational approach of Pontryagin's minimum principle leads to a nonlinear two-point boundary-value problem that must be solved by numerical methods. Three applicable techniques are the method of steepest descent, the variation of extremals, and quasi linearization. A fourth numerical technique known as gradient projection minimizes a function of several variables subject to constraints. In contrast to dynamic programming the inequality constraints complicate the solutions using the variational approach. Furthermore, the numerical solutions for the optimal control are in open-loop form.

Open-loop optimal control may be acceptable, particularly when unknown or unpredictable disturbances are absent or small. If closed-loop control is needed or desired, the open-loop optimal solution obtained by the variational

---

* Originally known as Pontryagin's maximum principle; to maximize a quantity, minimize its negative.

† An alternate approach for continuous plants is the Hamilton-Bellman-Jacobi partial differential equation.

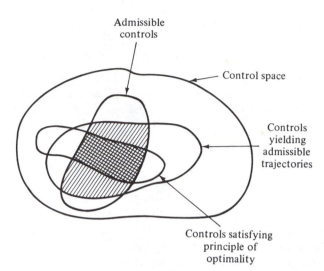

Admissible
controls

Control space

Controls
yielding
admissible
trajectories

Controls satisfying
principle of
optimality

Figure 12–3–1. A two-dimensional representation of control space showing effect of principle of optimality.

approach can be used to limit the area to be searched with dynamic programming. This combination of the two approaches makes it possible to obtain an optimal control law with an appreciable reduction in the requirements for computer storage and in computation time.

When the designer has selected the best optimal control, he or she still faces the problem of constructing a practical controller to generate this particular control vector. It may be impossible to do so, or the designer may be willing to accept a lower quality of performance in exchange for a simpler and less expensive controller. A controller that generates a control force other than the optimal control is called a *suboptimal controller*. In designing a suboptimal controller, the designer can use the optimal control solution to the performance index as a standard of best performance in evaluating different configurations.

To illustrate both optimal and suboptimal control, let us consider the simple problem of controlling a pure inertia plant defined by the equation

$$\ddot{x} = a \qquad (12\text{–}3\text{–}7)$$

where a is an applied specific force with the dimensions of an acceleration. If the state and control variables are defined by $x_1 = x$, $x_2 = \dot{x}$, and $u = a$, then Eq. (12–3–7) can be written in the form $\dot{\mathbf{x}} = \mathbf{Ax} + \mathbf{Bu}$ as

$$\begin{bmatrix} \dot{x}_1 \\ \dot{x}_2 \end{bmatrix} = \begin{bmatrix} 0 & 1 \\ 0 & 0 \end{bmatrix} \begin{bmatrix} x_1 \\ x_2 \end{bmatrix} + \begin{bmatrix} 0 \\ 1 \end{bmatrix} [u] \qquad (12\text{–}3\text{–}8)$$

The plant matrix $\mathbf{A}$ has two eigenvalues equal to zero and is in the Jordan canonical form. Since the second row in $\mathbf{B}$, which corresponds to the last row in the single Jordan block, is not zero, the plant is completely controllable.

The problem is to transfer the plant from any initial state to the origin in the minimum time. The state constraints are equality constraints only and are

$$\mathbf{x}(0) = \begin{bmatrix} x_{10} \\ x_{20} \end{bmatrix} \qquad \mathbf{x}(t_f) = \begin{bmatrix} 0 \\ 0 \end{bmatrix} \qquad (12\text{–}3\text{–}9)$$

where t_0 is taken equal to zero and t_f is unspecified. The control will be limited by the inequality constraint

$$-1 \leq u(t) \leq +1 \qquad (12\text{--}3\text{--}10)$$

In this case application of Pontryagin's minimum principle does yield analytical results. These results show that the optimal control is bang-bang with $u^*(t) = \pm 1$ and that there is only one switching at the curve defined by the nonlinear equation

$$x_1(t) = -\tfrac{1}{2}x_2(t)|x_2(t)| \qquad (12\text{--}3\text{--}11)$$

    Parabolic trajectories for $u = +1$ and $u = -1$ are shown in Fig. 12–3–2a. Notice that only the segments AO and BO satisfy the final state constraint, that is, pass through the origin. Therefore, one of them must be the terminal segment of any optimal trajectory and together they form the switching curve described by Eq. (12–3–11). This optimal switching curve along with the optimal trajectories for several initial states is sketched in Fig. 12–3–2b. For states above the switching curve the optimal control is $u^* = -1$, and for those below it is $u^* = +1$. This is an optimal control law (closed loop) inasmuch as the optimal control at any instant of time is determined by the state at the time. Notice that the origin is a singular point and is a stable node. Mathematically, therefore, the system is asymptotically stable and the plant will remain at the origin if $u^*(t)$ is set and kept equal to zero upon arrival. In actuality, the plant will go into a limit cycle about the origin.

    To instrument this optimal control law requires, among other hardware, a nonlinear function generator to produce the switching curve. A linear switching curve would be easier to generate. Such a suboptimal switching line and a few typical trajectories are sketched in Fig. 12–3–2c. Notice that the number of switchings has increased, leading to an oscillatory response. Further analysis is required to determine the best suboptimal switching line. That there is a best suboptimal controller may be deduced from the fact that with a vertical switching line the origin becomes a center and the system will be marginally stable.

    Much has been left unsaid about optimal control theory and its application. It is a direct synthesis approach that has many apparent drawbacks. Analytical solutions are limited in number. Numerical solutions generally require a large computational effort and are presently limited to certain classes of problems. Optimal solutions are difficult, if not impossible, and expensive to implement with hardware. Finally, physical understanding and sound judgment are necessary to a useful interpretation of solutions. In spite of these disadvantages, optimal control theory has great promise for complex and difficult problems* that are not responsive to any other approach. For any problem it can provide an insight into the best system behavior that cannot be obtained elsewhere and that can be of great value in the design of better suboptimal controllers.

---

* The theory is not limited to control problems alone.

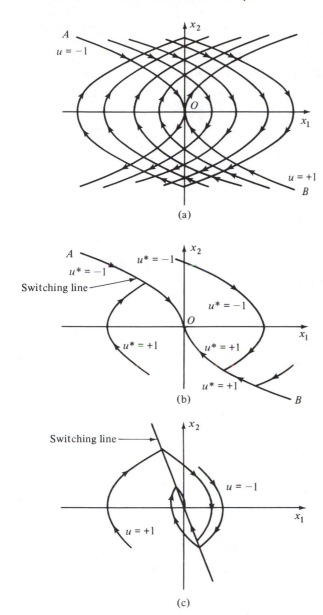

Figure 12-3-2. Minimum time transfer of an inertia plant: (a) some trajectories; (b) optimal switching; (c) suboptimal switching.

12-4 STOCHASTIC CONTROL SYSTEMS

A system is termed *stochastic* if any of the variables are random or if the plant or any component is nondeterministic. The descriptive words *stochastic, random,* and *nondeterministic* are similar in meaning in that whatever they describe is unpredictable, that is, contains uncertainties.

Random variables may appear as random inputs, either command or disturbance, noise in command inputs, or noise in state and output measurements. With a deterministic plant, determination of the state of the plant in the presence of random variables is known as the *state estimation problem,* and control of the plant is known as the *stochastic control problem.*

If the plant is nondeterministic, determination of the best representation is referred to as the *parameter estimation* or *plant identification problem.* Control of a nondeterministic plant with random variables is known as *adaptive control,* which can be open loop or closed loop. If the adaptive control is a function of the state or output vector, it is closed loop and the system may be called a *self-adaptive* or *self-learning control system.*

The analysis and design of stochastic control systems are based on a statistical representation of the uncertainties and the application of probability theory. The approach, techniques, and vocabulary differ sufficiently from anything previously discussed so as to be considered completely beyond the scope of this book. Just remember that there are methods for handling any random processes and uncertainties that might affect the performance of your otherwise acceptable deterministic control system.

Appendix A

THE COMPLEX VARIABLE

Any complex number (variable) can be written in the form

$$z = a + jb \qquad \text{(A–1)}$$

where a is called the real part of z and b is called the imaginary part of z; both a and b are real numbers. The symbol j denotes the imaginary unit, the square root of -1, which has the relationships

$$j^2 = -1 \qquad j^3 = -j \qquad j^4 = +1 \qquad \ldots \qquad \text{(A–2)}$$

The complex plane is a rectangular coordinate system whose abscissa represents the real part and whose ordinate represents the imaginary part of any complex number. The complex number z can now be represented geometrically either by the point P or by the vector represented by the line OP as shown in Fig. A–1a. With the vector representation, the complex number z is specified by its magnitude $|z|$ and its angle ϕ, the latter being also referred to as the argument of z with the symbol $\underline{/z}$. The complex number can now be written as

$$z = |z|e^{j\phi} \qquad \text{(A–3)}$$

where

$$|z| = \sqrt{a^2 + b^2} \qquad \text{(A–4)}$$

and

$$\phi = \underline{/z} = \tan^{-1}\frac{b}{a} \qquad \text{(A–5)}$$

The conjugate of a complex number differs only in the sign of the imaginary

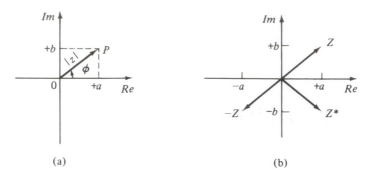

Figure A–1. (a) Geometric representation of the complex number z; (b) geometric relationships among z, z^*, and $-z$.

part. For example the conjugate of z, which is given the symbol z^*, is

$$z^* = a - jb \qquad (A–6)$$

The product of z and z^* is a real positive number as can be seen by performing the multiplication

$$zz^* = (a + jb)(a - jb) = a^2 - j^2b^2 = a^2 + b^2 = |z|^2 \qquad (A–7)$$

The geometric relationships of z to its conjugate z^* and to its negative $-z$ are shown in Fig. A–1b. The conjugate is symmetric with respect to the real axis, and the negative is symmetric with respect to the origin.

Vector representation provides a convenient graphical technique for adding and subtracting complex numbers, as shown in Fig. A–2. In Fig. A–2a the complex variable s is added to the complex number a by completing the parallelogram to obtain $(s + a)$, a vector originating at the origin and having the angle ϕ. In Fig. A–2b, the negative of a is subtracted vectorially from s, a simpler process, to obtain $(s + a)$. This time the vector representing $(s + a)$ does not pass through the origin, and the angle ϕ is measured from the horizontal. Notice that the complex plane in Fig. A–2 is labeled the s plane, implying that the complex Laplace variable, $s = \sigma + j\omega$, is the variable of primary interest.

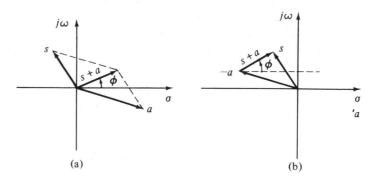

Figure A–2. Vector addition and subtraction of complex numbers; (a) addition; (b) subtraction.

Two complex numbers or functions are equal only if their real and imaginary parts are equal or if their magnitudes and angles are equal. If

$$z_1 = a_1 + jb_1 = |z_1|e^{j\phi_1}$$

and

$$z_2 = a_2 + jb_2 = |z_2|e^{j\phi_2} \tag{A-8}$$

then z_1 and z_2 are equal only if $a_1 = a_2$ and $b_1 = b_2$ or if $|z_1| = |z_2|$ and $\phi_1 = \phi_2$.

The vector representation simplifies the multiplication and division of complex numbers and the evaluation of complex functions. Consider the complex function represented by the equation

$$F(s) = \frac{z_1 z_2}{z_3} \tag{A-9}$$

where $z_1 = s + a$, $z_2 = s + b$, and $z_3 = s + c$; a, b, and c may also be complex numbers. The operations of the right side yield

$$F(s) = \frac{|z_1||z_2|}{|z_3|}e^{j(\phi_1 + \phi_2 - \phi_3)} \tag{A-10}$$

The left side can be expressed in vector form as

$$F(s) = |F(s)|e^{j\phi} \tag{A-11}$$

Therefore, for the equality of Eq. (A-9) to be valid,

$$|F(s)| = \frac{|z_1||z_2|}{|z_3|} \tag{A-12}$$

and

$$\phi = \phi_1 + \phi_2 - \phi_3 \tag{A-13}$$

Equation (A-10) and the subsequent relationships can be used to evaluate graphically the coefficients in partial fraction expansions and thus the relative importance of individual transient modes. Consider first the case of all real denominator factors where

$$c(s) = \frac{K(s + 2)}{s(s + 3)(s + 4)} = \frac{C_1}{s} + \frac{C_2}{s + 3} + \frac{C_3}{s + 4} \tag{A-14}$$

and thus

$$c(t) = C_1 + C_2 e^{-3t} + C_3 e^{-4t} \tag{A-15}$$

From the Heaviside expansion theorems,

$$C_2 = \left[\frac{K(s + 2)}{s(s + 4)}\right]_{s=-3} \tag{A-16}$$

and

$$C_3 = \left[\frac{K(s + 2)}{s(s + 3)}\right]_{s=-4} \tag{A-17}$$

The graphical evaluations of these coefficients are illustrated in Fig. A-3.

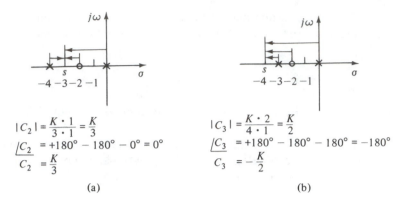

$$|C_2| = \frac{K \cdot 1}{3 \cdot 1} = \frac{K}{3}$$
$$\underline{/C_2} = +180° - 180° - 0° = 0°$$
$$C_2 = \frac{K}{3}$$

(a)

$$|C_3| = \frac{K \cdot 2}{4 \cdot 1} = \frac{K}{2}$$
$$\underline{/C_3} = +180° - 180° - 180° = -180°$$
$$C_3 = -\frac{K}{2}$$

(b)

Figure A–3. Graphical evaluation of the coefficients of a partial fraction expansion:

(a) $C_2 = \left[\dfrac{K(s + 2)}{s(s + 4)} \right]_{s = -3}$;   (b) $C_3 = \left[\dfrac{K(s + 2)}{s(s + 3)} \right]_{s = -4}$.

If any of the denominator factors are complex, such as

$$c(s) = \frac{K}{s(s^2 + 4s + 13)} = \frac{C_1}{s} + \frac{C_2}{s + 2 + j3} + \frac{C_2^*}{s + 2 - j3} \qquad (A–18)$$

it can be shown that C_2 and C_2^* are complex conjugates and that

$$c(t) = C_1 + 2|C_2|e^{-2t} \sin (3t + \phi) \qquad (A–19)$$

where

$$\phi = \underline{/C_2} + 90° \qquad (A–20)$$

The magnitude and argument of C_2 are graphically determined in Fig. A–4.

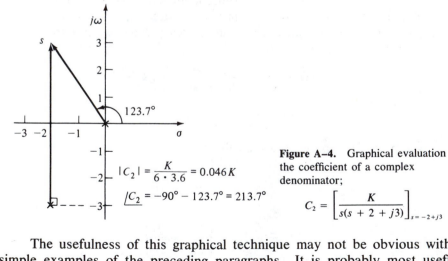

$$|C_2| = \frac{K}{6 \cdot 3.6} = 0.046\,K$$
$$\underline{/C_2} = -90° - 123.7° = 213.7°$$

Figure A–4. Graphical evaluation of the coefficient of a complex denominator;

$$C_2 = \left[\frac{K}{s(s + 2 + j3)} \right]_{s = -2+j3}$$

The usefulness of this graphical technique may not be obvious with the simple examples of the preceding paragraphs. It is probably most useful in approximating the effects upon the transient modes of adding poles and closed-loop zeros during system compensation by the root locus method.

Appendix B

POLYNOMIAL FACTORING

There may be times when it is necessary or desirable to factor polynomials by hand in order to locate the roots of a characteristic equation. Descartes' rule of signs may be helpful. It states that the number of positive real roots is no more than the number of sign changes among the coefficients of the characteristic equation, $D(s) = 0$; similarly, the maximum number of negative real roots is no more than the number of sign changes for $D(-s) = 0$. A characteristic equation of odd order must have at least one real root since complex roots must appear in conjugate pairs.

Of the many methods for factoring polynomials, only the methods of Newton and Lin will be described. Newton's method is useful in determining the real root when the order is odd and low. Lin's method is good for finding a quadratic factor of higher-order polynomials.

Let us apply Newton's method to the third-order characteristic equation

$$D(s) = s^3 + 6s^2 + 20s + 24 = 0 \qquad \text{(B–1)}$$

The general procedure is to make a first best guess as to the value of the real root, denoted s_1, and evaluate $D(s_1)$. Then for a second choice let

$$s_2 = s_1 - \frac{D(s_1)}{D'(s_1)} \qquad \text{(B–2)}$$

where

$$D'(s_1) = \left. \frac{dD(s)}{ds} \right|_{s = s_1} \qquad \text{(B–3)}$$

For this example,

$$D'(s) = 3s^2 + 12s + 20 \qquad \text{(B–4)}$$

From Descartes' rule we know that the real root must be negative. The first trial value of s is obtained by dividing the constant term in Eq. (B–1) by the coefficient of the s term. Therefore, s_1 is -1.2, $D(s_1) = 6.91$, and $D'(s_1) = 9.92$. From Eq. (B–2) we find that

$$s_2 = -1.2 - \frac{6.91}{9.92} \cong 1.9 \qquad \text{(B–5)}$$

and continue on; $D(s_2)$ is equal to $+0.801$ and $D'(s_2)$ is equal to $+9.03$, so that $s_3 = -1.99$, which is close to the actual value of -2.0. The other two roots are found from the quadratic that remains when $(s + 2.0)$ is divided into $D(s)$.

Lin's method is one of successive trial divisions by a quadratic factor. Consider the fourth-order characteristic polynomial

$$D(s) = s^4 + 9s^3 + 37s^2 + 81s + 52 \qquad \text{(B–6)}$$

The first quadratic trial divisor is obtained from the three lowest-order terms by dividing by the coefficient of the second-order term and is

$$s^2 + \frac{81}{37}s + \frac{52}{37} = s^2 + 2.1s + 1.4 \qquad \text{(B–7)}$$

Then we divide Eq. (B–6) by this trial divisor:

$$
\begin{array}{r}
s^2 + 6.9s + 21.1 \\
s^2 + 2.1s + 1.4 \,\overline{)\, s^4 + 9.0s^3 + 37.0s^2 + 81.0s + 52.0} \\
\underline{s^4 + 2.1s^3 + 1.4s^2} \\
6.9s^3 + 35.6s^2 + 81.0s \\
\underline{6.9s^3 + 14.5s^2 + 9.7s} \\
21.1s^2 + 72.3s + 52 \\
\underline{21.1s^2 + 44.3s + 29.5} \\
28.0s + 22.5
\end{array}
$$

The remainder is much too large. The second trial divisor is

$$21.1s^2 + 72.3s + 52 = s^2 + 3.9s + 2.5 \qquad \text{(B–8)}$$

The second trial division is

$$
\begin{array}{r}
s^2 + 5.1s + 14.6 \\
s^2 + 3.9s + 2.5 \,\overline{)\, s^4 + 9.0s^3 + 37.0s^2 + 81.0s + 52.0} \\
\underline{s^4 + 3.9s^3 + 2.5s^2} \\
5.1s^3 + 34.5s^2 + 81.0s \\
\underline{5.1s^3 + 19.9s^2 + 12.8s} \\
14.6s^2 + 68.2s + 52.0 \\
\underline{14.6s^2 + 56.9s + 36.5} \\
11.3s + 15.5
\end{array}
$$

Although smaller, the remainder is still too large. The third trial divisor is

$$14.6s^2 + 68.2s + 52.0 = s^2 + 4.67s + 2.56 \qquad \text{(B-9)}$$

The procedure continues until the remainder is sufficiently small for the accuracy desired. As the remainder becomes small, the number of significant figures in the trial divisor should be increased for greater accuracy.

If a person is handy with the Spirule, the root locus can be used to factor polynomials. To do so, rewrite the characteristic equation of Eq. (B–6) in the form

$$1 + \frac{81(s + 0.642)}{s^2(s^2 + 9s + 37)} = 0 \qquad \text{(B-10)}$$

Now plot the root locus for a variable K, taking

$$K \frac{Z(s)}{P(s)} = \frac{K(s + 0.642)}{s^2(s^2 + 9s + 37)} \qquad \text{(B-11)}$$

Then the four roots of the characteristic equation for $K = 81$ are found graphically from the root locus.

Appendix C

MATRICES AND DETERMINANTS

A matrix is an ordered array of elements such as the *rectangular* matrix

$$
\mathbf{A} = [a_{mn}] = \begin{bmatrix} a_{11} & a_{12} & \cdots & a_{1n} \\ a_{21} & a_{22} & \cdots & a_{2n} \\ \cdot & \cdot & \cdots & \cdot \\ \cdot & \cdot & \cdots & \cdot \\ \cdot & \cdot & \cdots & \cdot \\ a_{m1} & a_{m2} & \cdots & a_{mn} \end{bmatrix}
$$

which is known as an $m \times n$ matrix since it has m rows and n columns. If $m = n$, the matrix becomes a *square* matrix of order n. A square matrix whose elements, other than those on the main diagonal, are zero is a *diagonal* matrix; if the diagonal elements are all unity, the diagonal matrix is known as the *identity* matrix and is given the symbol $\mathbf{I}$. Examples of the diagonal and identity matrices are

$$
\mathbf{A} = \begin{bmatrix} 3 & 0 & 0 \\ 0 & 2 & 0 \\ 0 & 0 & 4 \end{bmatrix} \qquad \mathbf{I} = \begin{bmatrix} 1 & 0 & 0 \\ 0 & 1 & 0 \\ 0 & 0 & 1 \end{bmatrix}
$$

A matrix with only one column (an $n \times 1$ matrix) such as

$$
\mathbf{x} = \begin{bmatrix} x_1 \\ x_2 \\ \cdot \\ \cdot \\ \cdot \\ x_n \end{bmatrix}
$$

is a *column* matrix and is often referred to as an *n*-dimensional *vector*. A matrix with one row (a $1 \times m$ matrix) such as

$$\mathbf{c} = [c_1 \; c_2 \ldots c_m]$$

is a *row* matrix. The *transpose* of the matrix $\mathbf{A}$ is the matrix formed by interchanging the rows and columns of $\mathbf{A}$ and is given the symbol $\mathbf{A}^T$. If

$$\mathbf{A} = \begin{bmatrix} 1 & 0 & 2 \\ 3 & 1 & 3 \\ 2 & 2 & 4 \end{bmatrix} \qquad (C–1)$$

then

$$\mathbf{A}^T = \begin{bmatrix} 1 & 3 & 2 \\ 0 & 1 & 2 \\ 2 & 3 & 4 \end{bmatrix} \qquad (C–2)$$

There are other types and classifications of matrices as well.

Two matrices of the *same* size can be added or subtracted by adding or subtracting the corresponding elements of each matrix. For example,

$$\begin{bmatrix} 1 & 2 & 3 \\ 0 & 1 & 2 \\ 1 & 1 & 2 \end{bmatrix} + \begin{bmatrix} 1 & 2 & 0 \\ 3 & 1 & 2 \\ 1 & 3 & 4 \end{bmatrix} = \begin{bmatrix} 2 & 4 & 3 \\ 3 & 2 & 4 \\ 2 & 4 & 6 \end{bmatrix}$$

$$\begin{bmatrix} 1 & 2 & 3 \\ 0 & 1 & 2 \\ 1 & 1 & 2 \end{bmatrix} - \begin{bmatrix} 1 & 2 & 0 \\ 3 & 1 & 2 \\ 1 & 3 & 4 \end{bmatrix} = \begin{bmatrix} 0 & 0 & 3 \\ -3 & 0 & 0 \\ 0 & -2 & -2 \end{bmatrix}$$

To multiply a matrix by a scalar quantity, multiply each element of the matrix by the scalar. Multiplication of one matrix by another requires that the matrices be conformable; i.e., the number of the columns in the first matrix must be equal to the number of rows in the second matrix. If the first matrix is $m \times n$ and the second is $n \times p$, they are conformable and their product will be an $m \times p$ matrix. The multiplication process, which can be confusing, is illustrated by example.

$$\underset{(3 \times 3)}{\begin{bmatrix} 1 & 2 & 3 \\ 0 & 1 & 1 \\ 1 & 1 & 2 \end{bmatrix}} \cdot \underset{(3 \times 2)}{\begin{bmatrix} 1 & 2 \\ 3 & 1 \\ 3 & 2 \end{bmatrix}} = \underset{(3 \times 2)}{\begin{bmatrix} 1+6+9 & 2+2+6 \\ 0+3+3 & 0+1+2 \\ 1+3+6 & 2+1+4 \end{bmatrix}} = \underset{(3 \times 2)}{\begin{bmatrix} 16 & 10 \\ 6 & 3 \\ 10 & 7 \end{bmatrix}}$$

The general rule for multiplication is that the element in the *i*th row and *j*th column of the product matrix is equal to the sum of the products of the *i*th row of the first matrix and the *j*th column of the second matrix. Generally, $\mathbf{A} \cdot \mathbf{B} \neq \mathbf{B} \cdot \mathbf{A}$, so that we must be careful to observe the sequence of multiplication.

A square matrix has a *determinant* which is the sum of all the products that can be formed from the elements of the matrix. The second-order determinant is

$$\begin{vmatrix} a_{11} & a_{12} \\ a_{21} & a_{22} \end{vmatrix} = a_{11}a_{22} - a_{21}a_{12} \qquad (C–3)$$

or

$$\begin{vmatrix} 1 & 3 \\ 2 & 4 \end{vmatrix} = 4 - 6 = -2 \tag{C-4}$$

A third-order determinant can be evaluated directly as

$$\begin{vmatrix} 1 & 0 & 2 \\ 3 & 1 & 3 \\ 2 & 2 & 4 \end{vmatrix} = 1 \cdot 1 \cdot 4 + 0 \cdot 3 \cdot 2 + 2 \cdot 3 \cdot 2$$

$$-2 \cdot 1 \cdot 2 - 2 \cdot 3 \cdot 1 - 4 \cdot 3 \cdot 0 = +6 \tag{C-5}$$

or by using minors and cofactors.

A *minor* of an element is the determinant formed by crossing out the row and column containing that particular element. In the determinant of Eq. (C–5) the minor of 1, the element in the first row and first column, is equal to the second-order determinant of Eq. (C–4). The *cofactor* of an element a_{ij} is equal to $(-1)^{i+j}$ times the minor of the element.

With the definition of a cofactor, a *determinant* is equal to the sum of the products of the elements of any row or column and their respective cofactors. Applying this statement to the third-order determinant of Eq. (C–5) and using the elements of the first column, we find that

$$\begin{vmatrix} 1 & 0 & 2 \\ 3 & 1 & 3 \\ 2 & 2 & 4 \end{vmatrix} = (-1)^{1+1} \cdot 1 \begin{vmatrix} 1 & 3 \\ 2 & 4 \end{vmatrix} + (-1)^{2+1} \cdot 3 \begin{vmatrix} 0 & 2 \\ 2 & 4 \end{vmatrix} + (-1)^{3+1} \cdot 2 \begin{vmatrix} 0 & 2 \\ 1 & 3 \end{vmatrix}$$

$$= 1(4 - 6) - 3(0 - 4) + 2(0 - 2) = +6 \tag{C-6}$$

Higher-order determinants are similarly evaluated by successive expansions by cofactors until third- or second-order determinants occur.

If a square matrix $\mathbf{A}$ is nonsingular (its determinant is not zero), there exists an *inverse* matrix $\mathbf{A}^{-1}$ such that

$$\mathbf{A} \cdot \mathbf{A}^{-1} = \mathbf{A}^{-1} \cdot \mathbf{A} = \mathbf{I} \tag{C-7}$$

It can be shown that $\mathbf{A}^{-1}$, the inverse of $\mathbf{A}$, can be obtained with the expression

$$\mathbf{A}^{-1} = \frac{\text{adj } \mathbf{A}}{|\mathbf{A}|} \tag{C-8}$$

where adj $\mathbf{A}$ is the symbol for the adjoint of $\mathbf{A}$. The *adjoint* of $\mathbf{A}$ is the transpose of the square matrix whose elements are the cofactors of the elements of $\mathbf{A}$. The adjoint can also be obtained by replacing the elements of the transpose of $\mathbf{A}$ by their cofactors.

To obtain the inverse of the matrix $\mathbf{A}$ of Eq. (C–1), we first find the adj $\mathbf{A}$ to be

$$\text{adj } \mathbf{A} = \begin{bmatrix} -2 & 4 & -2 \\ -6 & 0 & 3 \\ 4 & -2 & 1 \end{bmatrix} \tag{C-9}$$

and then divide each element by $|\mathbf{A}|$, which is $+6$, to obtain

$$\mathbf{A}^{-1} = \begin{bmatrix} -\frac{1}{3} & \frac{2}{3} & -\frac{1}{3} \\ -1 & 0 & \frac{1}{2} \\ \frac{2}{3} & -\frac{1}{3} & \frac{1}{6} \end{bmatrix} \tag{C-10}$$

If Eq. (C–10) is the inverse of $\mathbf{A}$, then $\mathbf{A}^{-1} \cdot \mathbf{A}$ should be equal to $\mathbf{I}$. If we perform the matrix multiplication, the product matrix is indeed the identity matrix.

Cramer's rule is a convenient method for solving a set of linear algebraic equations and can be helpful in obtaining transfer functions. Consider the set of equations

$$a_{11}x_1 + a_{12}x_2 + a_{13}x_3 = c_1$$
$$a_{21}x_1 + a_{22}x_2 + a_{23}x_3 = c_2 \tag{C-11}$$
$$a_{31}x_1 + a_{32}x_2 + a_{33}x_3 = c_3$$

Cramer's rule states that

$$x_1 = \frac{\begin{vmatrix} c_1 & a_{12} & a_{13} \\ c_2 & a_{22} & a_{23} \\ c_3 & a_{32} & a_{33} \end{vmatrix}}{\begin{vmatrix} a_{11} & a_{12} & a_{13} \\ a_{21} & a_{22} & a_{23} \\ a_{31} & a_{32} & a_{33} \end{vmatrix}} \tag{C-12}$$

We see that the solution for x_1 is the ratio of two determinants. The denominator is the determinant of the matrix formed from the coefficients of x_1, x_2, and x_3 and must not be zero. The numerator is the determinant of the matrix formed by replacing the column of the x_1 coefficients by the column of the right-hand terms. Solutions for x_2 and x_3 can be obtained by appropriately changing the numerator determinant. The denominator determinant remains unchanged.

If the equations of Eq. (C–11) are in the Laplace domain where a_{11}, a_{12}, ... may be functions of s and x_1, x_2, and x_3 are the transforms of x_1, x_2, and x_3, then the denominator determinant set equal to zero is the characteristic equation of the plant or system represented by these transformed equations.

References

1. Anderson, B. D., and J. B. Moore, *Linear Optimal Control,* Englewood Cliffs, N.J.: Prentice-Hall, Inc., 1971.

2. Aoki, M., *Optimization of Stochastic Systems,* New York: Academic Press, 1967.

3. Aström, K. J., *Introduction to Stochastic Control Theory,* New York: Academic Press, 1970.

4. Aström, K. J., and B. Wittenmark, *Computer-Controlled Systems—Theory and Design,* Englewood Cliffs, N.J.: Prentice-Hall, Inc., 1984.

5. Atherton, D. P., *Nonlinear Control Engineering,* London: Van Nostrand Reinhold Company, 1982.

6. Bellman, R. E., and R. E. Kalaba, *Dynamic Programming and Modern Control Theory,* New York: Academic Press, 1965.

7. Brown, G. S., and D. P. Campbell, *Principles of Servomechanisms,* New York: John Wiley & Sons, Inc., 1948.

8. Bryson, A. E., Jr., and Y. C. Ho, *Applied Optimal Control,* Waltham, Mass.: Blaisdell Publishing Company, 1969.

9. Chen, C. F., and I. J. Haas, *Elements of Control Systems Analysis: Classical and Modern Approaches,* Englewood Cliffs, N.J.: Prentice-Hall, Inc., 1968.

10. Citron, S. J., *Elements of Optimal Control,* New York: Holt, Rinehart and Winston, Inc., 1969.

11. D'Azzo, J. J., and C. H. Houpis, *Linear Control System Analysis and Design,* New York: McGraw-Hill Book Company, 1975.

12. Deshpande, P. B., and R. H. Ash, *Elements of Computer Process Control,* Research Triangle Park, N.C.: Instrument Society of America, 1981.

13. Dobelin, E. O., *System Modeling and Response,* New York: John Wiley & Sons, Inc., 1980.

14. Dorf, R. C., *Modern Control Systems,* 3d ed., Reading, Mass.: Addison-Wesley Publishing Company, Inc., 1980.

15. Elgerd, O. I., *Control Systems Theory,* New York: McGraw-Hill Book Company, 1967.

16. Evans, W. R., *Control System Dynamics,* New York: McGraw-Hill Book Company, 1954.

17. Eveleigh, V. W., *Adaptive Control and Optimization Techniques,* New York: McGraw-Hill Book Company, 1967.

18. Flügge-Lotz, I., *Discontinuous and Optimal Control,* New York: McGraw-Hill Book Company, 1968.

19. Franklin, G. F., and J. D. Powell, *Digital Control of Dynamic Systems,* Reading, Mass.: Addison-Wesley Publishing Company, Inc., 1981.

20. Gelb, A., and W. VanderVelde, *Multiple Input Describing Functions and Nonlinear System Design,* New York: McGraw-Hill Book Company, 1968.

21. Gibson, J. E., *Nonlinear Automatic Control,* New York: McGraw-Hill Book Company, 1963.

22. Green, W. L., *Aircraft Hydraulic Systems,* Chichester, England: John Wiley and Sons, 1985.

23. Hougen, J. O., *Measurements and Control Applications,* Research Triangle Park, N.C.: Instrument Society of America, 1979.

24. Hsu, J. C., and A. V. Meyer, *Modern Control Principles and Applications,* New York: McGraw-Hill Book Company, 1968.

25. Kirk, D. E., *Optimal Control Theory,* Englewood Cliffs, N.J.: Prentice-Hall, Inc., 1970.

26. Koppel, L. B., *Introduction to Control Theory with Applications to Process Control,* Englewood Cliffs, N.J.: Prentice-Hall, Inc., 1968.

27. Kuo, B. C., *Analysis and Synthesis of Sampled-Data Control Systems,* Englewood Cliffs, N.J.: Prentice-Hall, Inc., 1963.

28. Kuo, B. C., *Automatic Control Systems,* 5th ed., Englewood Cliffs, N.J.: Prentice-Hall, Inc., 1987.

29. Kushner, H. J., *Introduction to Stochastic Control,* New York: Holt, Rinehart and Winston, Inc., 1971.

30. Lago, G., and L. M. Benningfield, *Control System Theory,* New York: The Ronald Press Company, 1962.

31. Lefschetz, S., *Stability of Nonlinear Control Systems,* New York: Academic Press, 1965.

32. Leondes, C. T., *Modern Control Systems Theory,* New York: McGraw-Hill Book Company, 1965.

33. Lindorff, D. P., *Theory of Sampled-Data Control Systems,* New York: John Wiley & Sons, Inc., 1965.

34. Lynch, W. A., and J. G. Truxal, *Introductory System Analysis,* New York: McGraw-Hill Book Company, 1961.

35. Lynch, W. A., and J. G. Truxal, *Principles of Electronic Instrumentation,* New York: McGraw-Hill Book Company, 1962.

36. Mason, S. J., and H. J. Zimmerman, *Electronic Circuits, Signals, and Systems,* New York: John Wiley & Sons, Inc., 1960.

37. Melsa, J. L., and D. G. Schultz, *Linear Control Systems,* New York: McGraw-Hill Book Company, 1969.

38. Minorsky, N., *Theory of Nonlinear Control Systems,* New York: McGraw-Hill Book Company, 1969.

39. Moore, J. A., *Digital Control Devices,* Research Triangle Park, N.C.: Instrument Society of America, 1986.

40. Murphy, G. J., *Basic Automatic Control Theory,* Princeton, N.J.: D. Van Nostrand Company, Inc., 1966.

41. Ogata, K., *Modern Control Engineering*, Englewood Cliffs, N.J.: Prentice-Hall, Inc., 1970.

42. Ogata, K., *State Space Analysis of Control Systems*, Englewood Cliffs, N.J.: Prentice-Hall, Inc., 1967.

43. Pugh, A. H., *Robotic Technology,* London: Peter Peregrinus Ltd., 1983.

44. Raven, F. H., *Automatic Control Engineering,* 3d ed., New York: McGraw-Hill Book Company, 1978.

45. Sage, A. P., *Optimum Systems Control,* Englewood Cliffs, N.J.: Prentice-Hall, Inc., 1968.

46. Sage, A. P., *Linear Systems Control,* Champaign, Ill.: Matrix Publishers, Inc., 1978.

47. Saucedo, R., and E. E. Schiring, *Introduction to Continuous and Digital Control Systems,* New York: The Macmillan Company, 1968.

48. Schwarzenback, J., and K. F. Gill, *System Modelling and Control,* London: Edward Arnold Ltd., 1978.

49. Takahashi, Y., M. J. Rabins, and D. M. Auslander, *Control and Dynamic Systems,* Reading, Mass.: Addison-Wesley Publishing Company, 1970.

50. Van de Vegte, J. *Feedback Control Systems,* Englewood Cliffs, N.J.: Prentice-Hall, Inc., 1986.

51. Watkins, B. O., *Introduction to Control Systems,* New York: The Macmillan Company, 1969.

Index